TETRAHEDRON ORGANIC CHEMISTRY SERIES
Series Editors: J E Baldwin, FRS & P D Magnus, FRS

VOLUME 13

The Chemistry

of *C*-Glycosides

Related Pergamon Titles of Interest

BOOKS

Tetrahedron Organic Chemistry Series:
CARRUTHERS: Cycloaddition Reactions in Organic Synthesis
DEROME: Modern NMR Techniques for Chemistry Research
GAWLEY: Asymmetric Synthesis*
HASSNER & STUMER: Organic Syntheses based on Name Reactions and Unnamed Reactions
PAULMIER: Selenium Reagents & Intermediates in Organic Synthesis
PERLMUTTER: Conjugate Addition Reactions in Organic Synthesis
SIMPKINS: Sulphones in Organic Synthesis
WILLIAMS: Synthesis of Optically Active Alpha-Amino Acids
WONG & WHITESIDES: Enzymes in Synthetic Organic Chemistry

JOURNALS

BIOORGANIC & MEDICINAL CHEMISTRY
BIOORGANIC & MEDICINAL CHEMISTRY LETTERS
JOURNAL OF PHARMACEUTICAL AND BIOMEDICAL ANALYSIS
TETRAHEDRON
TETRAHEDRON: ASYMMETRY
TETRAHEDRON LETTERS

Full details of all Elsevier Science publications/free specimen copy of any Elsevier Science journal are available on request from your nearest Elsevier Science office

* In Preparation

The Chemistry of *C*-Glycosides

DANIEL E. LEVY

Glycomed Inc., California

and

CHO TANG

Sugen Inc., California

PERGAMON

U.K.	Elsevier Science Ltd, The Boulevard, Langford Lane, Kidlington, Oxford OX5 1GB, U.K.
U.S.A.	Elsevier Science Inc., 660 White Plains Road, Tarrytown, New York 10591-5153, U.S.A.
JAPAN	Elsevier Science Japan, Tsunashima Building Annex, 3-20-12 Yushima, Bunkyo-ku, Tokyo 113, Japan

---

First Edition 1995

Transferred to digital printing 2007

Library of Congress Cataloging in Publication Data

A catalog record for this book is available from the Library of Congress

British Library Cataloguing in Publication Data

A catalogue record for this book is available from the British Library

ISBN: 9780080420813

Transferred to Digital Printing 2007

CONTENTS

ACKNOWLEDGEMENTS

Brock, W. de Lappe (Information Services, Glycomed, Inc.) was instrumental in performing the necessary literature database searches for this work.

John H. Musser (Vice President of Medicinal Chemistry, Glycomed, Inc.) reviewed this work in preparation for publicatiion clearance by Glycomed's intellectual property committee.

Glycomed, Inc. (860 Atlantic Ave. Alameda, CA 94501) supported ongoing research efforts in the area of *C*-glycosides.

The following figures were reproduced with the kind permission of the American Chemical Society:

> Figures 1.2.1 and 1.2.3 (Tse-Chang Wu, Peter G. Goekjian, Yoshito Kishi "Preferred Conformations of *C*-Glycosides. 1. Conformational Similarity of Glycosides and Corresponding *C*-Glycosides" *The Journal of Organic Chemistry* **1987**, *52*, 4819).

> Figures 1.2.2 (Yuan Wang, Peter G. Goekjian, David M. Ryckman, William H. Miller, Stefan A. Babirad, Yoshito Kishi "Preferred Conformations of *C*-Glycosides. 9. Conformational Analysis of 1, 4-Linked Carbon Disaccharides" *The Journal of Organic Chemistry* **1992**, *57*, 482)

> Figure 8.11.2 (Toru Haneda, Peter G. Goekjian, Sung H Kim, Yoshito Kishi "Preferred Conformation of *C*-Glycosides. 10. Conformational Analysis of Carbon Trisaccharides" *The Journal of Organic Chemistry* **1992**, *57*, 482).

The following figure was reproduced with the kind permission of the International Union of Pure and Applied Chemistry:

> Figure 1.2.4 (Yoshito Kishi "Preferred Solution Conformations of Marine Natural Product Palytoxin and of *C*-Glycosides and their Parent Glycosides" *Pure and Applied Chemistry* **1993**, *65*, 771).

The following figure was reproduced with the kind permission of Academic Press:

> Figure 1.4.3 (Daniel E. Levy, Peng Cho Tang, John H. Musser "Cell Adhesion and Carbohydrates" *Annual Reports in Medicinal Chemistry* **1994**, *29*, 215).

The following diagrams and the text from the bottom of page 53 through the middle of page 54 were reproduced with the kind permission of Elsevier Science Ltd, The Boulevard, Langford Lane, Kidlington OX5 1GB, UK:

> Scheme 2.3.17, Scheme 2.3.18, and Table 2.3.1 (Daniel E. Levy, Falguni Dasgupta, Peng Cho Tang "Synthesis of Novel Fused Ring *C*-Glycosides" *Tetrahedron Asymmetry* **1994**, *5*, 1937).

Abbreviations

Ac	acetyl
AIBN	2,2'-azobisisobutyronitrile
allyl-TMS	allyl trimethylsilane
9-BBN	9-borabicyclo[3.3.1]nonane
Bn	benzyl
BSA	bis-trimethylsilylacetimidate
Bz	benzoyl
CBZ	benzyl carbamoyl
CN	cyano
CSA	camphorsulfonic acid
DAHP	3-deoxy-D-arabino-heptulosonic acid-7-phosphate
DBU	1,8-diazabicyclo[5.4.0]undec-7-ene
DHQ	dehydroquinate
DIBAL	diisobutyl aluminumhydride
DMF	dimethyl formamide
GDP	guanidine diphosphate
LDA	lithium diisopropylamide
LTD_4	leukotriene D_4
m	meta
m-CPBA	meta-chloroperbenzoic acid
Me	methyl
mMPM	meta-methoxyphenylmethyl
MOM	methoxymethyl
Ms-Cl	methanesulfonyl chloride
n	normal
NC	isocyano
NHS	N-hydroxysuccinimide
NMR	nuclear magnetic resonance
nOe	nuclear Overhauser effect
NPhth	phthalimido
o	ortho
p	para
PMB	para-methoxybenzyl
PNB	para-nitrobenzoyl
p-Tol	para-tolyl
p-TolSCl	para-toluene chlorosulfide
p-TsOH	para-toluenesulfonic acid
SAr	arylsulfide
sec	secondary
SPh	phenylsulfido
TBDMS	tert-butyl dimethylsilyl

TBDPS	tert-butyl diphenylsilyl
tert	tertiary
THP	tetrahydropyranyl
TIPS	triisopropylsilyl
TMS	trimethylsilyl
TMSCN	trimethylsilyl cyanide
TMSI	trimethylsilyl iodide
TMSOTf	trimethylsilyl triflate
Tr	triphenylmethyl, trityl

Chapter One   Introduction

1   Organization of the book

This book is divided into eight chapters. Each of the first seven chapters centers around a different class of reactions. The sections within each chapter define subsets of the reaction classes. For example, Chapter 2 deals extensively with electrophilic substitutions as a means of forming C-glycosides. However, each section within the chapter deals with methods used for the formation or introduction of specific groups at the C-glycosidic linkage.

Chapter 8 is different from the first seven chapters. Instead of being divided into sections, it discusses the chronology of the evolution of C-di and trisaccharide syntheses from 1983 through 1994. The final section speculates on the future of C-glycoside chemistry with respect to new synthetic methodologies. It is the hope of the authors that this book will serve as a comprehensive review of C-glycoside chemistry and a valuable reference tool.

1.1   Definition and Nomenclature of C-Glycosides

Carbohydrates have long been a source of scientific interest due to their abundance in nature and the synthetic challenges posed by their polyhydroxylated structures. However, the commercial use of carbohydrates has been greatly limited by the hydrolytic lability of the glycosidic bond. With the advent of C-glycosides, this limitation promises to be overcome thus paving the way for a new generation of carbohydrate-based products.

C-glycosides occur when the oxygen atom of the exo-glycosidic bond is replaced by a carbon atom for any given O-glycoside. Additionally, for any furanose or pyranose, any exocyclic atoms linked to one or two etheral carbons tetrahedrally will be only carbon or hydrogen and there will be at least one carbon and one OH group on the ring. In accordance with these statements, structural definitions of C and O-glycosides are illustrated in Figure 1.1.1.

With the relatively minor structural modification involved in the transition from O-glycosides to C-glycosides comes the more involved question of how to name these compounds. Figure 1.1.2 shows a series of C-glycosides with their corresponding names. Compound $\underline{1}$ derives its name from the longest continuous chain of carbons. Since this is a seven carbon sugar, the name will be derived from heptitol. As the heptitol is cyclized between C_2 and

C_6 it is designated as 2,6-anhydro. Finally, carbons 2-5 and carbons 6-7 bear the D-gulo and D-glycero configurations, respectively. By comparison, compound **4** is a 2,6-anhydro heptitol bearing the D-glycero configuration at carbons 6-7. However, carbons 2-5 bear the D-ido configuration and the molecule is now designated as such.

Figure 1.1.1   **Definition of *C*-Glycosides**

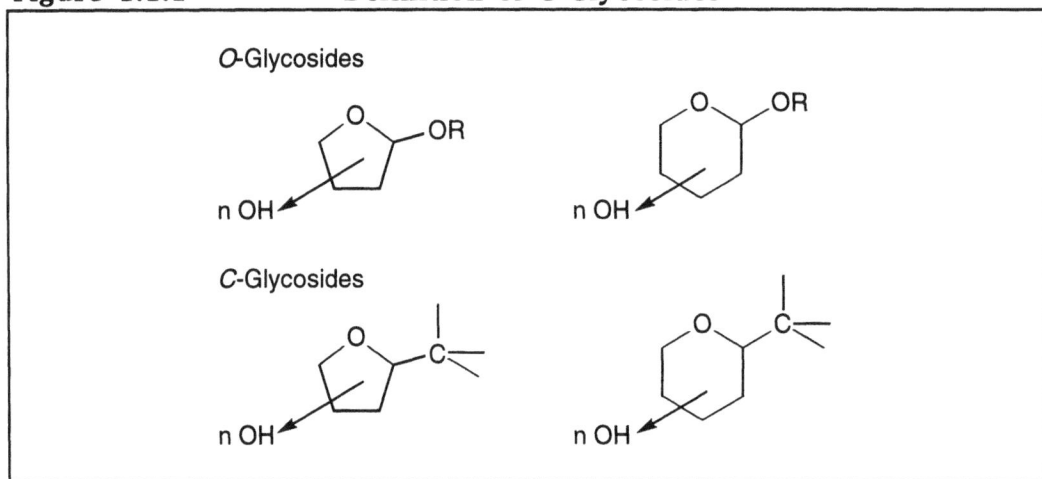

O-Glycosides

C-Glycosides

Figure 1.1.2   **Representative *C*-Glycosides and Their Nomenclature**

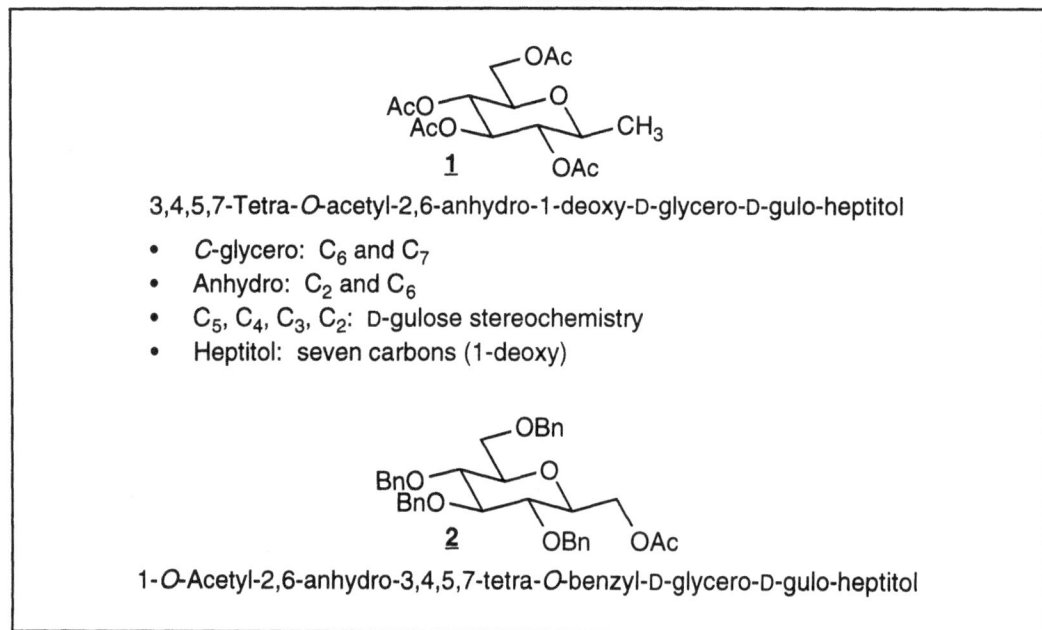

1

3,4,5,7-Tetra-*O*-acetyl-2,6-anhydro-1-deoxy-D-glycero-D-gulo-heptitol

- *C*-glycero: C_6 and C_7
- Anhydro: C_2 and C_6
- C_5, C_4, C_3, C_2: D-gulose stereochemistry
- Heptitol: seven carbons (1-deoxy)

2

1-*O*-Acetyl-2,6-anhydro-3,4,5,7-tetra-*O*-benzyl-D-glycero-D-gulo-heptitol

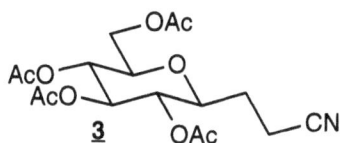

3

5,6,7,9-Tetra-*O*-acetyl-4,8-anhydro-2,3-dideoxy-D-glycer-D-gulo-nonanitrile

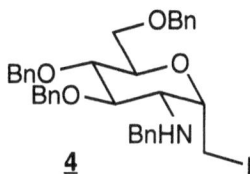

4

2,6-Anhydro-4,5,7-tri-*O*-benzyl-3-(N-benzylamino)-1,3-dideoxy-1-iodo-
D-glycero-D-ido-heptitol

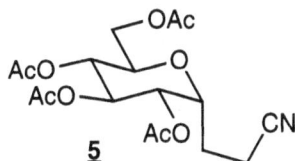

5

5,6,7,9-Tetra-*O*-acetyl-4,8-anhydro-2,3-dideoxy-D-glycer-D-ido-nonanitrile

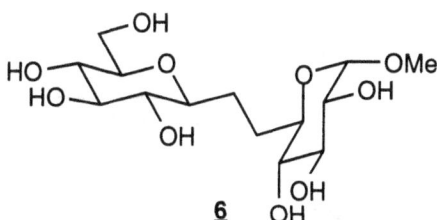

6

Methyl-8,12-anhydro-6,7-dideoxy-D-glycero-D-gulo-α-D-gluco-tridecapyranoside

or   D-Glc-*C*-β-(1-6)-D-GlcOMe

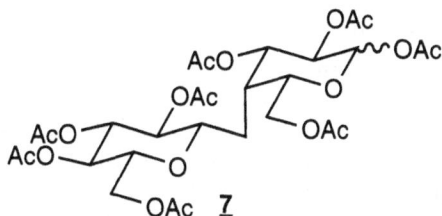

7

1,2,3,6-Tetra-*O*-acetyl-4-deoxy-4-(3,4,5,7-tetra-*O*-acetyl-2,6-anhydro-1-deoxy-D-
glycero-D-gulo-heptitol-1-yl)-α,β-D-galactopyranose

1.2 *O*-Glycosides vs. *C*-Glycosides: Comparisons of Physical Properties, Anomeric Effects, H-Bonding Abilities, Stabilities and Conformations

The structural and chemical similarities prevalent between *C* and *O*-glycosides are illustrated in Table 1.2.1 and are summarized as follows. In light of the bond lengths, Van der Waal radii, electonegativities and bond rotational barriers being very similar between *O* and *C*-glycosides, we find the largest difference between physical constants to be in the dipole moments. However, as minor differences do exist, the conformations of both *O* and *C*-glycosides are represented by similar antiperiplanar arrangements. These conformational similarities are illustrated in Figure 1.2.1.

Table 1.2.1   Physical Properties of *O* and *C*-Glycosides

	O-Glycosides	*C*-Glycosides
Bond Length	O-C = 1.43 Angstroms	O-C = 1.54 Angstroms
Van der Waal	O = 1.52 Angstroms	O = 2.0 Angstroms
Electronegativity	O = 3.51	C = 2.35
Dipole Moment	C-O = 0.74D	C-C = 0.3D
Bond Rotational Barrier	CH_3-O-CH_3 = 2.7 kcals/mole	CH_3-CH_3 = 2.88 kcals/mole
H-Bonding	Two	None
Anomeric Effect	Yes	No
Exoanomeric Effect	Yes	No
Stability	Cleaved by Acid and Enzymes	Stable to Acid and Enzymes
Conformation	$C_{1'}$-$C_{2'}$ antiperiplanar to O_1-C_1	$C_{1'}$-$C_{2'}$ antiperiplanar to C_1-C_2

Perhaps the major difference between *C* and *O*-glycosides is found within chemical reactivities. Not only are *C*-glycosides absent of anomeric effects, they are also stable to acid hydrolysis and are incapable of forming hydrogen bonds.

Within the realm of physical properties, *O* and *C*-glycosides exhibit similar coupling constants in their ^{1}H NMR spectra. A summary of the respective average coupling constants[1] is presented in Figure 1.2.2 and specific illustrations[2,3] are shown in Figure 1.2.3.

Regarding the study summarized in Figure 1.2.3, it was determined that as with *O*-glycosides, the $C_{1'}$-$C_{2'}$ bond is antiperiplanar to the C_1-C_2 bond in the corresponding *C*-glycosides. Furthermore, C_2 substituents were found to have no effect on *C*-glycosidic bond conformations and, considering 1,3-diaxial-like interactions, the *C*-aglyconic region will distort first. In summary, *C*-glycosides were found to be not conformationally rigid with conformations predictable based on steric effects. These conformations were shown to parallel those of corresponding *O*-glycosides thus suggesting potential uses for *C*-glycosides.

Figure 1.2.1   Conformations of *O* and *C*-Glycosides

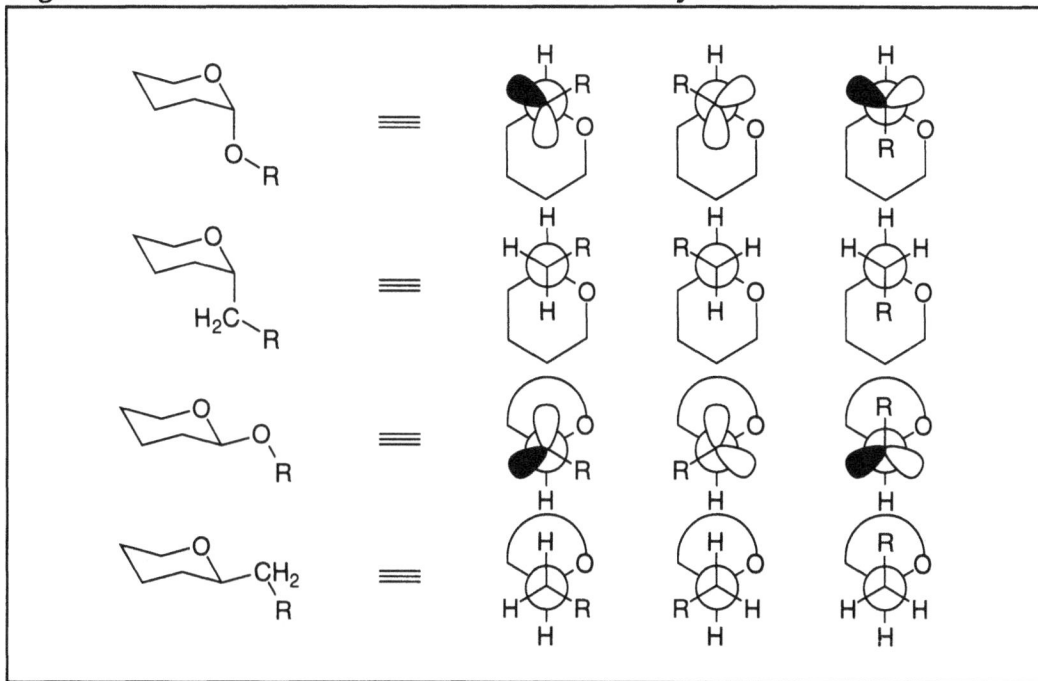

Figure 1.2.2   Conformations and Coupling Constants

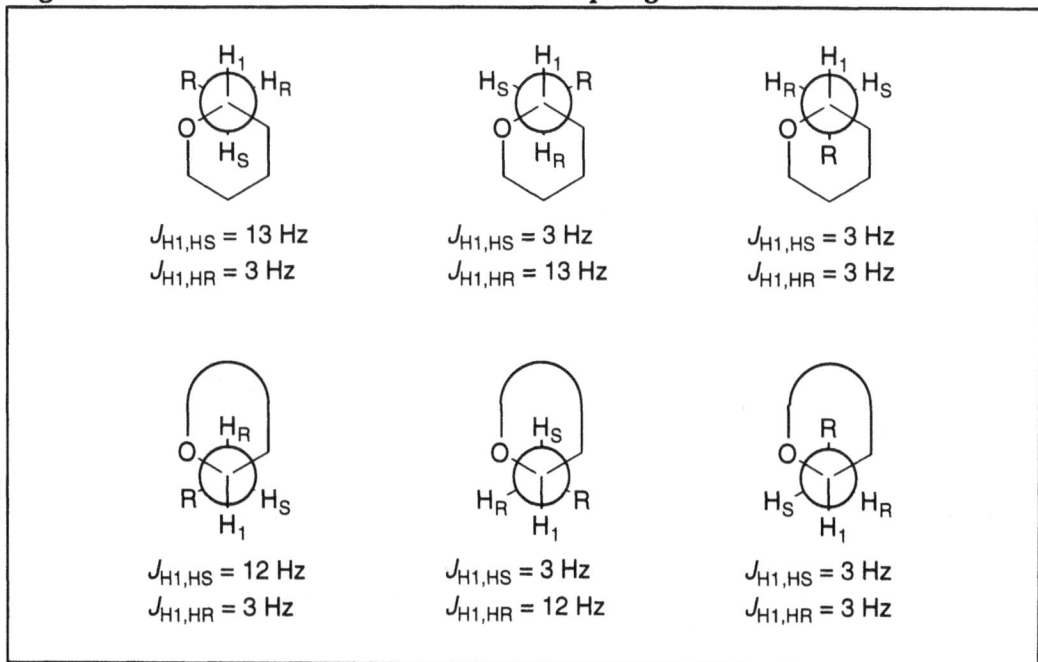

$J_{H1,HS} = 13$ Hz
$J_{H1,HR} = 3$ Hz

$J_{H1,HS} = 3$ Hz
$J_{H1,HR} = 13$ Hz

$J_{H1,HS} = 3$ Hz
$J_{H1,HR} = 3$ Hz

$J_{H1,HS} = 12$ Hz
$J_{H1,HR} = 3$ Hz

$J_{H1,HS} = 3$ Hz
$J_{H1,HR} = 12$ Hz

$J_{H1,HS} = 3$ Hz
$J_{H1,HR} = 3$ Hz

Kishi, *et al.*,[4] demonstrated that "it is possible to predict the conformational behavior around the glycosidic bond of given *C*-saccharides by placing the $C_{1'}$-$C_{2'}$ bond antiperiplanar to the C_α-C_n bond and then focusing principally on the steric interaction around the non-glycosidic bond." This postulate was demonstrated, as shown in Figure 1.2.4, utilizing a diamond lattice to arrange the glycosidic conformations of the *C*-trisaccharide analogs of the blood group determinants.[5,6]

Figure 1.2.3   Literature Examples of Coupling Constants

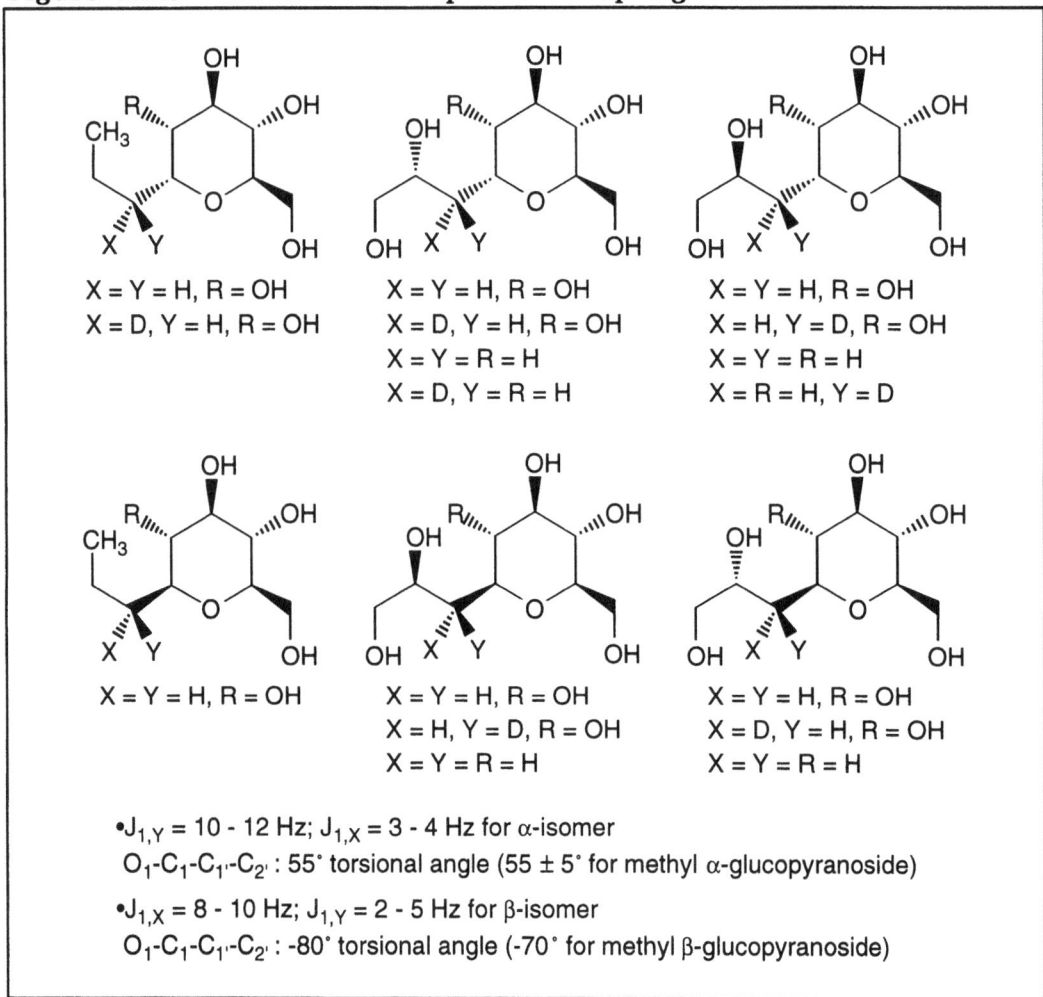

X = Y = H, R = OH
X = D, Y = H, R = OH

X = Y = H, R = OH
X = D, Y = H, R = OH
X = Y = R = H
X = D, Y = R = H

X = Y = H, R = OH
X = H, Y = D, R = OH
X = Y = R = H
X = R = H, Y = D

X = Y = H, R = OH

X = Y = H, R = OH
X = H, Y = D, R = OH
X = Y = R = H

X = Y = H, R = OH
X = D, Y = H, R = OH
X = Y = R = H

- $J_{1,Y}$ = 10 - 12 Hz; $J_{1,X}$ = 3 - 4 Hz for α-isomer
  O_1-C_1-$C_{1'}$-$C_{2'}$: 55° torsional angle (55 ± 5° for methyl α-glucopyranoside)

- $J_{1,X}$ = 8 - 10 Hz; $J_{1,Y}$ = 2 - 5 Hz for β-isomer
  O_1-C_1-$C_{1'}$-$C_{2'}$: -80° torsional angle (-70° for methyl β-glucopyranoside)

In an NMR study of the conformations of *C*-glycofuranosides and *C*-glycopyranosides, Brakta, *et al.*,[7] demonstrated that the $^1H_{1'-2'}$ coupling constants were limited in their abilities to define conformations. However, the

$H_{1'}$ chemical shifts were shown to be more down field for the α anomers while the $C_{1'}$, and $C_{5'}$ chemical shifts were found to be more up field. Furthermore, the ^1H-^{13}C coupling constants for the anomeric protons were measured as larger for the α isomer. In light of all of this information, the nOe is the most diagnostic. When applied to $H_{1'}$ and $H_{4'}$ or $H_{5'}$, no nOe exists for the α isomer while that for the β isomer can be readily determined.

Figure 1.2.4   *C*-**Trisaccharide of Blood Group Determinants in Diamond Lattice**

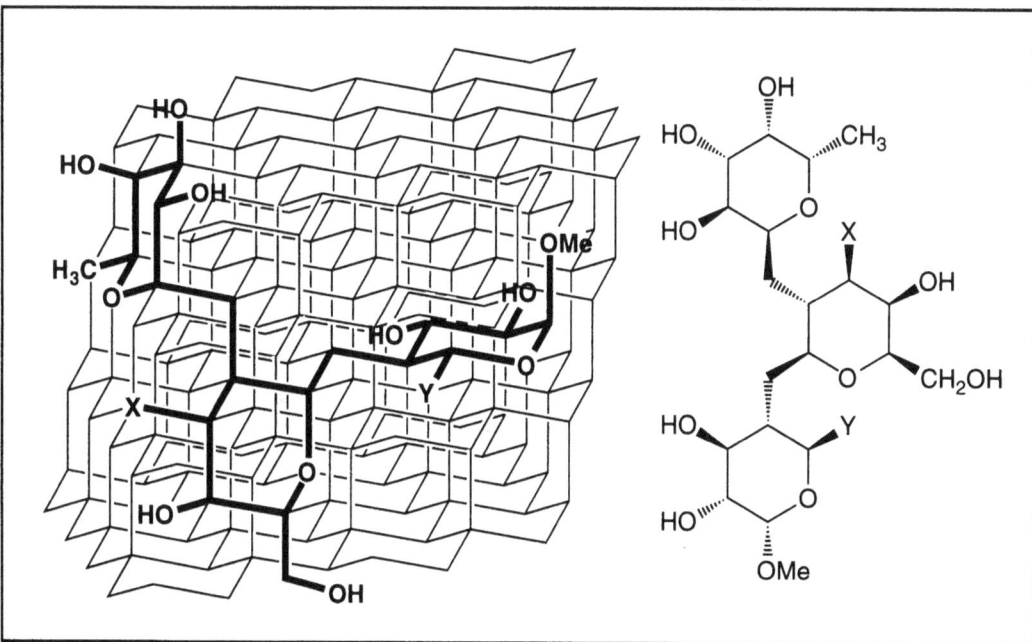

1.3   Naturally Occurring *C*-Glycosides

Figure 1.3.1   **Aloins A and B**

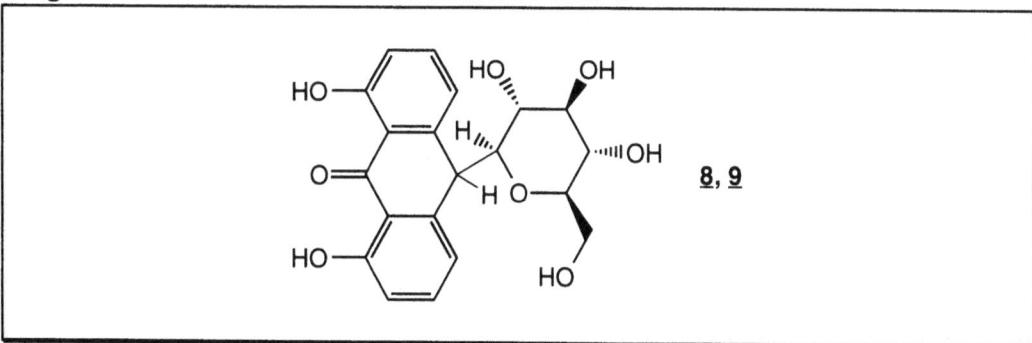

While *C*-glycoside chemistry has recently attracted much attention, they were not invented by chemists. Many examples of *C*-glycosides have been isolated from nature. For example, in the structures represented in Figure 1.3.1 were isolated from Barbados aloe. These compounds, aloin A and aloin B, are collectively known as barbaloin make up the bitter and purgative principle of aloe.[8-12]

Figure 1.3.2   *Carthamus tinctorius*

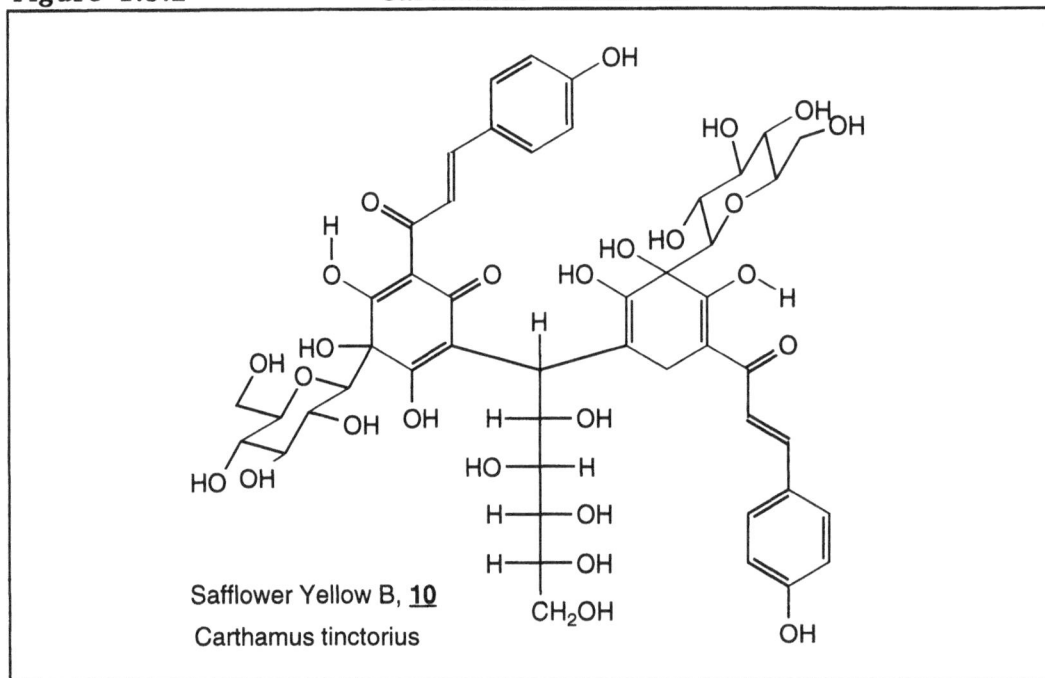

Safflower Yellow B, **10**

Carthamus tinctorius

Figure 1.3.3   *Trachelospermum asiaticum*

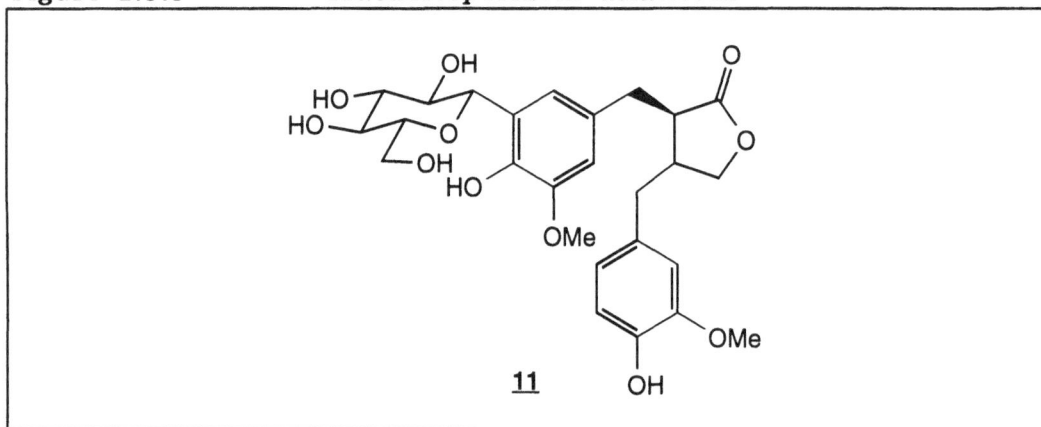

11

In 1984, a yellow pigment was isolated from the petals of *Carthamus tinctorius*.[13] The structure assigned to the isolate is shown in Figure 1.3.2 and contains an open-chain *C*-glycoside.

Figure 1.3.4   ***Asphodelus ramosus***

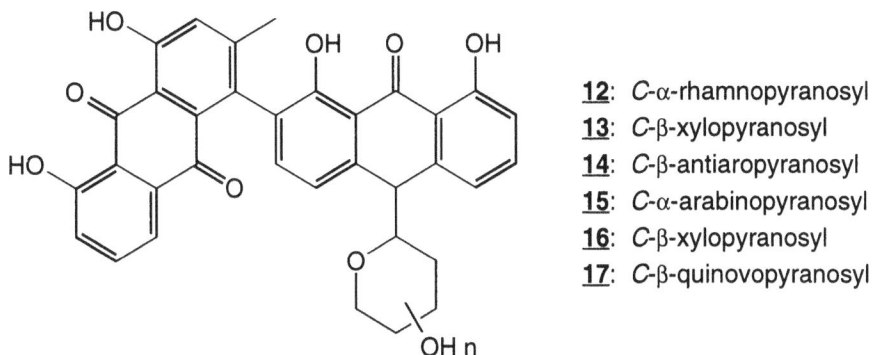

12: *C*-α-rhamnopyranosyl
13: *C*-β-xylopyranosyl
14: *C*-β-antiaropyranosyl
15: *C*-α-arabinopyranosyl
16: *C*-β-xylopyranosyl
17: *C*-β-quinovopyranosyl

Figure 1.3.5   ***Chrysopogon aciculatis***

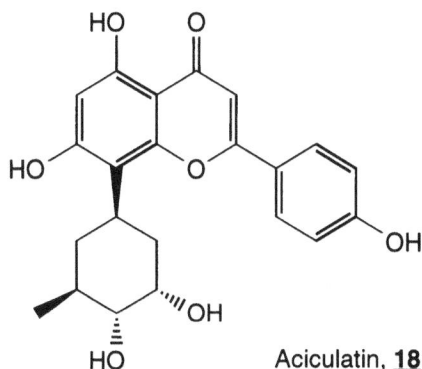

Aciculatin, **18**

In 1986, the first naturally occurring lignan *C*-glycoside was isolated from *Trachelospermum asiaticum* (Figure 1.3.3).[14] Another example of naturally occurring *C*-glycosides was reported in 1991 when extracts of *Asphodelus ramosus* tubers were shown to posses IC_{50} values of 4.5 ppm in the *Artemia salina* bioassay. Analysis of the extracts yielded the family of six new anthraquinone-anthrone-*C*-glycosides shown in Figure 1.3.4.[15] All six compounds share the same aglycone unit and differ only in the specific sugar attached.

Again, in 1991, a flavone derived *C*-glycoside was isolated from *Chrysopogon aciculatis* and shown to be cytotoxic to KB cells. Interestingly, this compound, shown in Figure 1.3.5, was also shown to bind to DNA with a K_d of about 15 μM.[16]

Considering the number of naturally occurring *C*-glycosides known, we cannot ignore the specific class of *C*-nucleosides. This class of compounds is particularly important due to the antibacterial, antiviral, and antitumor properties of many of these molecules. Some examples of naturally occurring *C*-nucleosides are shown in Figure 1.3.6 and include pyrazofurin, showdomycin, oxazinomycin, formycin, and formycin B.[17,18]

Figure 1.3.6   Naturally Occurring *C*-Nucleosides

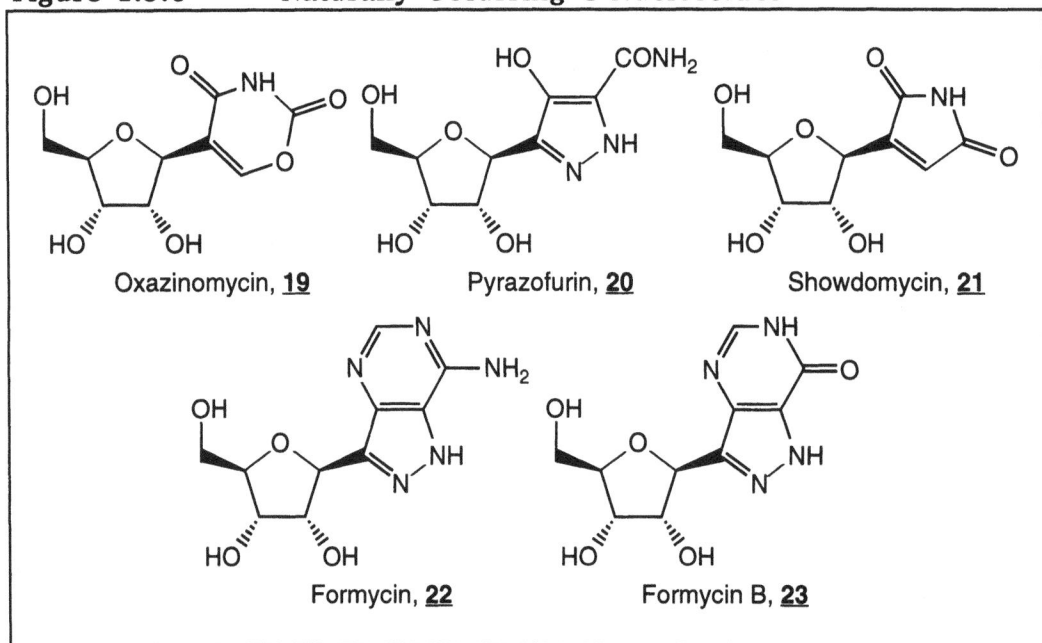

Oxazinomycin, **19**   Pyrazofurin, **20**   Showdomycin, **21**

Formycin, **22**   Formycin B, **23**

1.4. *C*-Glycosides as Stable Pharmacophores

With the fast paced development of the chemistry of *C*-glycosides, the pharmaceutical and biotechnology industries have recognized that *C*-glycoside analogs of biologically active carbohydrates may be studied as stable pharmacophores. Currently, glycosides and saccharides are used in a variety of applications ranging from foodstuffs to components of nucleic acids and cell surface glycoconjugates.[19-24] Realizing the extent to which these types of compounds are involved in everyday life, the advantages of using *C*-glycosides as drug candidates become apparent. Included among these advantages are the facts that *C*-glycosides are stereochemically stable and that chiral starting materials are readily available (sugar units). Furthermore, *C*-glycosides as pharmacophores may yield novel enzyme inhibitors with structures that may be difficult or impossible to construct utilizing standard carbohydrate chemistry. The following Figures and Schemes illustrate the use of *C*-glycosides in biological systems.

The biosyntheses and biodegradations of glycosides are accomplished by enzymes called glycosidases. In cases where the product of a glycosidation is mediating an undesired biological function, the utility of glycosidase inhibitors becomes apparent. Figure 1.4.1 shows a series of molecules designed to inhibit the specific activity of β-glucosidase (cellobiase).[25] The assay used involves the cleavage of an *O*-nitrophenyl-β-D-glucoside with the above mentioned enzyme isolated from sweet almonds.[26] The biological activities of the illustrated synthetic compounds were compared to the activity of 1-deoxynojirimycin and the relevant K_is accompany the structures in Figure 1.4.1.

Figure 1.4.1   *C*-**Glycosidic β-Glucosidase Inhibitors**

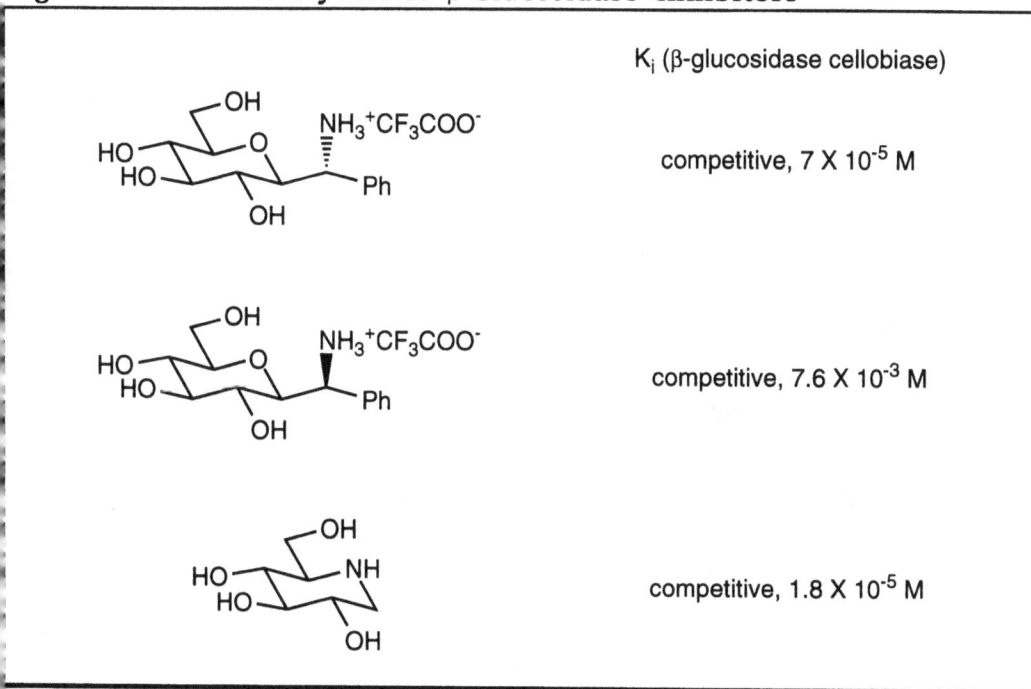

K_i (β-glucosidase cellobiase)

competitive, 7×10^{-5} M

competitive, 7.6×10^{-3} M

competitive, 1.8×10^{-5} M

With the interest in glycosidase inhibitors, compounds lending insight to the structures and mechanisms of these enzymes have received much attention. Recently, Lehmann, *et al.*,[27] prepared a series of diastereotopic *C*-glycosides designed to be substrates for β-D-galactosidase. The premise was that in order to confirm this enzyme's mechanism of action, proposed from an extrapolation of lysozyme activity,[28-31] the structural analysis of products resulting from the enzyme's effect on its substrates would be helpful. Therefore, the structures shown in Figure 1.4.2 were prepared and, as illustrated in Scheme 1.4.1, were shown to be substrates for β-D-galactosidase.[27]

Figure 1.4.2   *C*-Glycosidic Substrates
                               for β-D-Galactosidase

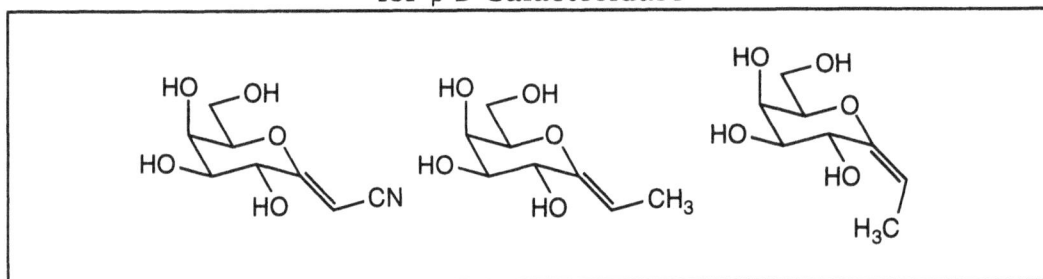

Scheme 1.4.1   *C*-Glycosidic Substrates
                              for β-D-Galactosidase

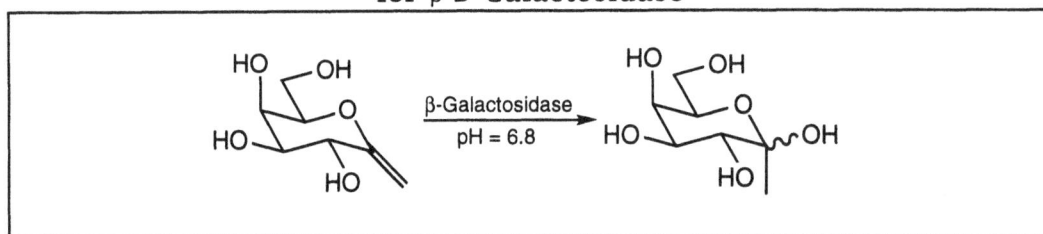

With the encouraging results with synthetic β-D-galactosidase inhibitors, Lehmann, *et al.*,[32] prepared deuterated analogs. Utilizing the stereochemical outcome, the face of the substrate serving as the recipient of proton donation by the enzyme was determined. As shown in Scheme 1.4.2, a proton donating substituent in the binding site was found to induce reaction at the *re* face of the bound substrate. These findings were consistent with the previously mentioned lysozyme model.

Considering the rapid advancement of our understanding of the inhibition of carbohydrate binding proteins, there is still a tremendous need for more selective molecules. In the area of affinity labeling, diazoketones have found much use.[33] With the irreversible nature of their binding to respective enzymes, Myers, *et al.*,[34] designed diazoketones into sugar derivatives. The general approach, shown in Scheme 1.4.3, was designed to create irreversible glycosidase inhibitors or suicide substrates.

Inhibitors of the enzyme glycogen phosphorylase have been studied for their ability to lower blood glucose levels.[35-37] In order to study the nature of the binding of glucose derived substrates to glycogen phosphorylase, the product shown in Scheme 1.4.4[38] was complexed to the enzyme and studied *via* x-ray crystallography.[39] Aside from demonstrating reasonable inhibitors, the conformations of *C*-glucosides bound to this particular enzyme were shown to exist in a skew boat conformation.

Scheme 1.4.2   ***C*-Glycosidic Substrates**
for β-Galactosidase

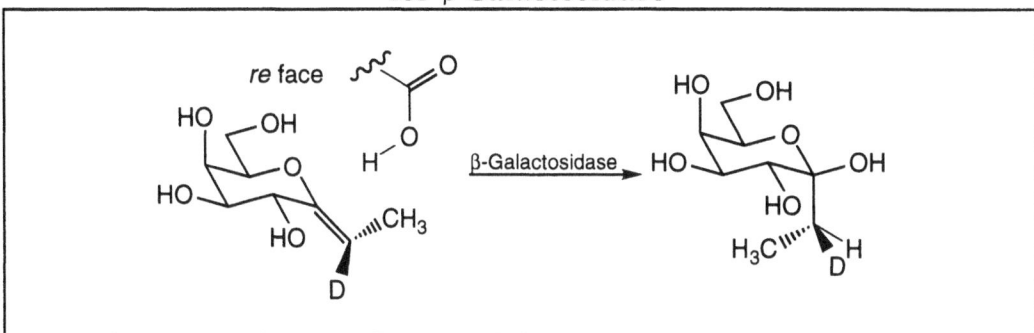

Scheme 1.4.3   ***C*-Glycosidic Affinity Labels for**
Carbohydrate Binding Proteins

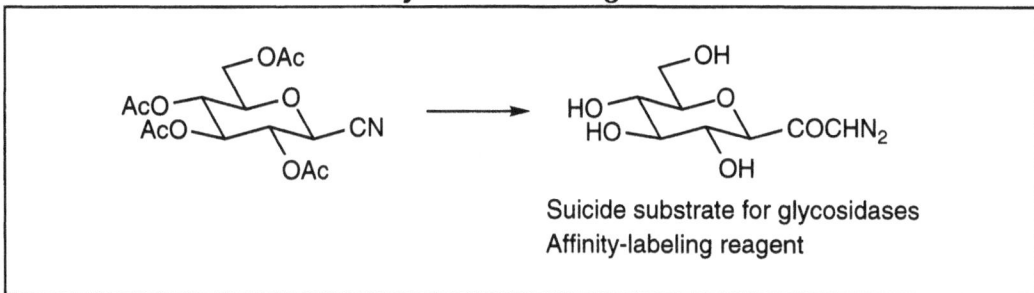

Suicide substrate for glycosidases
Affinity-labeling reagent

Scheme 1.4.4   ***C*-Glycosidic Glycogen**
Phosphorylase Inhibitors

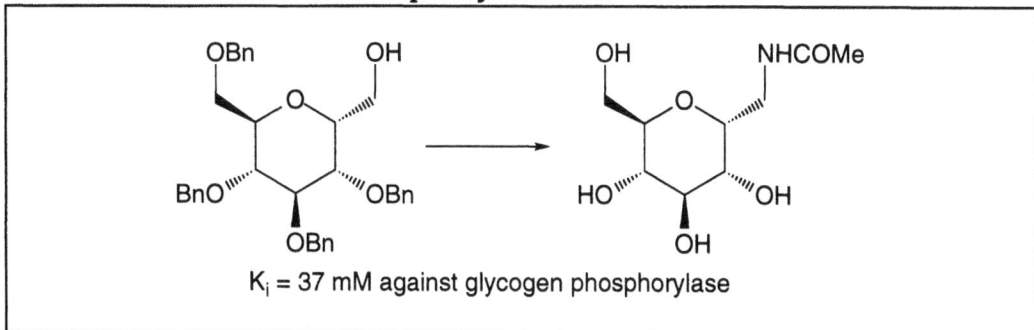

K_i = 37 mM against glycogen phosphorylase

Glycosyltransferases assemble polysaccharides from their monomeric units by adding, to the growing carbohydrate chain, activated nucleotide sugars.[40] The chemistry of glycosyltransferases is well known[41-43] and is illustrated in the biosynthesis of sialyl Lewis[x], shown in Figure 1.4.3.[44] First, β-1,4-galactosyltransferase forms the lactose scaffolding. Second, α-2,3-sialyltransferase adds the sialic acid. Finally, the addition of fucose by α-1,3-fucosyltransferase completes the synthesis.

Figure 1.4.3   The Biosynthesis of Sialyl Lewis[x]

The glycosidic bonds are formed by the corresponding glycosyltransferases in sequence and where specified:

  a: β-1,4-Galactosyltransferase
  b: α-2,3-Sialyltransferase
  c: α-1,3-Fucosyltransferase

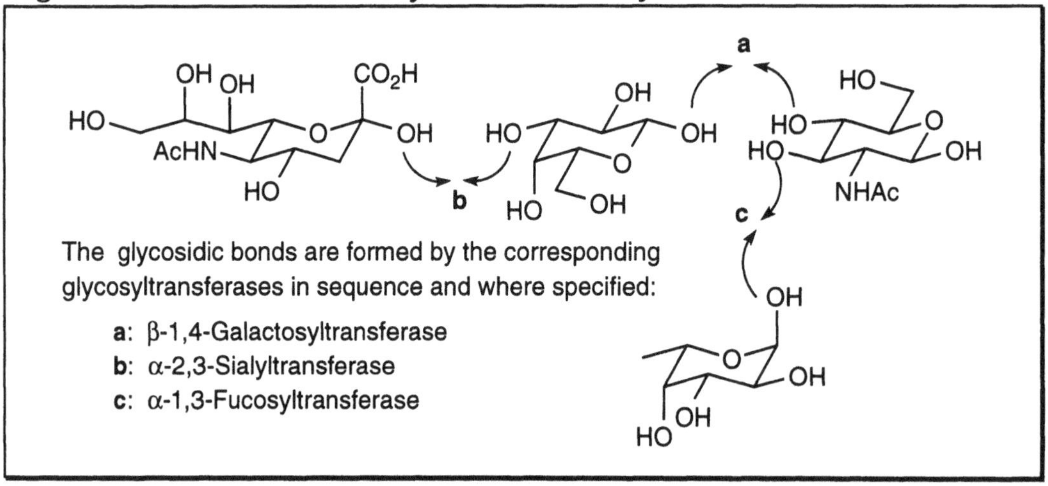

Figure 1.4.4   *C*-Glycosidic Fucosyltransferase Inhibitors

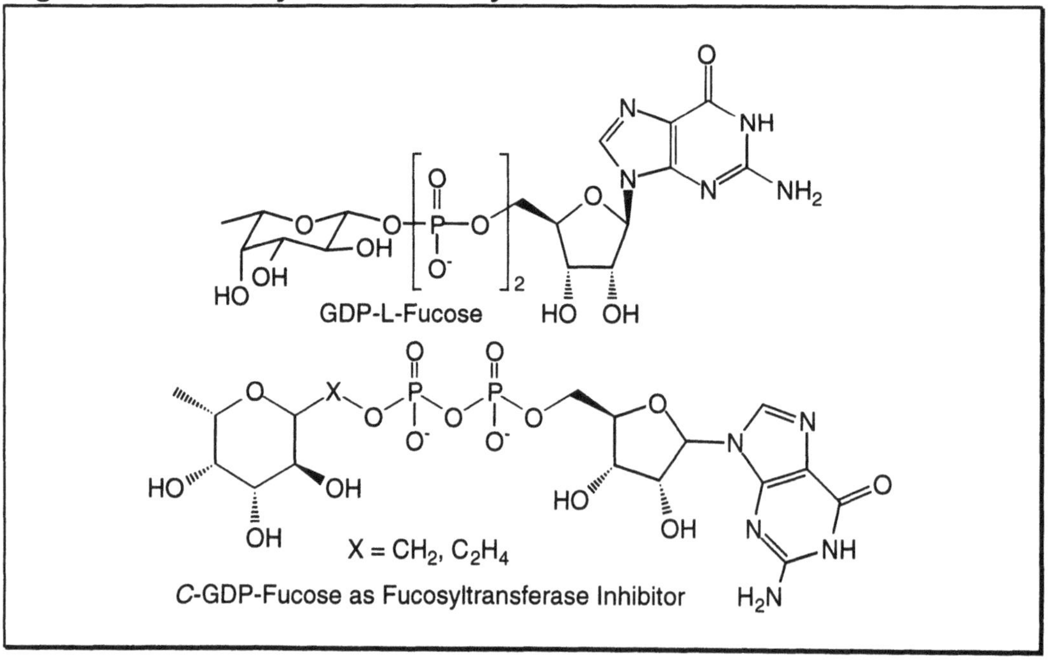

GDP-L-Fucose

$X = CH_2, C_2H_4$

C-GDP-Fucose as Fucosyltransferase Inhibitor

Since sialic acid and fucose are important in the binding interaction of sialyl Lewis[x] to the selectins,[45-50] a reasonable conjecture is that removal of fucose from sialyl Lewis[x] will completely stop the binding of leukocytes to the endothelium. Furthermore, this may be accomplished by inhibiting the activity of α-1,3-fucosyltransferase thus preventing the formation of sialyl Lewis[x].

Human α-1,3-fucosyltransferases are found in milk,[51-53] saliva,[54] and other specific cell lines.[55-57] Through cloning studies, a role for an α-1,3-fucosyltransferase in cell adhesion was identified for the biosynthesis of sialyl Lewis[x] on neutrophils.[58,59] As these studies link α-1,3-fucosyltransferase to cell adhesion and since the activated fucose species used by this enzyme is GDP-L-fucose, Luengo, *et al.,*[60] prepared the *C*-glycoside analog of GDP-fucose shown in Figure 1.4.4.

Scheme 1.4.5   *C*-**Glycosidic Analogs of**
Lipid A and Lipid X

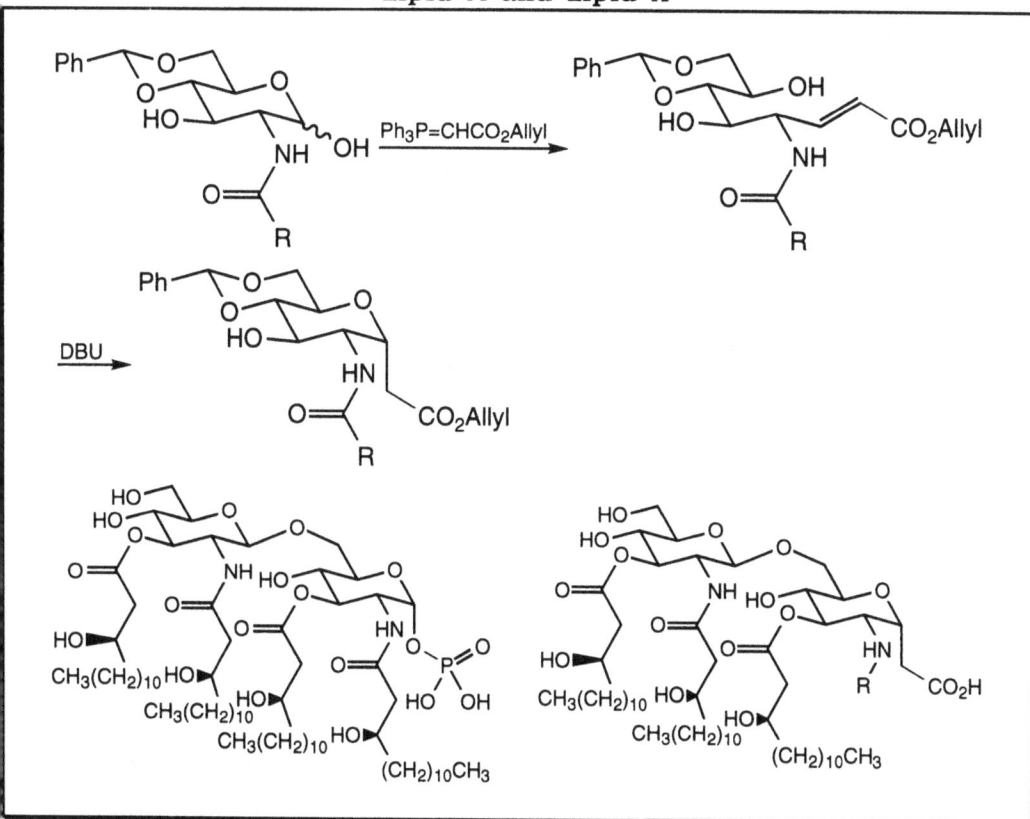

Lipopolysaccharides or endotoxins are present in the cell walls of Gram-negative bacteria and are immunostimulants responsible for many pathological effects.[61-69] Lipid A is the lipophilic terminal of endotoxins and lipid X is the reducing monosaccharide of lipid A. As the immunostimulatory effects of endotoxins are of interest, lipid A has been studied with the intent of eliminating the deleterious pathological effects.[70-72] In fact, it was not until recently that lipid X was shown to posses some of the immunostimulatory effects present in lipid A.[73-75] In 1991, Vyplel, *et al.*,[76] reported the synthesis of *C*-glycosidic analogs of these two lipopolysaccharide subunits. The syntheses, summarized in Scheme 1.4.5, involve the use of Wittig methodology in order to effect the *C*-glycosidation.   The results of this study demonstrated the feasibility of replacing the anomeric phosphate groups on these lipids with *C*-glycosidic carboxylic acid units as well as retaining at least three lipid units on any given analog in order to retain immunostimulatory activity. Endotoxic potential appeared to not be changed in the transition to *C*-glycosides.

Scheme 1.4.4   *C*-Glycosidic Ligands for *e-coli* Receptors

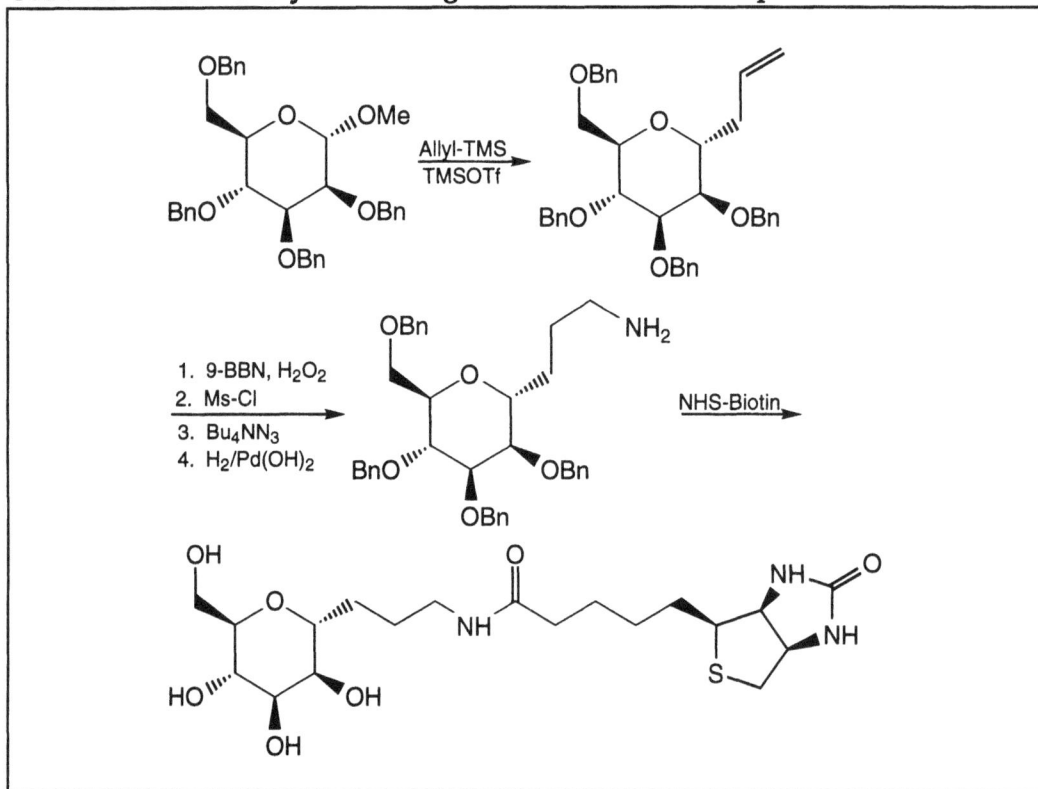

Pathogenic organisms are known to use carbohydrate receptors such as lectins to bind to cell surfaces.[77-79] Type I pili, a type of proteinaceous appendage, are found on the surface of *Escherichia coli*[80] and contain receptors capable of binding to α-linked mannosides.[81-83] In 1992, Bertozzi, *et al.*,[84] Prepared new *C*-mannosides that bind to *Escherichia coli* receptors and are able to target proteins to this organism. A specific synthesis, shown in Scheme 1.4.6, began with the per-benzylated *O*-methylmannoside shown. The initial *C*-glycosidation was accomplished on treatment with allyltrimethylsilane and trimethylsilyl triflate giving the *C*-mannoside in high yield with the α anomer favored at a ratio of 15 : 1. Elaboration of this compound to the aminopropyl mannoside was accomplished *via* conversion of the allyl group to an alcohol on treatment with 9-BBN followed by an oxidative workup. The alcohol was then mesylated and the mesylate was displaced with azide. Simultaneous reduction of the azide to the amine and removal of the benzyl groups was accomplished utilizing a palladium hydroxide mediated catalytic hydrogenation. Once formed, the aminopropylmannoside was coupled with biotin. This product was shown to totally block the bacterial mediated agglutination of yeast cells at a concentration of 7 mM. It is interesting that this compound is 9.6 times more active than the parent methyl-α-D-mannopyranoside.[85,86]

Scheme 1.4.5   ***C*-Glycosidic Polymers as Inhibitors of Influenza Virus**

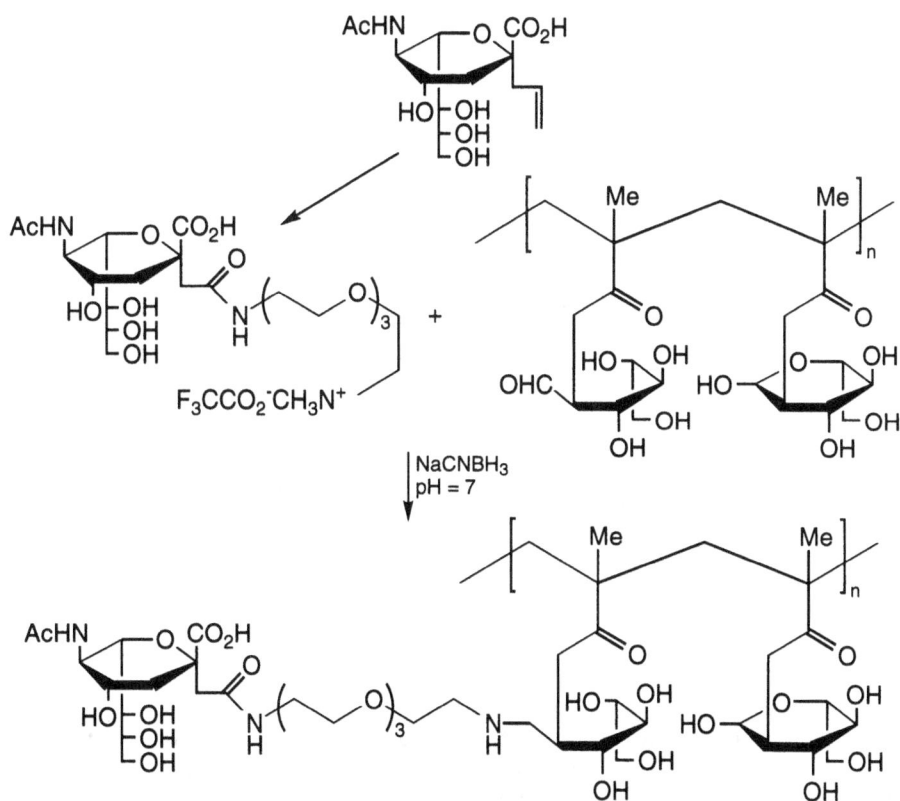

Sialic acids are found terminating cell-surface glycoproteins and glycolipids.[87-89] Pathogens use sialic acid α-glycosides in order to attach to cells prior to infection. This process is mediated by lectins, a class of glycoproteins.[77,90,91] Among the most studied of these glycoproteins is influenza hemagglutinin.[92-94] Recently, the ability of this protein to attach to erythrocytes has been inhibited by polyvalent sialic acid derivatives.[95-100] In 1992, Nagy, *et al.*,[101] reported the preparation of new carbohydrate materials resistant to neuraminidase, the enzyme present on the surface of influenza virus[92] responsible for cleavage of the hydrolyzable *O*-sialoside linkage of previously tested compounds. The fact that the newly reported compounds, prepared as shown in Scheme 1.4.5, are *C*-glycosides and, thus, resistant to neuraminidase explains their ability to inhibit the *in vitro* infectivity of the influenza virus.

Scheme 1.4.6   *C*-Glycosidic Hemagglutination Inhibitors

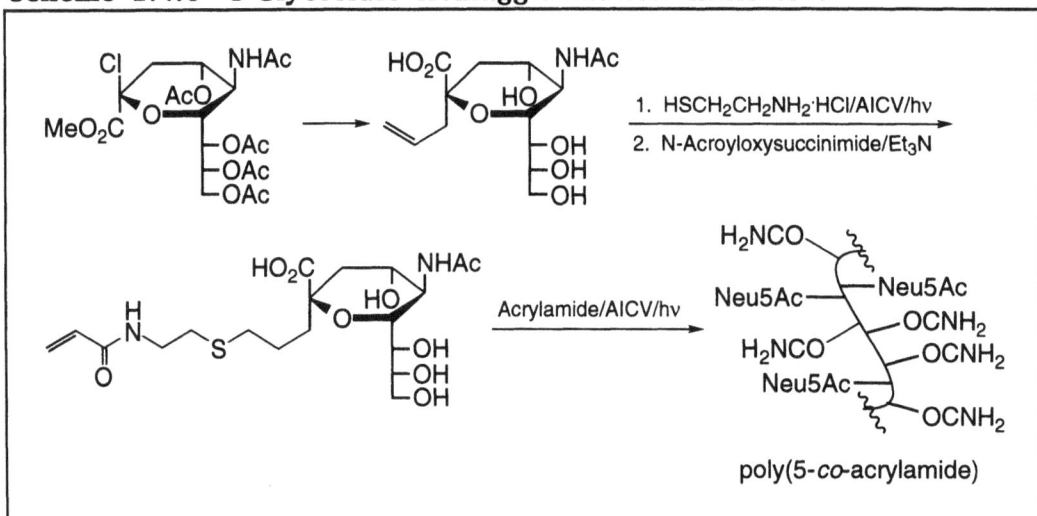

The compounds of the present study were prepared from the *C*-allyl sialic acid derivative described by Paulsen *et al.*[102] and Vasella, *et al.*[103] The ethyl ester of this compound was converted to the amide, shown. Completion of the synthesis involved reductive amination to the illustrated glucose based polymer thus giving the product. In one particular instance, the starting materials were combined to yield a product containing approximately 30% of the sialic acid analog. This polyvalent *C*-sialoside inhibited infection by influenza virus with an IC_{50} of 0.2μM.

The polymeric nature of the above example is advantageous in that many pathogen-host cell interactions are of a polyvalent nature.[104] The rational behind this relates to the much stronger binding observed when a series of weaker interactions are acting in unison. In 1993, Whitesides, *et al.*,[105] took advantage of this phenomenon and reported a study complimentary to Nagy's work utilizing a glucose-based polymer. The approach of the Whitesides study is shown in Scheme 1.4.6 and involves the preparation of sialic acid substituted acrylamide polymers.

Figure 1.4.5   ***C*-Glycosides as LTD$_4$ Antagonists**

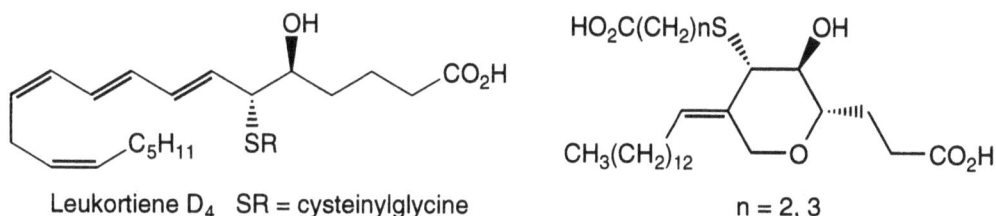

Leukortiene D$_4$ SR = cysteinylglycine

n = 2, 3

Scheme 1.4.7   **Preparation of *C*-Glycosides as LTD$_4$ Antagonists**

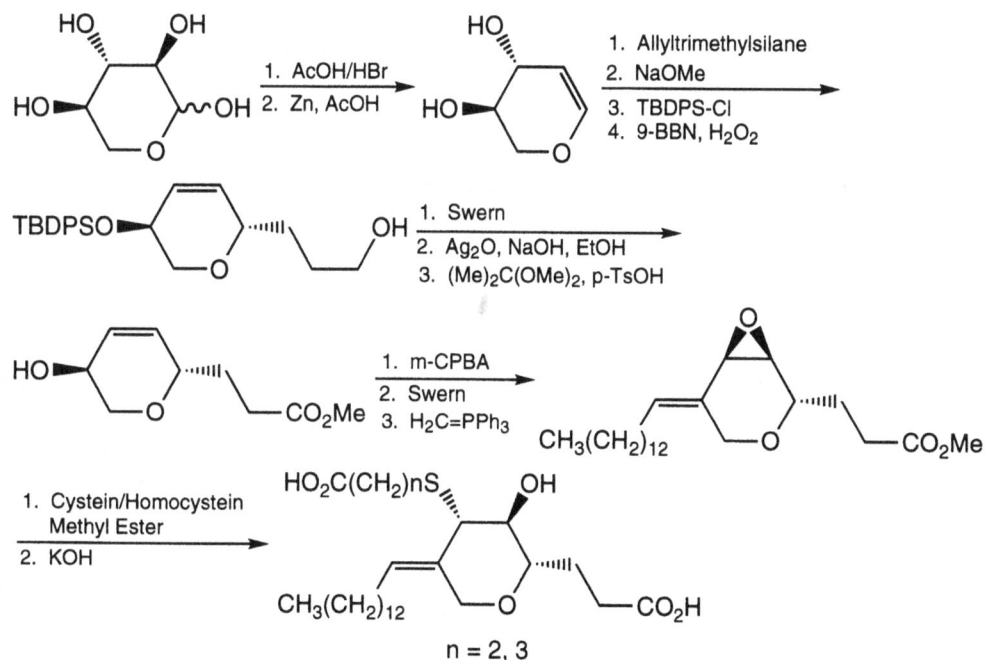

Beginning with the *C*-allyl sialoside described in Scheme 1.4.5, elaboration to the extended chain analog was accomplished by first treating with aminoethanethiol under radical conditions and then acylating the resulting amine with *N*-acroyloxysuccinimide. This compound was then added to polyacrylamide under radical forming conditions to obtain the final product. These new compounds, compared to the corresponding O-glycosidic analogs, showed approximately the same activities in inhibiting influenza induced agglutination with K_is ranging from 0.2 to 0.4 µM.

Many other approaches have been reported for the preparation of stable C-glycosidic pharmacophores. Among these is the synthesis of the leukotriene antagonists shown in Figure 1.4.5.[106] This work, shown in Scheme 1.4.7, involves the initial acetylation and dehydration of D-xylose giving the illustrated glycal. Treatment with allyltrimethylsilane gave a C-allyl glycoside. Further elaboration *via* deacetylation, silylation, and hydroxylation utilizing 9-BBN/H_2O_2 yielded the alcohol, shown. The alcohol was then oxidized to the corresponding methyl ester and the silyl group was removed. Conversion of the double bond to an epoxide on treatment with *m*-CPBA was followed by a Swern oxidation of the secondary alcohol followed by a Wittig reaction, thus providing the illustrated olefin. The synthesis was completed by opening the epoxide with cysteine or homocysteine esters followed by hydrolysis to the corresponding carboxylic acid. Once synthesized, these compounds were shown to inhibit the leukotriene D_4 induced contractions of guinea pig ileum *in vitro*.

Scheme 1.4.8   Early Steps in the Shikimate Pathway

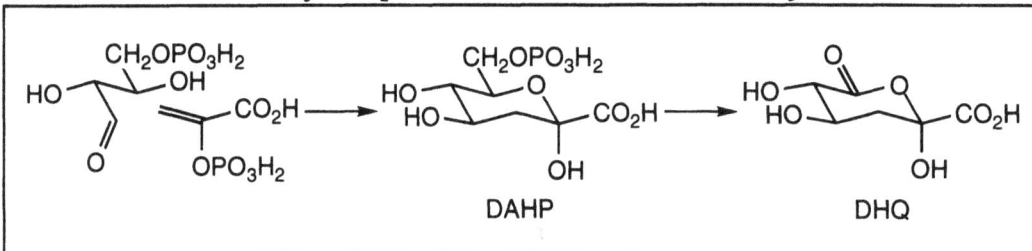

The shikimate pathway is utilized by plants to form aromatic amino acids.[107-109] In this bioprocess, shown in Scheme 1.4.8, D-erythrose-4-phosphate is combined with phosphophenylpyruvate giving 3-deoxy-D-arabino-heptulosonic acid-7-phosphate (DAHP). The next step utilizes DHQ synthase to convert DAHP to dehydroquinate (DHQ).

Figure 1.4.6   *C*-Glycosidic Tetrazole DAHP Analogs

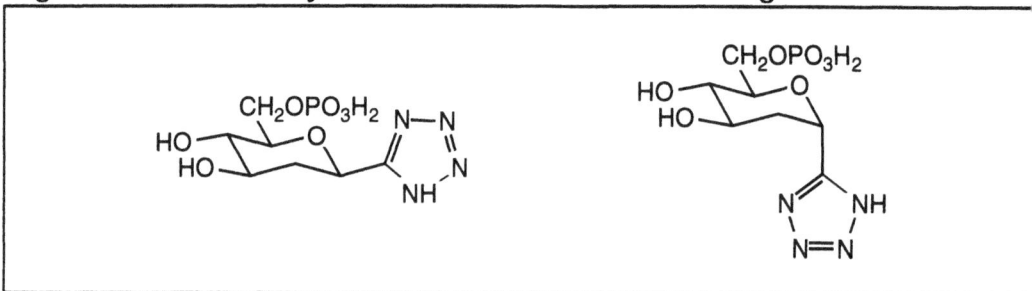

Scheme 1.4.9   *C*-Glycosidic Tetrazoles

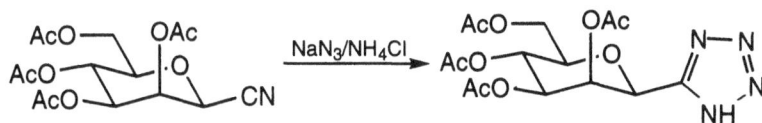

Figure 1.4.7   **Peptidoglycan Lipid-Bearing Monomer**

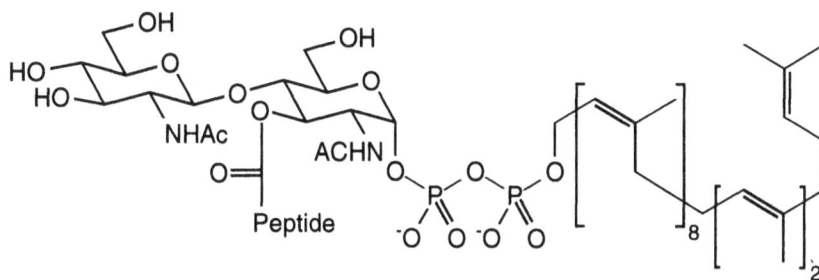

Scheme 1.4.10   *C*-Glycosidic
                         Transglycosylase Inhibitor

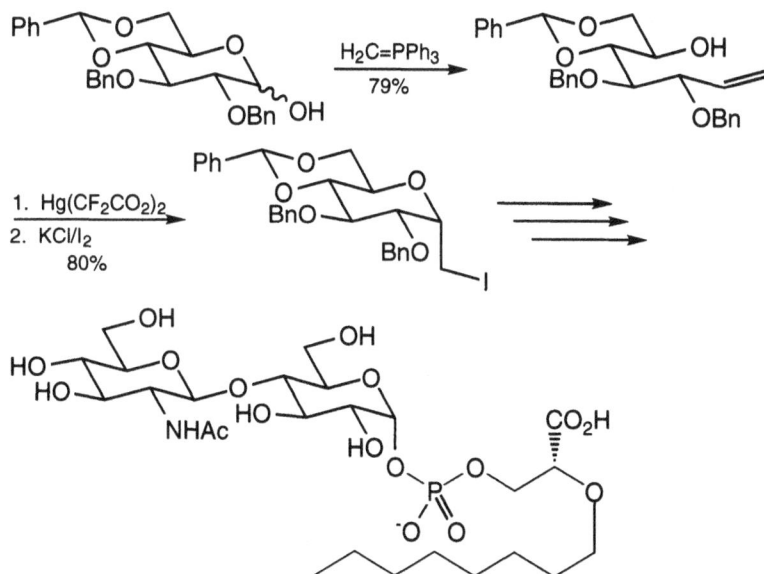

As the shikimate pathway is essential to the biology of plants, it has been identified as a potential target for herbicides.[110] As potential inhibitors of DHQ synthase, Wightman, *et al.*,[111] prepared the *C*-glycosidic tetrazole DAHP analogs shown in Figure 1.4.6. The key synthetic step, represented in Scheme 1.4.9, involved the conversion of cyanoglycosides to tetrazoles on treatment with sodium azide and ammonium chloride.[112]

Transglycosylase[113] is a penicillin-binding protein responsible for the connection of lipid-bearing carbohydrate-based monomers to peptidoglycan. This process is important in that the proliferation of bacteria requires polymerization of the peptidoglycan carbohydrate chain[114] in addition to the processing of the peptide backbone. Referring to the structure of the peptidoglycan monomer unit, represented in Figure 1.4.7, Vederas, *et al.*,[115] prepared a *C*-phosphonate disaccharide as a potential inhibitor of the transglycosylase induced peptidoglycan polymerization.

The key step in the preparation of the inhibitor is shown in Scheme 1.4.10 and involves the elaboration of the glucose derivative, shown, to the olefin on reaction with methylenetriphenylphosphorane. Once the olefin was isolated, the *C*-glycoside was obtained *via* mercury mediated cyclization followed by oxidation of the mercury with iodine. Further elaboration of the resulting iodide gave the desired *C*-phosphate disaccharide shown.

1.5 Further Reading

This chapter introduced the nomenclature, physical properties, chemistry, and potential uses of *C*-glycosides. Through specific examples and detailed analyses of synthetic strategies, this book will endeavor to explore the various chemical reactions associated with *C*-glycosides as well as methods for their preparation. In addition to the present compilation, numerous review articles[116-122] have been written that cover a wide range of subjects regarding *C*-glycosides. Although this book is meant to be extensive in its treatment of the subject, it is not exhaustive. Therefore, for more detailed treatments of the topics presented herein, the reader is referred to the cited references.

1.6 References

1.   Wang, Y.; Goekjian, P. G.; Ryckman, D. M.; Miller, W. H.; Babirad, S. A.; Kishi, Y. *J. Org. Chem.* **1992**, *57*, 482.
2.   Wu, T. C.; Goekjian, P. G.; Kishi, Y. *J. Org. Chem.* **1987**, *52*, 4819.
3.   Goekjian, P. G.; Wu, T.-C.; Kishi, Y. *J. Org. Chem.* **1991**, *56*, 6412.
4.   Wang, Y.; Goekjian, P. G.; Ryckman, D. M.; Kishi, Y. *J. Org. Chem.* **1988**, *53*, 4151.
5.   Haneda, T.; Goekjian, P. G.; Kim, S. H.; Kishi, Y. *J. Org. Chem.* **1992**, *57*, 490.
6.   Kishi, Y. *Pure & Appl. Chem.* **1993**, *65*, 771.

7.   Brakta, M.; Farr, R. N.; Chaguir, B.; Massiot, G.; Lavaud, C.; Anderson, W. R.; Sinou, D.; Daves, G. D. Jr. *J. Org. Chem.* **1993**, *58*, 2992.

8.   Grun, M.; Franz, G. *Pharmazie* **1979**, *34*, H. 10, 669.

9.   Auterhoff, H.; Graf, E.; Eurisch, G.; Alexa, M. *Arch. Pharm. (Weinheim, Ger.)* **1980**, *313*, 113.

10. Graf, E.; Alexa, M. *Planta Med.* **1980**, *38*, 121.

11. Rauwald, H.-W. *Arch. Pharm. (Weinheim, Ger.)* **1982**, *315*, 769.

12. Rauwald, H.-W.; Roth, K. *Arch. Pharm. (Weinheim, Ger.)* **1984**, *317*, 362.

13. Takahashi, Y.; Saito, K.; Yanagiya, M.; Ikura, M.; Hikichi, K.; Matsumoto, T.; Wada, M. *Tetrahedron Lett.* **1984**, *25*, 2471.

14. Abe, F.; Yamauchi, T. *Chem. Pharm. Bull.* **1986**, *34*, 4340.

15. Adinolfi, M.; Lanzetta, R.; Marciano, C. E.; Parrilli, M.; Giuliu, A. D. *Tetrahedron* **1991**, *47*, 4435.

16. Carte, B. K.; Carr, S.; DeBrosse, C.; Hemling, M. E.; Mackenzie, L.; Offen, P.; Berry, D. *Tetrahedron* **1991**, *47*, 1815.

17. Buchanan, J. G.; Wightman, R. H. *Prog. Chem. Org. Nat. Prod.* **1984**, *44*, 243.

18. Humber, D. C.; Mulholland, K. R.; Stoodley, R. J. *J. Chem. Soc. Perkin Trans. 1* **1990**, 283.

19. Schmidt, R. R. *Angew. chem.* **1986**, *98*, 213.

20. Schmidt, R. R. *Angew. chem. Int. Ed. Engl.* **1986**, *25*, 212.

21. Schmidt, R. R. *Pure Appl. Chem.* **1989**, *61*, 1257.

22. Paulson, H. *Angew. chem.* **1990**, *102*, 851.

23. Paulson, H. *Angew. chem. Int. Ed. Engl.* **1990**, *29*, 823.

24. Hakomori, S. *J. Biol. Chem.* **1990**, *265*, 823.

25. Schmidt, R. R.; Dietrich, H. *Angew. Chem. Int. Ed. Engl.* **1991**, *30*, 1328.

26. Lehmann, J.; Ziser, L. *Carbohydr. Res.* **1989**, *188*, 45.

27. Fritz, H.; Lehmann, J.; Schlesslmann, P. *Carbohydrate Res.* **1983**, *113*, 71.

28. Hehre, E. J.; Genghoff, D. S.; Sternlicht, H.; Brewer, C. F. *Biochemistry* **1977**, *16*, 1780.

29. Flowers, H. M.; Sharon, N. *Adv. Enzymol.* **1979**, *48*, 30.

30. Blake, C. C. F.; Johnson, L. N.; Meir, G. A.; North, A. T. C.; Phillips, D. C.; Sarma, V. R. *Proc. R. Soc. (London), Ser. B* **1967**, *167*, 378.

31. Phillips, D. C. *Sci. Am.* **1966**, *215*, 78.

32. Lehmann, J.; Schlesselmann, P. *Carbohydrate Res.* **1983**, *113*, 93.

33. Sinott, M. L. *CRC Crit. Rev. Biochem.* **1982**, *12*, 327.

34. Myers, R. W.; Lee, Y. C. *Carbohydrate Res.* **1986**, *152*, 143.

35. Martin, J. L.; Veluraja, K.; Ross, K.; Johnson, L. N.; Fleet, G. W. J.; Ramsden, N. G.; Bruce, I.; Orchard, M. G.; Oikonomakos, N. G.; Papageorgiou, A. C.; Leonidas, D. D.; Tsitoura, H. S. *Biochemistry* **1991**, *30*, 10101.

36. Acharya, K. R.; Stuart, D. I.; Varvil, K. M.; Johnson, L. N. *Glycogen Phosphorylase b*, World Scientific Press, Singapore, 1st ed. **1991**.

37. Martin, J. L.; Johnson, L. N.; Withers S. G. *Biochemistry* **1990**, *29*, 10745.

38. Pougny, J.-P.; Nassr, M. A. M.; Sinnay, P. *J. Chem. Soc., Chem. Commun.* **1981**, 375.
39. Watson, K. A.; Mitchell, E. P.; Johnson, L. N.; Son, J. C.; Bichard, C. J.; Fleet, G. W. J.; Ford, P.; Watkin, D. J.; Oikonomakos. *J. Chem. Soc. Chem. Comm.* **1993**, 654.
40. Paulson, J. C.; Colley, K. J. *J. Biol. Chem.* **1989**, *264*, 17615.
41. Kornfeld, R.; Kornfeld, S. *Annu. Rev. Biochem.* **1985**, *54*, 631.
42. Sadler, J. E. *Biology of Carbohydrates* **1984**, *2*, 87.
43. Basu, S.; Basu, M. *Glycoconjugates* **1982**, *3*, 265.
44. Lowe, J. B.; Stoolman, L. M.; Nair, R. P.; Larsen, R. D.; T.L. Berhend, T. L.; Marks, R. M. *Cell* **1990**, *63*, 475.
45. Berg, E. L.; Robinson, M. K.; Mansson, O.; Butcher, E. C.; Magnani, J. L. *J. Biol. Chem.* **1991**, *265*, 14869.
46. Tyrrell, D.; James, P.; Rao, N.; Foxall, C.; Abbas, S.; Dasgupta, F.; Nashed, M.; Hasegawa, A.; Kiso, M.; Asa, D.; Kidd, J.; Brandley, B. K. *Proc. Natl. Acad. Sci. USA.* **1991**, *88*, 10372.
47. Polley, M. J.; Phillips, M. L.; Wayner, E.; Nucelman, E.; Kinghal, A. K.; Hakomori, S.; Paulson, J. C. *Proc. Natl. Acad. Sci. USA* **1991**, *88*, 6224.
48. Imai, Y.; Singer, M. S.; Fennie, C.; Lasky, L. A.; Rosen, S. D. *J. Cell Biol.* **1991**, *113*, 1213.
49. Stoolman, L. M.; Rosen, S. D. *J. Cell Biol.* **1983**, *96*, 722.
50. True, D. D.; Singer, M. S.; Lasky, L. A.; Rosen, S. D. *J. Cell Biol.* **1990**, *111*, 2757.
51. Prieels, J. P.; Monnom, D.; Dolmans, M.; Beyer, T. A.; Hill, R. L. *J. Biol. Chem.* **1981** *256*, 10456.
52. Johnson, P. H.; Watkins, W. M. *Biochem. Soc. Trans.* **1982**, *10*, 445.
53. Eppenberger-Castori, S.; Lotscher, H.; Finne, J. *Glycoconjugate J.* **1989**, *8* 264.
54. Johnson, P. H.; Yates, A. D.; Watkins, W. M. *Biochem. Biophys. Res Commun.* **1981**, *100*, 1611.
55. Holmes, E. H.; Levery, S. B. *Arch. Biochem. Biophys.* **1989**, *274*, 633.
56. Kukowska-Latallo, J. F.; Larsen, R. D.; Nair, R. P.; and Lowe, J. B. *Genes Dev* **1990**, *4*, 1288.
57. Stroup, G. B.; Anumula, K. R.; Kline, T. F.; Caltabiano, M. M. *Cancer Res.* **1990**, *50*, 6787.
58. Macher, B. A.; Holmes, E. H.; Swiedler, S. J.; Stults, C. L. M.; Srnka, C. A. *Glycobiology* **1991**, *1*,6.
59. Kuijpers, T. W. *Blood* **1993**, *81*, 873.
60. Luengo, J. I.; Gleason, J. G. *Tetrahedron Lett.* **1992**, *33*, 6911.
61. Galanos, C.; Lüderitz, O.; Rietschel, E. T.; Westphal, O.; Brade, H.; Brade, L.; Freudenberg, M.; Schade, U.; Imoto, M.; Yoshimura, H.; Kusumoto, S.; Shiba, T. *Eur. J. Biochem.* **1985**, *148*, 1.
62. Homma, J. Y.; Matsura, M.; Kanegasaki, S.; Kawakubo, Y.; Kojima, Y.; Shibukawa, N.; Kumazawa, Y.; Yamamoto, A.; Tanamoto, K.; Yasuda, T.;

Imoto, M.; Yoshimura, H.; Kusomoto, S.; Shiba, T. *J. Biochem.* **1985**, *98*, 395.

63.   Kotani, S.; Takada, H.; Tsujimoto, M.; Ogawa, T.; Harada, K.; Mori, Y.; Kawasaki, A.; Tanaka, A.; Nagao, S.; Tanaka, S.; Shiba, T.; Kusumoto, S.; Imoto, M.; Yoshimura, H.; Yamamoto, M.; Shimamoto, T. *Infect. Immun.* **1984**, *45*, 293.

64.   Kotani, S.; Takada, H.; Tsujimoto, M.; Ogawa, T.; Takahashi, I.; Ikeda, T.; Otsuka, K.; Shimauchi, H.; Mashimo, J.; Nagao, S.; Tanaka, A.; Harada, K.; Nagaki, K.; Kitamura, H.; Shiba, T.; Kusumoto, S.; Imoto, M.; Yoshimura, H. *Infec. Immun.* **1985**, *49*, 225.

65.   Kusumoro, S.; Yamamoto, M.; Shiba, T. *Tetrahedron Lett.* **1984**, *25*, 3727.

66.   Galanos, C.; Rietschel, E. T.; Lüderitz, O.; Westphal, O.; Kim, Y. B.; Watson, D. W. *Eur. J. Biochem.* **1975**, *54*, 603.

67.   Levin, J.; Poore, T. E.; Zauber, N. P.; Oser, R. S. *N. Engl. J. Med.* **1970**, *283*, 1313.

68.   Galanos, C.; Lüderitz, O.; Rietschel, E. T.; Westphal, O. In *International Review of Biochemistry, Biochemistry of Lipids, II*; Goodwin, T. W., Ed.; University Park Press: Baltimore, MD 1977; Vol. 14, pp239-335.

69.   Takahashi, I.; Kotani, S.; Takada, H.; Tsujimoto, M.; Ogawa, T.; Shiba, T.; Kusumoto, S.; Yamamoto, M.; Hasegawa, A.; Kiso, M.; Nishijima, M.; Harada, K.; Tanaka, S.; Okumura, H.; Tamura, T. *Infect. Immun.* **1987**, *65*, 57.

70.   Nakamoto, S.; Takahashi, T.; Ikeda, K.; Achiwa, K. *Chem. Pharm. Bull.* **1987**, *35*, 4517.

71.   Kumazawa, Y.; Matsura, M.; Maruyama, T.; Kiso, M.; Hasegawa, A. *Eur. J. Immunol.* **1986**, *16*, 1099.

72.   Lasfargues, A.; Charon, D.; Tarigalo, F.; Ledu, A. L.; Szabo, L.; Chaby, R. *Cell. Immunol.* **1986**, *98*, 8.

73.   Nishijima, M.; Amano, F.; Akamatsu, Y.; Akegawa, K.; Tokunaga, T.; Raetz, C. R. H. *Proc. Natl. Acad. Sci. USA* **1985**, *82*, 282.

74.   Amano, F.; Nishijima, M.; Akamatsu, Y. *J. Immunol.* **1986**, *136*, 4122.

75.   Sayers, T. J.; Macher, I.; Chung, J.; Kugler, E. *J. Immunol.* **1987**, *138*, 2935.

76.   Vyplel, H.; Scholz, D.; Macher, I.; Schindlmaier, K.; Schutze, E. *J. Med. Chem.* **1991**, *34*, 2759.

77.   Sharon, N.; Lis, H. *Science* **1989**, *246*, 227.

78.   Sharon, N.; Lis, H. *Lectins*, Chapman and Hall, London, 1989.

79.   Mirelman, D. *Microbial Lectins and Agglutins: Properties and Biological Activity*, D. Mirelman, Wiley-Interscience, Sons, New York, 1986.

80.   Clegg, S.; Gerlach, G. F. *J. Bacteriol.* **1987**, *169*, 934.

81.   Firon, N.; Ofek, I.; Sharon, N. *Infection and Immunity* **1984**, *43*, 1088.

82.   Firon, N.; Ofek, I.; Sharon, N. *Carbohydr. Res.* **1983**, *120*, 235.

83.   Ofek, I.; Mirelman, D.; Sharon, N. *Nature* **1977**, *265*, 623.

84.   Bertozzi, C.; Bednarski, M. *Carbohydrate Res.* **1992**, *223*, 243.

85.   Esdat, Y.; Ofek, I.; Yashouv-Gan, Y.; Sharon, N.; Mirelman, D. *Biochem. Biophys. Res. Commun.* **1978**, *85*, 1551.

86. Firon, N.; Ofek, I.; Sharon, N. *Biochem. Biophys. Res. Commun.* **1982**, *105*, 1426.

87. Cornfield, A. P.; Schauer, R. Occurrence of Sialic Acids. In *Sialic Acids; Chemistry, Metabolism and Function*; Schauer, R., Ed.; Springer-Verlag Publishing Co.: New York, 1982; Vol. 10, pp 5-39.

88. McGuire, E. J. *Biological Roles of Sialic Acid*; Rosenberg, A., Schengrund, C.-L., Eds.; Plenum Publishing Co.: New York, 1976.

89. Sharon, N. The Sialic Acids. In *Complex Carbohydrates*; Addison-Wesley Publishing Co.: London, 1975; pp142-154.

90. Ruigrok, R. W. H.; André, P. J.; Hoof Van Huysduynen, R. A. M.; Mellema, J. E. *J. Gen. Virol.* **1985**, *65*, 799.

91. Markwell, M. A. K. Viruses as Hemagglutinins and Lectins. In *Microbial Lectins and Agglutinins: Properties and Biological Activity*; Mirelman, E., Ed.; Wiley Series in Ecological and Applied Microbiology; Wiley-Interscience Publication, John Wiley and Sons Publishing Co.: New York, 1986; pp 21-53.

92. Wiley, D. C.; Skehel, J. J. *Annu. Rev. Biochem.* **1987**, *56*, 365.

93. Weis, W.; Brown, J. H.; Cusack, S.; Paulson, J. C.; Skehel, J. J.; Wiley, D. C. *Nature* **1988**, *333*, 426.

94. Sauter, N. K.; Bednarski, M. D.; Wurzburg, B. A.; Hanson, J. E.; Whitesides, G. M.; Skehel, J. J.; Wiley, D. C. *Biochemistry* **1989**, *28*, 8388.

95. Spaltenstein, A.; Whitesides, G. M. *J. Am. Chem. Soc.* **1991**, *113*, 686.

96. Glick, G. D.; Knowles, J. R. *J. Am. Chem. Soc.* **1991**, *113*, 4701.

97. Sabesan, S.; Duus, J. O.; Domaille, P.; Kelm, S.; Paulson, J. C. *J. Am. Chem. Soc.* **1991**, *113*, 5865.

98. Gamain, A.; Chomik, M.; Laferriere, C. A.; Roy, R. *Can. J. Microbiol.* **1991**, *37*, 233.

99. Byramova, N. E.; Mochalova, L. V.; Belyanchikov, I. M.; Matrosovich, M. N.; Bovin, M. V. *J. Carbohydr. Chem.* **1991**, *10*, 691.

100. Matrosovich, M. N.; Mochalova, L. V.; Marinina, V. P.; Byramova, N. E.; Bovin, M. V. *FEBS Lett.* **1990**, *272*, 209.

101. Nagy, J. O.; Wang, P.; Gilbert, J. H.; Schaefer, M. E.; Hill, T. G.; Gallstrom, M. R.; Bednarski, M. D. *J. Med. Chem.* **1992**, *35*, 4501.

102. Paulsen, H.; Matschulat, P. *Liebigs Ann. Chem.* **1991**, 487.

103. Wallimann, K.; Vasella, A. *Helv. Chim. Acta* **1991**, *74*, 1520.

104. Matrosovich, M. N. *FEBS Lett.* **1989**, *252*, 1.

105. Sparks, M. A.; Williams, K. W.; Whitesides, G. M. *J. Med. Chem.* **1993**, *36*, 778.

106. Sabol, J.; Cregge, R. J. *Tetrahedron Lett.* **1989**, *30*, 6271.

107. Weiss, U.; Edwards, J. M. *The Biosynthesis of Aromatic Compounds*, Wiley, New York, 1980.

108. Ganem, B. *Tetrahedron* **1978**, *34*, 3353.

109. Dewick, P. M. *Nat. Prod. Rep.* **1992**, *9*, 153.

110. Kishmore, G. M.; Shah, D. M. *Ann. Rev. Biochem.* **1988**, *57*, 627.

111. Buchanan, J. G.; Clelland, A. P. W.; Johnson, T.; Rennie, R. A. C.; Wightman, R. H. *J. Chem. Soc. Perkin Trans. 1* **1992**, 2593.
112. Farkas, I.; Szabó, I. F.; Bognár *Carbohydr. Res.* **1977**, *56*, 404.
113. Taku, A.; Stuckey, M.; Fan, D. P. *J. Biol. Chem.* **1982**, *257*, 5018.
114. Van Heijenoort, Y.; Leduc, M.; Singer, H.; Van Heijenoort, J. *J. Gen. Microbiol.* **1987**, *133*, 667.
115. Qiau, L.; Vederas, J. C. *J. Org. Chem.* **1993**, *58*, 3480.
116. Herscovici, J.; Antonakis, K. *Stud. Nat. Prod. Chem.* **1992**, *10*, 337.
117. Postema, M. H. D. *Tetrahedron* **1992**, *48*, 8545.
118. Daves, G. D. Jr. *Acc. Chem. Res.* **1990**, *23*, 201.
119. *Carbohydr. Res.* **1987**, *171*. Special Issue for C-Glycosides.
120. Hacksell, U.; Daves, G. D. Jr. *Progress in Medicinal Chemistry* **1985**, *22*, 1.
121. Hanessian, S.; Pernet, A. G. *Adv. Chem. Biochem.* **1976**, *33*, 111.
122. *Carbohydrate Chemistry, Specialist Periodical Reports, Royal Chemical Society* **1968-1990**, *Vol. 1-24*, Sections on C-Glycosides.

2   Introduction

One of the most widely used methods for the formation of *C*-glycosides involves electrophilic species derived from the sugar component. Some representative examples, shown in Figure 2.0.1, include oxonium compounds, lactones, enones, carbenes, and enol ethers. Once formed, nucleophiles can be induced to react with these reactive species thus providing tremendous diversity in the types of *C*-glycosides accessible. This chapter focuses on various approaches made use of in the formation of *C*-glycosides *via* the addition of nucleophiles to carbohydrate-derived electrophiles.

Figure 2.0.1   Examples of Electrophilic Substitutions

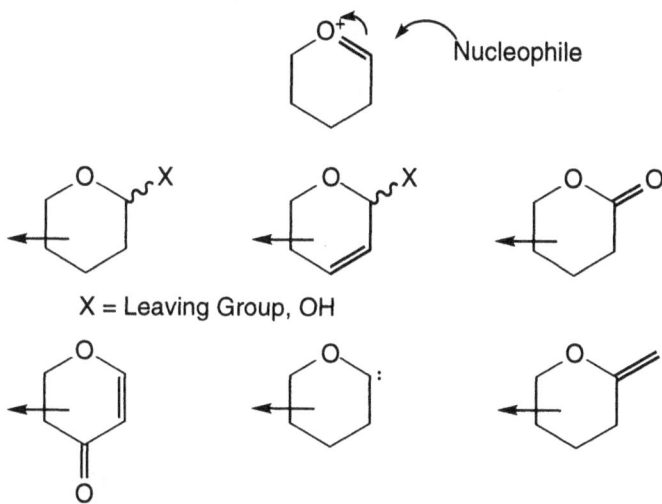

X = Leaving Group, OH

2.1   Anomeric Activating Groups and Stereoselectivity

When considering the formation of *C*-glycosides by electrophilic substitution, several direct comparisons can be made to the similar reactions utilized in the preparation of *O*-glycosides. Specifically, any activating group that is good enough for the formation of *O*-glycosides may be used in the preparation of *C*-glycosides. This rational arises from the fact that *C*-glycosidations *via* electrophilic substitution proceed through mechanisms similar to those observed in the corresponding *O*-glycosidations. Additionally, since *C*-glycoside technology generally utilizes slightly more stringent conditions, less reactive groups such as OAc and OTMS may also be used. One particular advantage to this chemistry is noted by the fact that heavy metal reagents are not required for the formation of *C*-glycosides. Furthermore, this chemistry is widely applicable to both benzyl and acetate protected sugars.

One concern applicable to all aspects of organic chemistry is the question of the stereochemical outcome of reactions. In the case of *C*-glycosides, since no anomeric effect exists, the stereochemistry is completely dictated by stereoelectronic effects. Additionally, it should be noted that neighboring group participation is not a predominant factor. The major consequence of these observations is that axially (α) substituted *C*-glycosides are far more accessible than the corresponding equatorial (β) isomers.

2.2 Cyanation Reactions

Direct cyanations to sugars or sugar derivatives, represented in Figure 2.2.1, effect one carbon extensions. The resulting nitriles are then available for a number of further reactions including cycloadditions, hydrolyses to acids, and reduction to aminomethyl groups. Unfortunately, there are no stereospecific methods available for these reactions.

Figure 2.2.1   Example of Cyanation Reactions

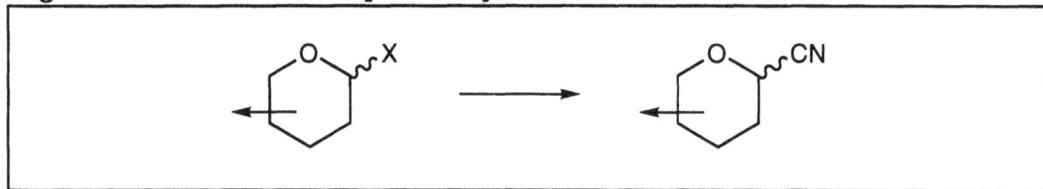

In addition to the formation of anomeric nitriles, similar chemistry allows for the formation of anomeric isocyanides. Requirements for these reactions include utilizing acetates, halides, and imidates as activating groups. Some Lewis acids used to effect these transformations include SnCl$_4$, TMSOTf, TMSCN, Et$_2$AlCN, and HgCN. Additionally, both benzylated and acetylated sugars may be used.

2.2.1   Cyanation Reactions on Activated Glycosides

Among the most basic of methods for the formation of *C*-cyanoglycosides involves the reaction of 1-*O*-acetoxyglycosides with TMSCN. Examples of this reaction, shown in Scheme 2.2.1,[1,2] incorporate the use of both benzyl and acetyl protected sugars. Where benzylated sugars were used, the results provided a 1 : 1 ratio of anomers. However, when mannose pentaacetate was used, the only isolated compound possessed the α configuration. This observation can be explained by the involvement of the adjacent acetate group as illustrated in figure 2.2.2. The result of this intermediate is apparent in that the approach of nucleophiles is only possible from the α face.

Scheme 2.2.1   Cyanations of 1-*O*-Acetoxyglycosides

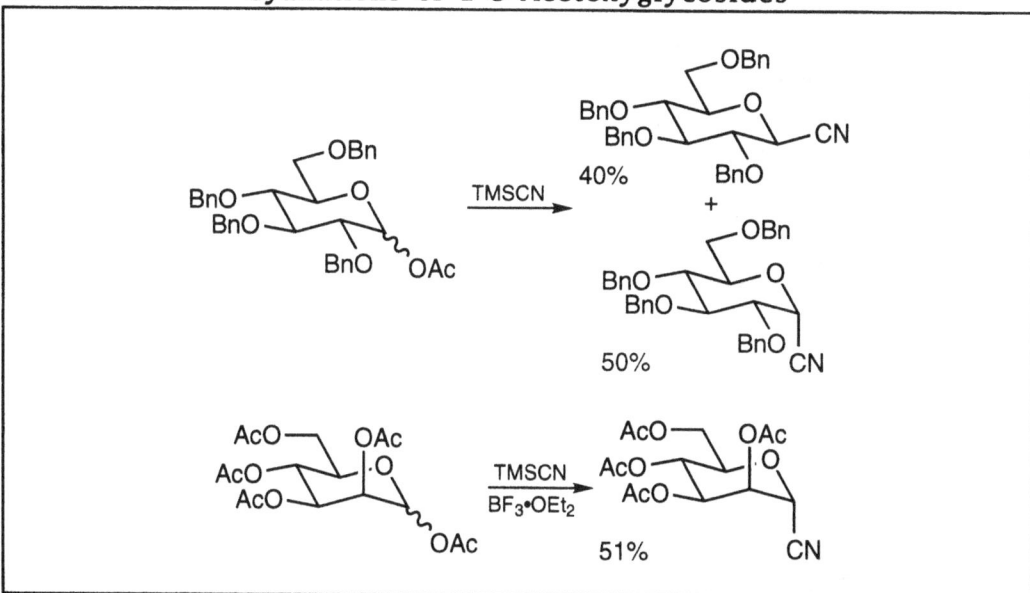

Figure 2.2.2   Cyanations with Acetate Participation

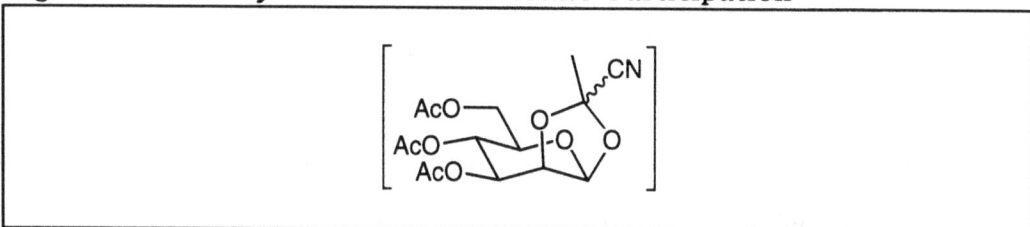

In addition to acetates as glycosidic activating groups, trichloroacetimidates are useful. This activating group is advantageous in the formation of *C*-cyanoglycosides as demonstrated by the near exclusive formation of the α anomer coupled with an 87% isolated yield. This specific example, shown in Scheme 2.2.2 was reported by Schmidt, *et al.*,[3] and utilized the α-trichloroacetimidate of 2,3,4,6-tri-*O*-benzyl glucose available from the reaction of trichloroacetonitrile with the protected glucose.

Scheme 2.2.2   Cyanations of Trichloroacetimidates

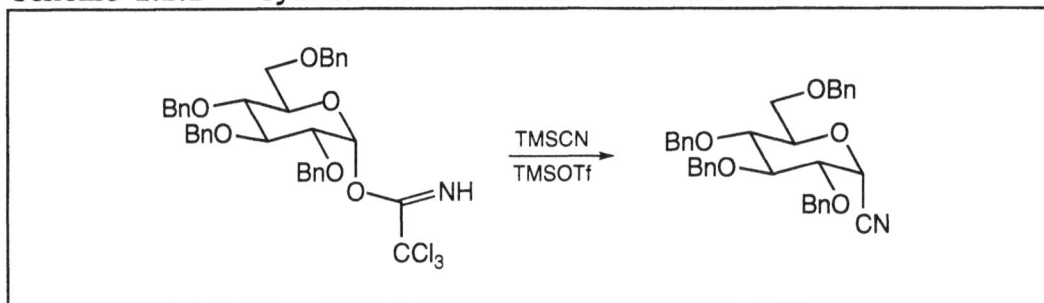

Scheme 2.2.3   Cyanations of Fluoroglycosides

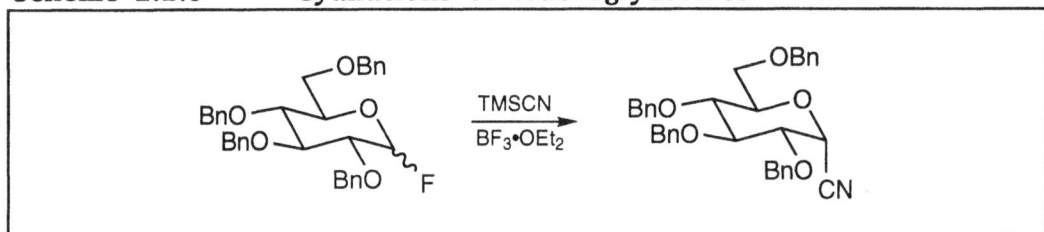

Scheme 2.2.4   Cyanations of Fluoroglycosides

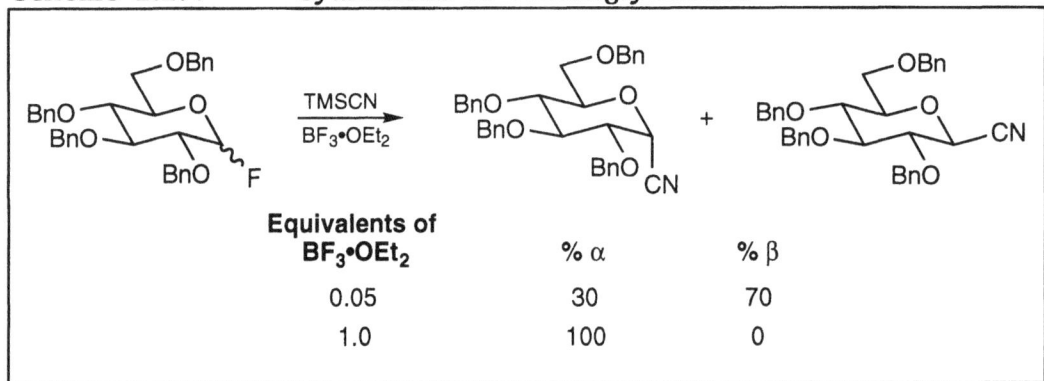

Equivalents of BF₃•OEt₂	% α	% β
0.05	30	70
1.0	100	0

As a final example of specific glycosidic activating groups utilized for the preparation of cyanoglycosides, fluoroglycosides are addressed. In the formation of *C*-glycosides, fluoride is a formidable activating groups. More examples utilizing this activating group are addressed later. For now, the reaction shown in Scheme 2.2.3 provides an excellent introductory example.

Utilizing an anomeric mixture of fluoroglucosides, Nicolau, *et al.*,[4] induced a reaction with TMSCN catalyzed by borontrifluoride etherate thus providing the desired cyanoglycosides. The reaction provided a 90% isolated yield containing a 3 : 1 ratio of anomers favoring the α configuration. This anomeric preference appears to be dependent on the amount of Lewis acid used in the reaction. Specifically, in an example reported by Araki, *et al.*,[5] shown in Scheme 2.2.4, 0.05 equivalents of borontrifluoride etherate effected a 3 : 7 ratio of anomers favoring the β configuration. However, when 1.0 equivalent of the catalyst was used the only identified product possessed the α configuration.

2.2.2   Cyanation Reactions on Glycals

Where the direct use of glycosides is widely useful, glycals have provide more complimentary and, in many cases, more stereoselective results. For example, as shown in Scheme 2.2.5, Grynkiewicz, *et al.*,[6] converted tri-*O*-acetyl glucal to the illustrated cyanoglycoside. The reaction proceeded in a 79% yield with only one prominent isomer. In a similar reaction, shown in Scheme 2.2.6, Nicolau, *et al.*,[7] accomplished a stereoselective cyanation reaction beginning with a 1-methyl substituted glycal. This reaction proceeded in an 82% yield with an additional 11% accounted for in the isolation of the corresponding isocyanide.

Scheme 2.2.5   **Cyanations of Glycals**

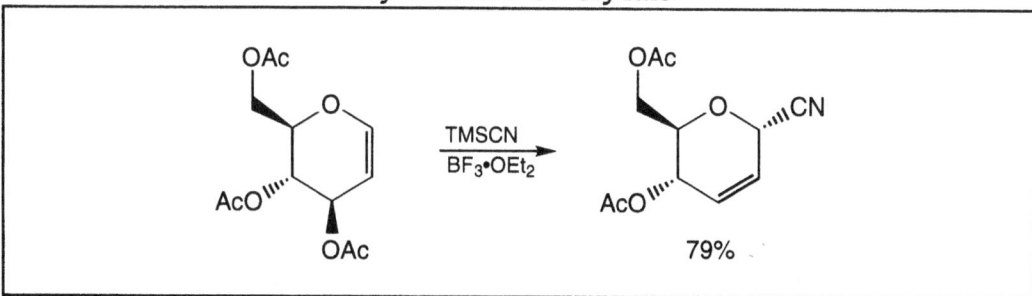

Scheme 2.2.6   **Cyanations of Glycals**

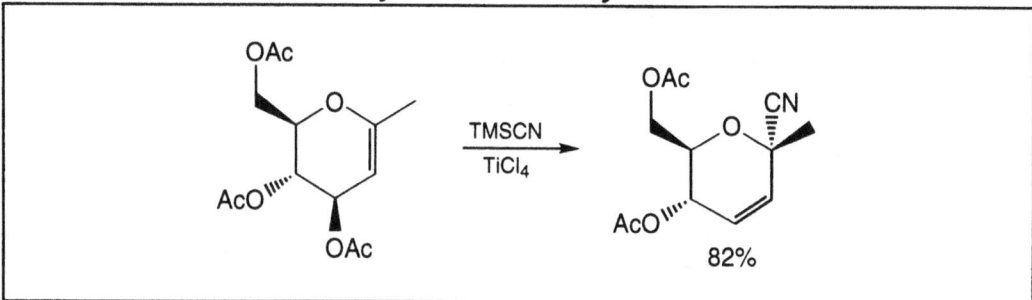

Scheme 2.2.7   **Arylsulfonium Glycals**

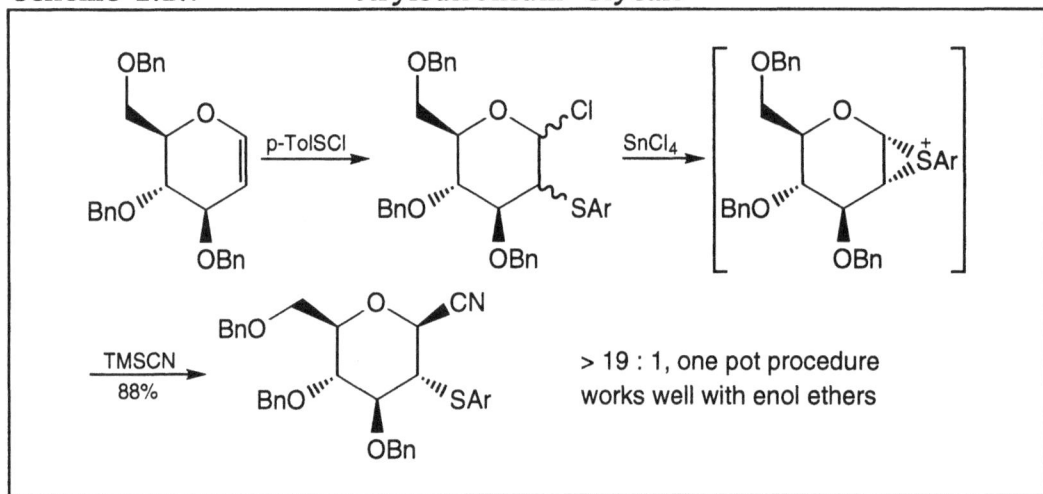

In one final example involving glycals, arylsulfonium compounds are shown to be useful intermediates. As illustrated in Scheme 2.2.7, Smolyadova, et al.,[8] formed arylsulfonium compounds on treatment of glycals with *p*-toluene chlorosulfide. The addition proceeded stereoselectively allowing for an S_N2 ring opening of the sulfonium ring yielding a *C*-1-cyano-2-arylsulfide glycoside in 88% yield. The stereochemical outcome resulted in a 19 : 1 bias towards the β isomer.

2.2.3   *Cyanation Reactions on Activated Furanosides*

Scheme 2.2.8   **Cyanations of Furanoses**

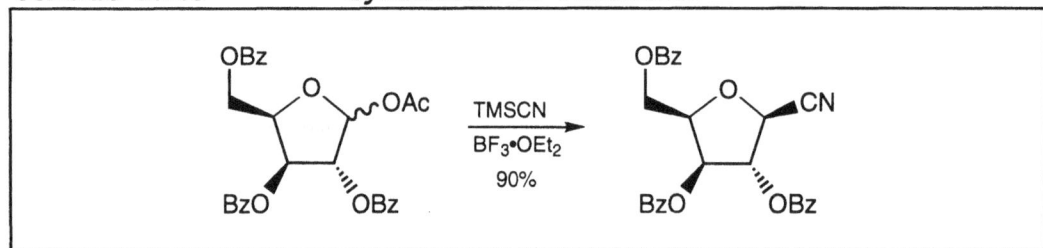

Until now, only the chemistry regarding the addition of TMSCN to pyranose sugars has been discussed. Such chemistry is, however, easily applicable to the corresponding furanose forms. These reactions have been demonstrated with benzoyl protected furanoses catalyzed by borontrifluoride etherate (Scheme 2.2.8)[2] and with benzyl protected sugars catalyzed by trityl perchlorate (Scheme 2.2.9).[9] In this second example, the anomeric ratio was shown to be solvent dependent with the α isomer favored by 93% when ether was used. Dimethoxyethane, on the other hand, allowed 37% of the product mixture to obtain the β configuration.

As with acetoxy furanosides, reactions with trimethylsilyl cyanide have also been applied to fluoro furanosides. These reactions, illustrated in Scheme 2.2.10,[5] unlike those mentioned above, produced high yields of 1 : 1 anomeric ratios even when run at -78°C.

Scheme 2.2.9   **Cyanations of Furanoses**

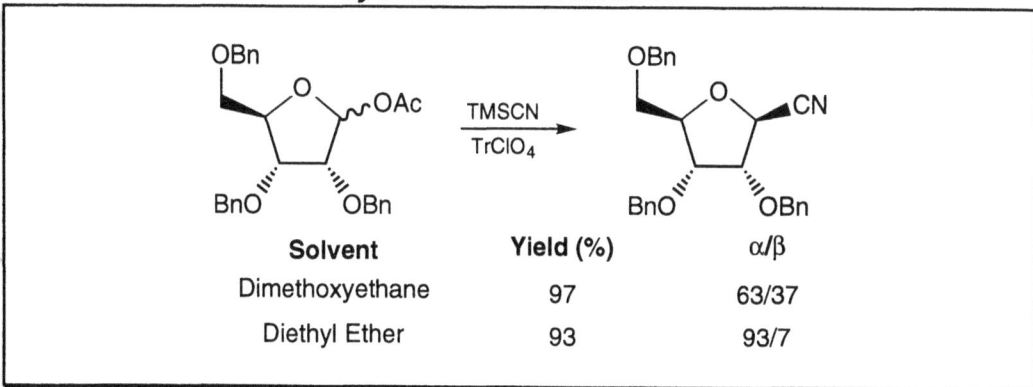

Solvent	Yield (%)	α/β
Dimethoxyethane	97	63/37
Diethyl Ether	93	93/7

Scheme 2.2.10   **Cyanations of Furanoses**

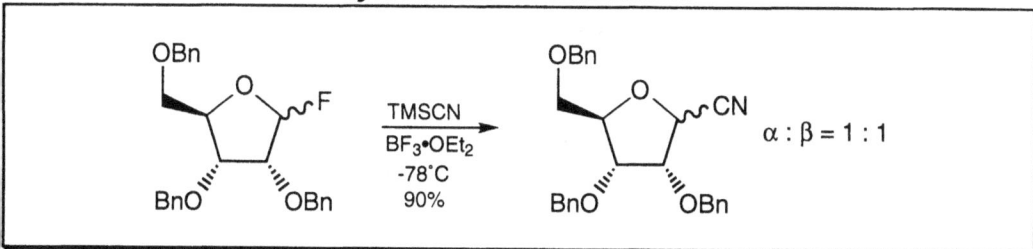

2.2.4   *Cyanation Reactions with Metallocyanide Reagents*

TMSCN is not the only reagent useful for the formation of cyanoglycosides. Complimentary and, perhaps, equally important is the use of aluminum reagents. As applied to pyranose sugars, Grierson, *et al.*,[10] demonstrated that the stereochemical outcome of the cyanation reaction with diethylaluminum cyanide was temperature dependent. Thus, as shown in Scheme 2.2.11, at room temperature, a mixture of α and β isomers were formed while, at reflux, the predominant isomer possessed the α configuration.

The use of aluminum reagents is also applicable to fluoroglycosides as demonstrated by the example shown in Scheme 2.2.12. Nicolau, *et al.*,[4] added dimethylaluminum cyanide to the illustrated fluoroglucoside.   When borontrifluoride etherate was used to catalyze the reaction, a 96% yield was obtained with a 10 : 1 ratio of anomers favoring the α configuration.

Scheme 2.2.11   Glycal Cyanations with Et₂AlCN

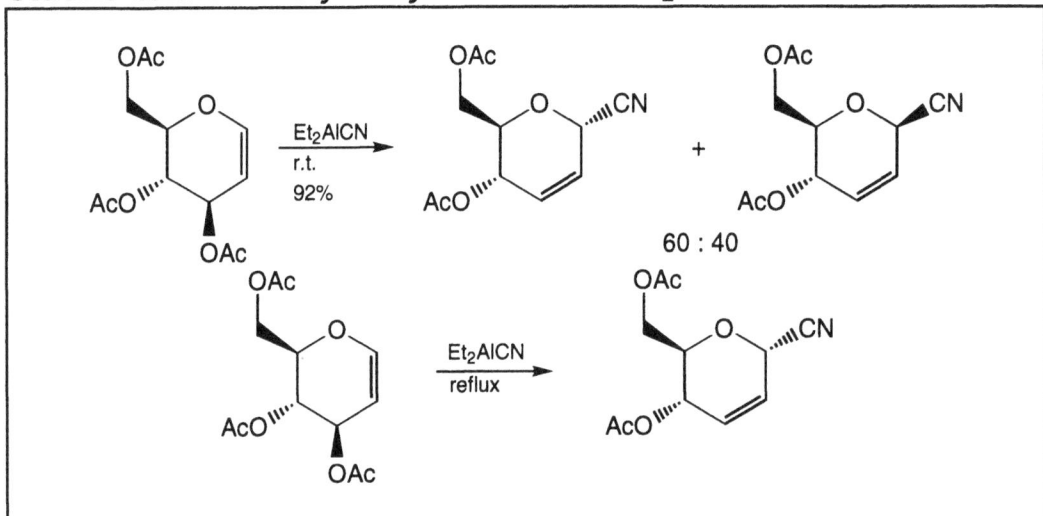

**Scheme 2.2.12   Fluoroglycoside Cyanations
with Me₂AlCN**

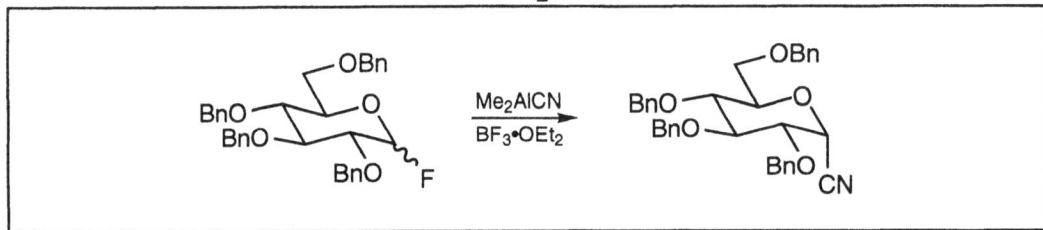

**Scheme 2.2.13   Fluoroglycoside Cyanations
with Et₂AlCN**

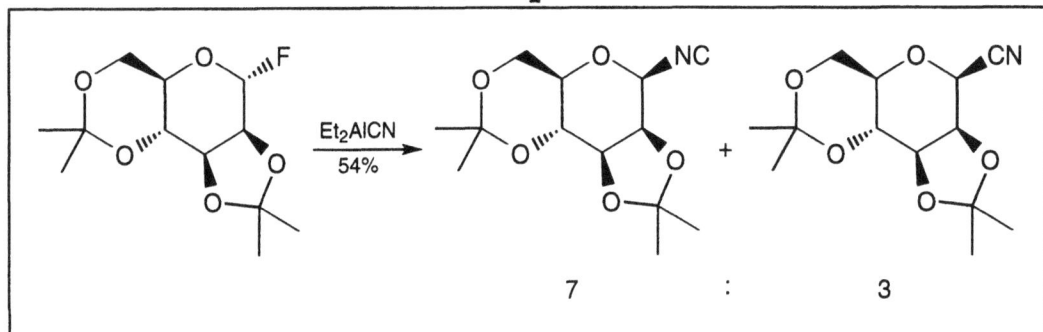

Scheme 2.2.14   **Fluoroglycoside Cyanations**
with Et₂AlCN

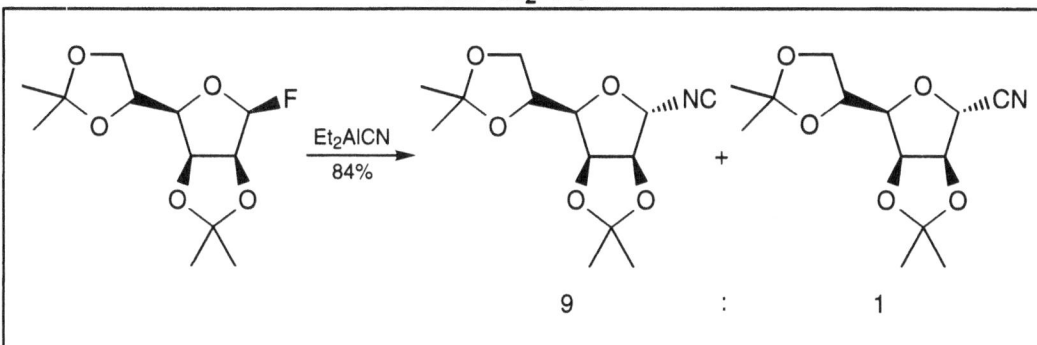

Regarding the use of aluminum reagents in the transformation of fluoroglycosides to cyanoglycosides, Drew, *et al.*,[11] demonstrated the compatibility of acetonide protecting groups with this technology. As shown in Scheme 2.2.13, utilizing diethylaluminum cyanide, the *bis*-acetonide of the illustrated α-fluoromannoside was converted to a 7 : 3 mixture of the β-isocyanoglycoside and the β-cyanoglycoside, respectively. Furthermore, as shown in Scheme 2.2.14, similar results were observed utilizing furanose analogs.

Scheme 2.2.15   **Mercuric Cyanide**
Cyanoglycosidations

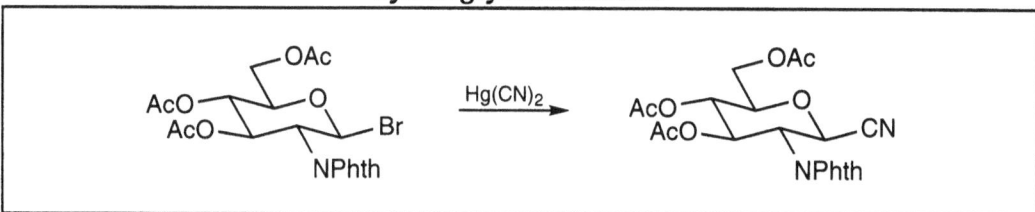

The previous two examples are useful if the desire is to form isocyanoglycosides. With respect to the formation of cyanoglycosides, the yields are poor; however, if no other methods apply, these examples are deemed useful. Another low yielding reaction which deserves mention is the conversion of bromoglycosides to cyanoglycosides *via* the *in situ* formation of mercuric cyanide.[12] The reaction, shown in Scheme 2.2.15, proceeded in only a 12% yield but, due to the ease of formation of bromoglycosides, provides a unique addition to the available methodology.

2.2.5   Other Cyanation Reactions

Scheme 2.2.16   Vinyligous Lactone Cyanations
with Acetone Cyanohydrin

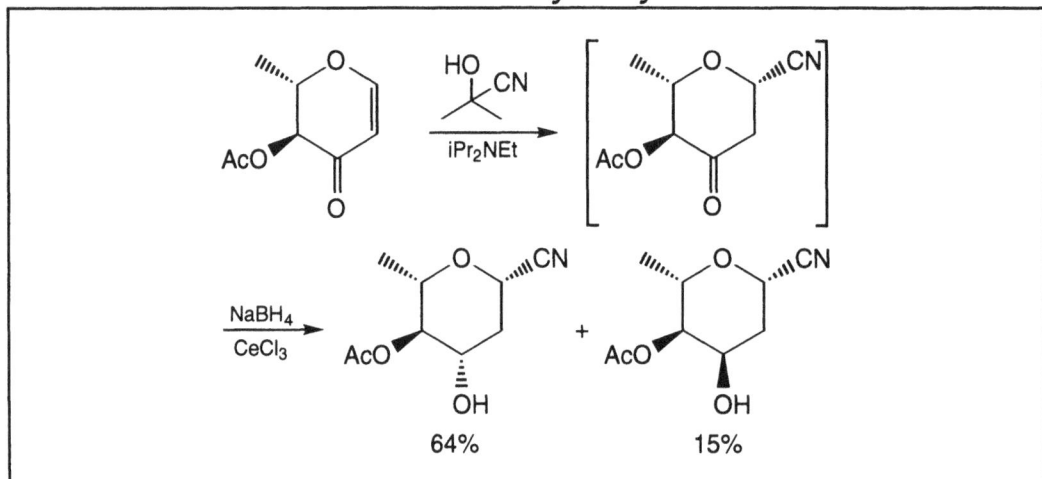

Scheme 2.2.17   Vinyligous Lactone Cyanations
with Acetone Cyanohydrin

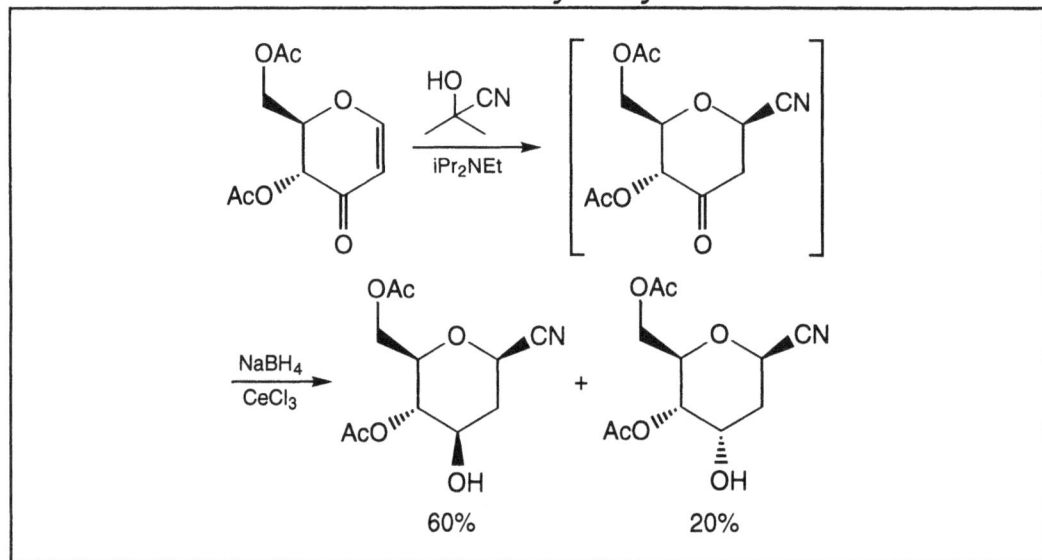

C-Cyanoglycosides are accessible from a number of starting materials other than glycosides and glycals. In concluding this section, the formation of cyanoglycosides from vinyligous lactones is explored. As shown in Scheme 2.2.16, Tatsuta, *et al.*,[13] formed 2-deoxy cyanoglycosides utilizing acetone cyanohydrin and Hünig's base. Following hydride reduction of the resulting ketone, a mixture of the 3-hydroxy isomers were isolated.

As illustrated by comparing the results shown in Scheme 2.2.16 with those illustrated in Scheme 2.2.17, the stereochemistry of the initial cyanation is dependent upon the stereochemistry of the starting material. Furthermore, the formation of the opposite 3-hydroxy configurations following hydride reduction is noted.

2.2.6   Cyanoglycoside Transformations

Scheme 2.2.18   **Hydride Reduction of Cyanoglycosides**

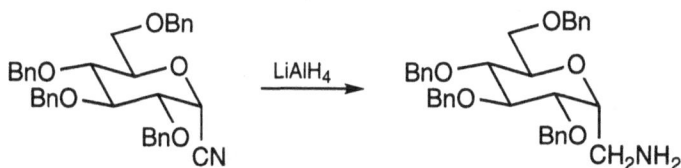

Scheme 2.2.19   *C*-**Glycosidic Acids and Amides from Nitriles**

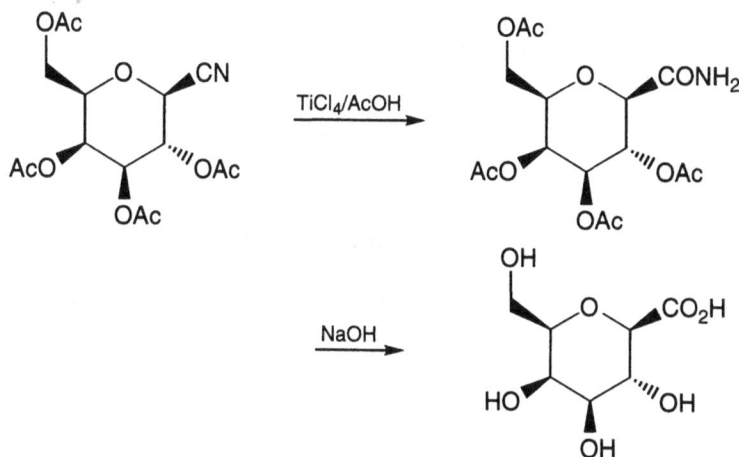

With the variety of methods available for the preparation of cyanoglycosides, a brief discussion of the versatility of these compound is warranted. To begin, as shown in Scheme 2.2.18, cyanoglycosides are easily reduced to the corresponding aminomethylglycosides on treatment with lithium aluminum hydride.[1] This reaction is effected without epimerization at the glycosidic position.

Scheme 2.2.20   Diazoketones from Cyanoglycosides

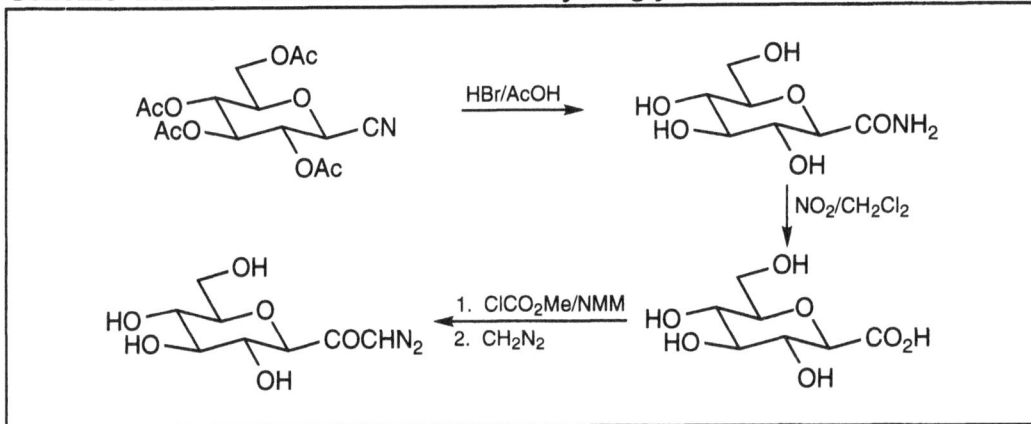

The hydrolysis of nitriles to amides is applicable to *C*-cyanoglycosides. As shown in Scheme 2.2.19, BeMiller, *et al.*,[14] utilized titanium tetrachloride in acetic acid to effect the conversion of a peracetylated *C*-cyanogalactoside to its corresponding primary amide. This transformation was effected in an 80% yield. Furthermore, base hydrolysis of the nitrile provides a clean conversion to the carboxylic acid.[15] Thus through the methods already mentioned, extensions with standard organic techniques allow the formation of a wide variety of useful functional groups.

Illustrating the versatility of *C*-cyanoglycoside nitrile derivatives, Myers, *et al.*,[15] formed a diazoketone from the peracetylated *C*-cyanoglucoside shown. The reaction sequence, illustrated in Scheme 2.2.20, involved the initial hydrolysis of the nitrile to the corresponding primary amide. Subsequent hydrolysis afforded the carboxylic acid. On conversion of the acid to a mixed carbonic anhydride followed by treatment with diazomethane, the desired diazoketone was obtained.

Among the most useful of synthetic transformation available to nitriles is the ability to form unnatural amino acids. An early example of the application of this technology to the formation of *C*-glycosyl amino acids was reported by Igolen, *et al.*[16] As shown in Scheme 2.2.21, the cyanofuranoside was converted to the corresponding aldehyde through initial formation of an *N,N'*-diphenylethylenediamine[17,18] followed by acid hydrolysis. Utilizing S-α-methylbenzylamine, benzoic acid, and *tert*-butyl isocyanate, application of the Ugi condensation[19] afforded the easily separable diastereomeric mixture, shown. Completion of the synthesis was then applied to the separated diastereomers *via* debenzylation on treatment with formic acid followed by acid hydrolysis of the resulting amides.

Scheme 2.2.21   *C*-Glycosyl Amino Acids

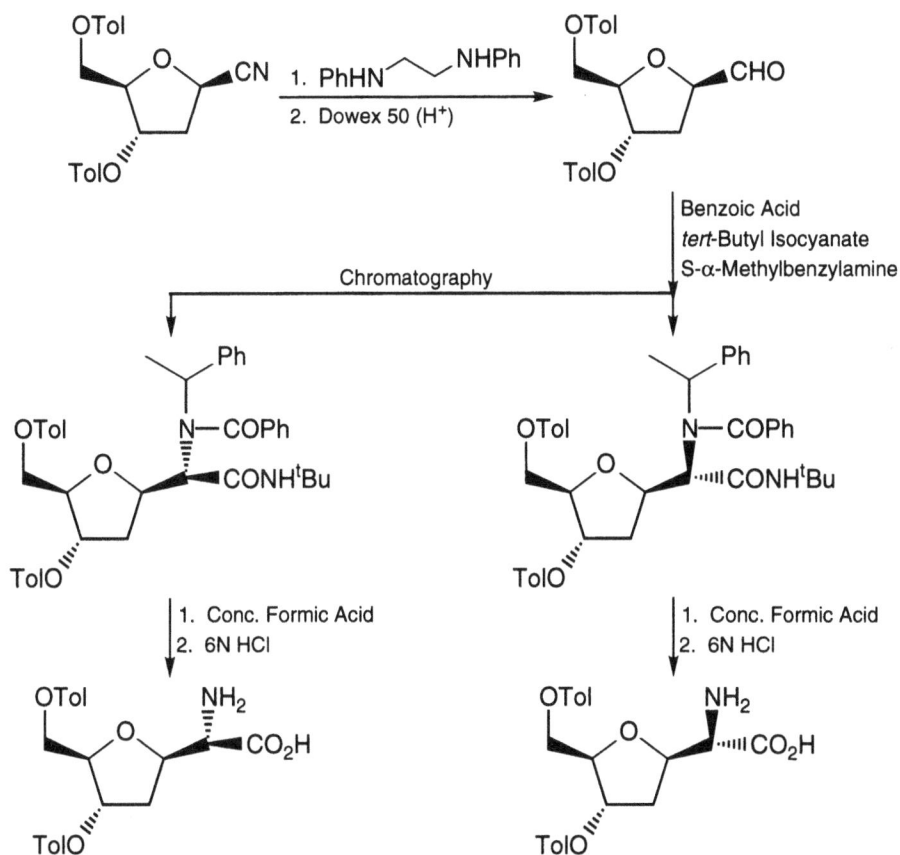

Figure 2.2.3   *C*-Glycosidic Heterocycles from Nitriles

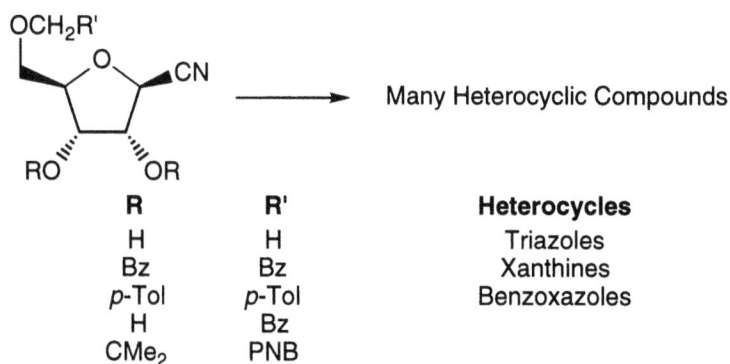

R	R'	Heterocycles
H	H	Triazoles
Bz	Bz	Xanthines
p-Tol	*p*-Tol	Benzoxazoles
H	Bz	
CMe$_2$	PNB	

As a final tribute to the versatility of cyanoglycosides, Poonian, *et al.*,[20] and El Khadem, *et al.*,[21] demonstrated that a wide variety of heterocycles may be derived from these compounds. Particularly interesting is the variety of compatible protecting groups including benzoyl, *p*-nitrobenzoyl, and acetonides. These observations are summarized in Figure 2.2.3 and the reader is referred to the original text for structural and experimental details.

2.3 Alkylation, Allenylation, Allylation and Alkynation Reactions

Concerning *C*-glycoside chemistry, in general, alkylation, allenylation, allylation, and alkynation reactions, generally illustrated in Figure 2.3.1, are among the most useful methods for preparative scales. Furthermore, they tend to yield the most valuable synthetic intermediates in that different functional groups of various lengths are easily obtained. As for the carbohydrate starting materials, many different glycosidic activating groups are useful in conjunction with a variety of Lewis acids. These reactions proceed well in many solvents including acetonitrile, nitromethane, and dichloromethane. Finally, both benzylated and acetylated sugars may be used and the predominant products bear the α configuration.

Figure 2.3.1   Examples of Alkylations, etc.

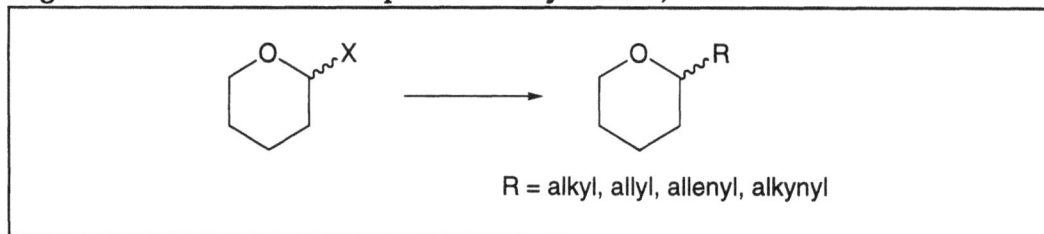

R = alkyl, allyl, allenyl, alkynyl

2.3.1   Use of Activated Glycosides and Anionic Nucleophiles

As early as 1974, Hanessian, *et al.*,[22] demonstrated the use of anion chemistry in the preparation of *C*-glycosides. As shown in Scheme 2.3.1, α-bromo-tetra-*O*-acetyl-D-glucose was treated with the anions of diethylmalonate and dibenzylmalonate yielding two distinct *C*-glycosides. In the case of the dibenzyl analog, debenzylation followed by treatment with the Meerwein reagent gave a product identical to that formed from diethylmalonate. Furthermore, decarboxylation of the diacid gave the corresponding carboxylic acid which, when subjected to a modified Hundsdiecker reaction[23] yielded the illustrated β-bromopyranoside. Subsequent treatment of this bromide with sodium acetate in dimethylformamide gave the known β-acetoxymethyl-*C*-glycoside thus demonstrating the accessibility of β anomers. One additional reaction demonstrating the feasibility of further functionalizing these *C*-glycosides involved the α bromination of the previously described monocarboxylic acid.

Scheme 2.3.1   *C*-Glycosides *via* Anionic Nucleophiles

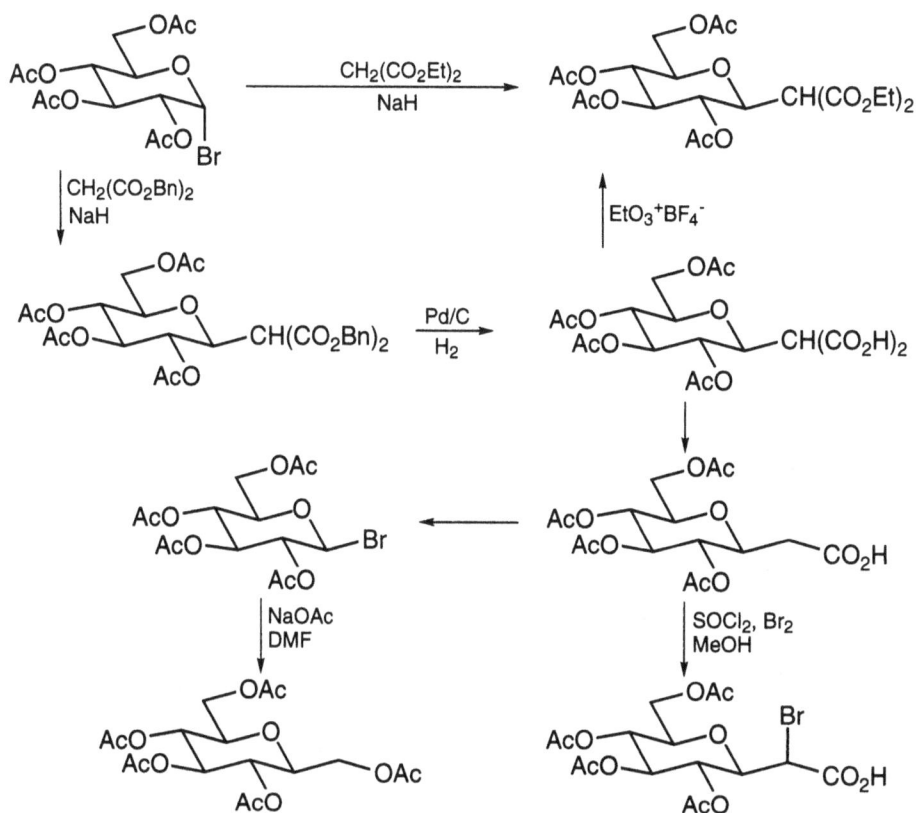

2.3.2   Use of Activated Glycosides and Cuprates

Similar *C*-glycosidations are available *via* cuprate chemistry. For example, as shown in Scheme 2.3.2, Bihovsky, *et al.*,[24] utilized a variety of cuprates in the preparation of α-*C*-methylglycosides. The yields were relatively poor when acetate protecting groups were used. However, substantial improvements in the yields were observed with methyl and benzyl protecting groups. In all cases, the α anomer was favored in ratios as high as 20 : 1.

Additional examples of the applicability of organocuprates to the preparation of *C*-glycosides were reported by Bellosta, *et al.*[25] His work, shown in Scheme 2.3.3, illustrated the selective formation of both α and β anomers through the appropriate epoxides.   Furthermore, an explanation of the regioselectivity of this reaction was addressed through the stereoelectronic interactions represented in Figure 2.3.2. In light of the promise demonstrated by this work, the difficulty in obtaining the required 1,2-epoxysugars is noted.

Scheme 2.3.2   ***C*-Glycosidations *via***
Cuprate Nucleophiles

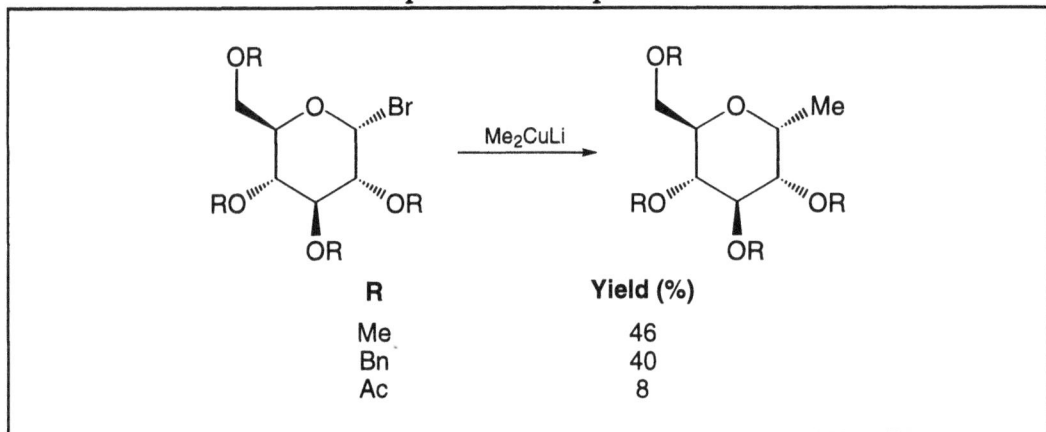

R	Yield (%)
Me	46
Bn	40
Ac	8

Scheme 2.3.3   ***C*-Glycosidations *via***
Cuprate Nucleophiles

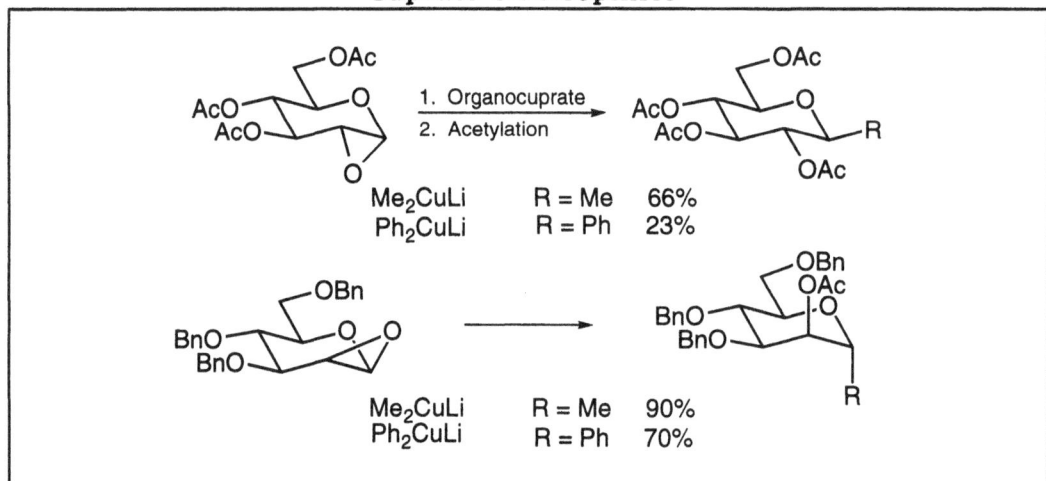

Me₂CuLi	R = Me	66%
Ph₂CuLi	R = Ph	23%

Me₂CuLi	R = Me	90%
Ph₂CuLi	R = Ph	70%

In an earlier example of the applicability of cuprate chemistry to *C*-glycosidations, Bellosta, *et al.*,[26] stereospecifically prepared epoxides from protected glucals (Scheme 2.3.4). Unfortunately, as shown in the previous example, the use of acetate protecting groups instead of benzyl protecting groups generally resulted in lower yields.   These observations were not improved through the use of mixed cuprates.

Figure 2.3.2   Regioselectivity of Cuprate Additions to Sugar-Derived Epoxides

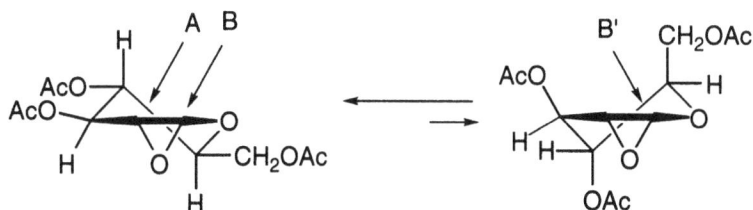

Scheme 2.3.4   *C*-Glycosidations *via* Cuprate Nucleophiles

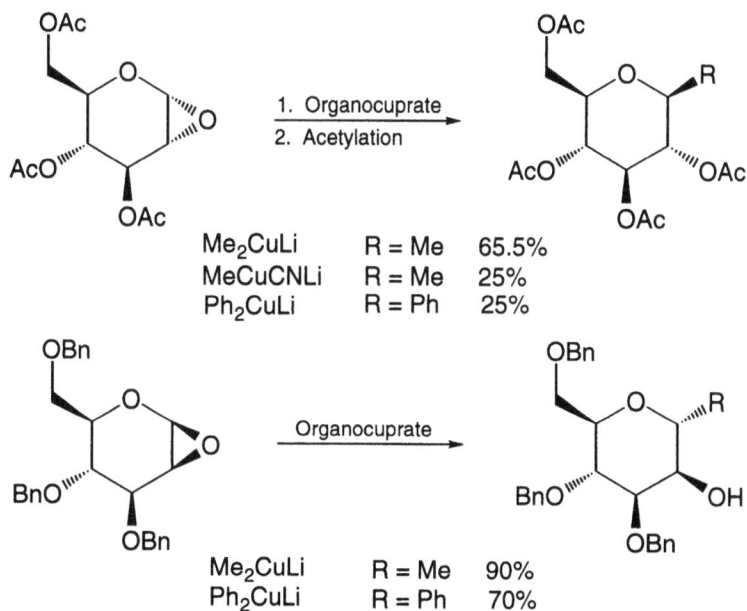

Me$_2$CuLi	R = Me	65.5%
MeCuCNLi	R = Me	25%
Ph$_2$CuLi	R = Ph	25%

Me$_2$CuLi	R = Me	90%
Ph$_2$CuLi	R = Ph	70%

2.3.3   Use of Activated Glycosides and Grignard Reagents

Complimentary to the use of cuprates, Grignard reagents have found a useful place in the technology surrounding *C*-glycosidations. As shown in Scheme 2.3.5, Shulman, *et al.*,[27] utilized allyl and vinyl Grignard reagents to give β-*C*-glycosides in modest yields. Subsequent epoxidation of these products provided potential glucosidase inhibitors. In a later report, Hanessian, *et al.*,[28] confirmed the above results and postulated the intermediate epoxide, shown in Scheme 2.3.6, as an explanation of the stereoselectivity of this chemistry.

Scheme 2.3.5   *C*-Glycosidations *via* Grignard Reagents

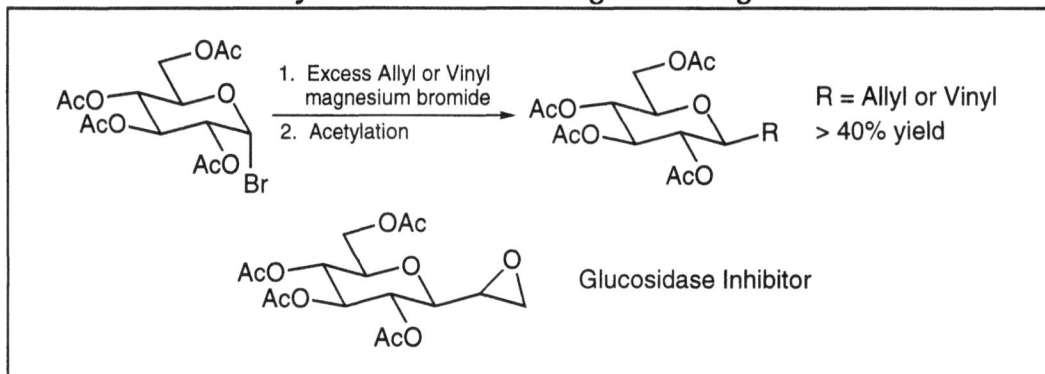

Scheme 2.3.6   *C*-Glycosidations *via* Grignard Reagents

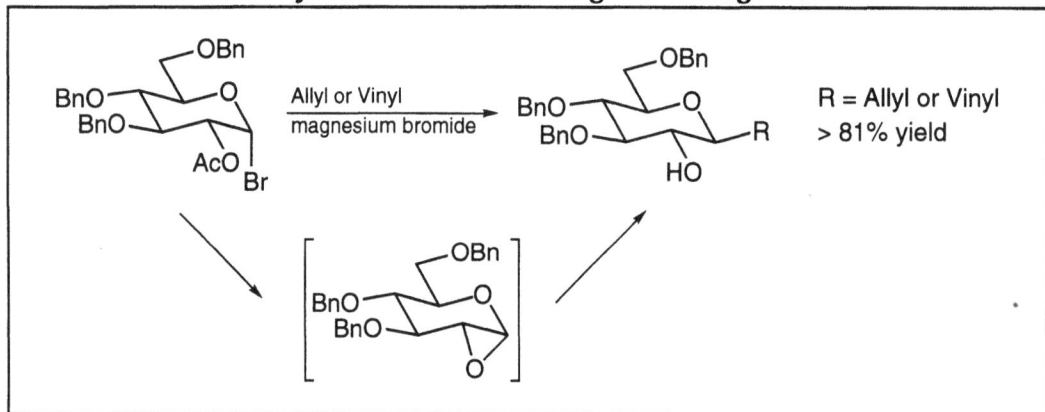

Scheme 2.3.7   *C*-Glycosidations *via* Grignard Reagents

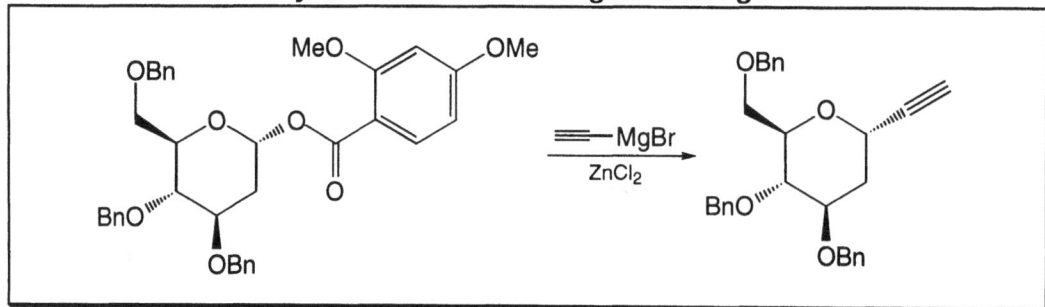

Grignard reagents are useful nucleophiles capable of effecting reactions on compounds stable to other types of nucleophiles. For example, the alkynation reaction shown in Scheme 2.3.7 proceeded in a 70% yield with complete retention of the α configuration.[29] However, when lithium and organocopper reagents were used, no reaction was observed. An additional feature of this reaction is the demonstration of the utility of glycosidic esters as viable leaving groups in the formation of C-glycosides *via* nucleophilic substitutions.

Scheme 2.3.8   *C*-Glycosidations *via* Grignard Reagents

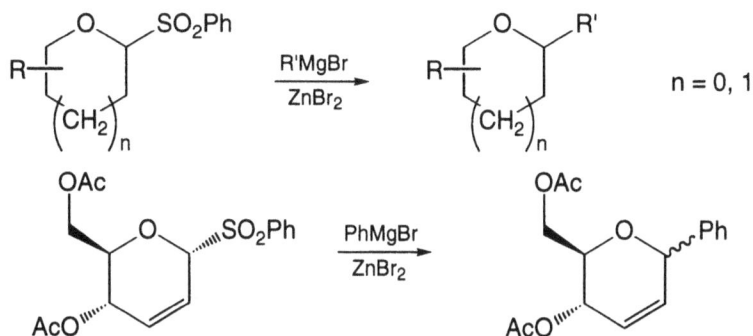

With the incorporation of new leaving groups in C-glycoside methodology, the versatility of the benzenesulphonyl group must be mentioned. As illustrated in Scheme 2.3.8, a variety of nucleophiles may be coupled with both furanose and pyranose compounds.[30] In all cases, the yields exceed 70%. This observation, coupled with the ready availability of glycosidic sulphones,[31-34] illustrates the general usefulness of this technology.

2.3.4   Use of Activated Glycosides and Organozinc Reagents

**Scheme 2.3.9   *C*-Glycosidations *via*
Organozinc Reagents**

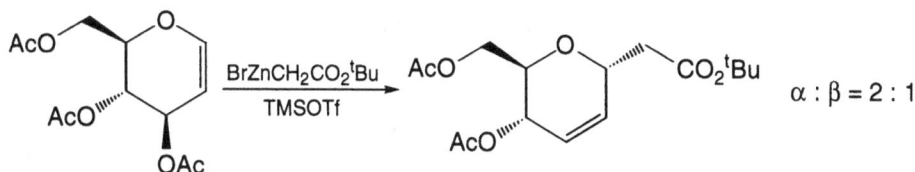

The previous two examples are the result of zinc modified Grignard reagents. However, the direct use of organozinc compounds is also known in the field of *C*-glycosidations. As shown in Scheme 2.3.9, Orsini, *et al.*,[35] reacted acetylated glycals with *tert*-butoxycarbonylmethylzinc bromide. In the case of tri-*O*-acetyl glucal, the reaction gave a 49% yield with a 2 : 1 ratio favoring the α anomer.

Scheme 2.3.10   *C*-Glycosidations *via* Organozinc Reagents

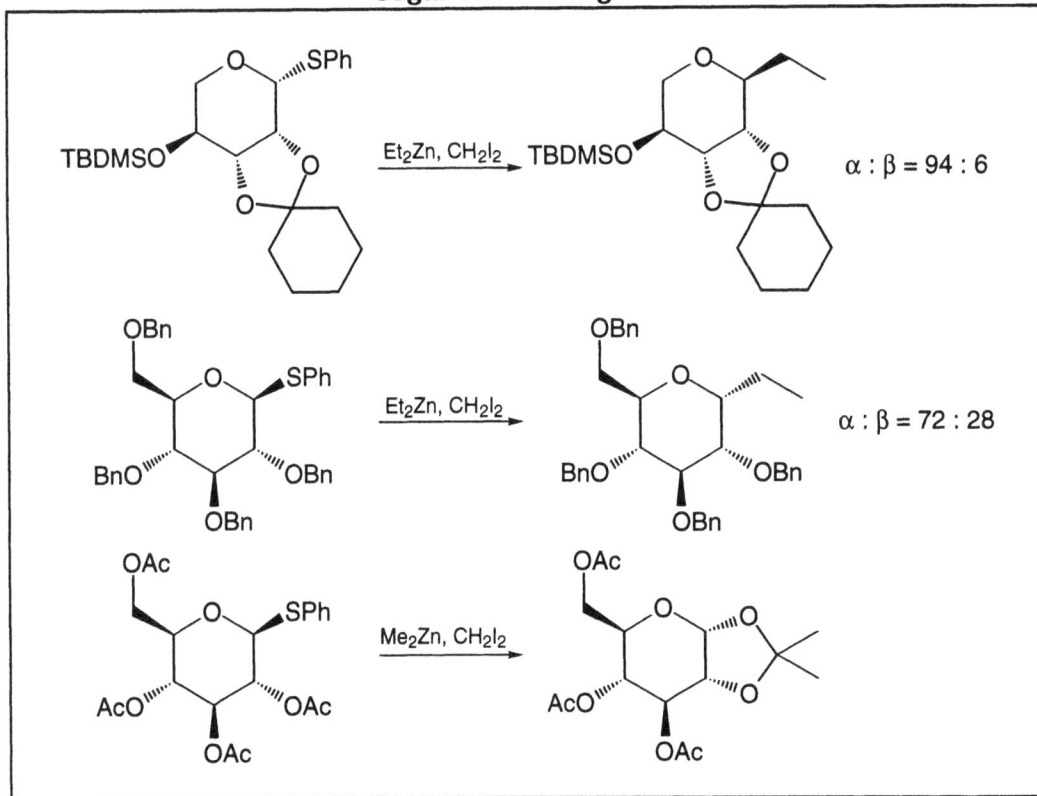

Although the use of organozinc reagents resembling Grignard reagents has been fruitful, the use of simpler reagents such as diethylzinc has, in many cases, effected higher yields. For example, as shown in Scheme 2.3.10, Kozikowski, *et al.*,[36] formed *C*-ethylglycosides from thioglycosides. In the cases of L-lyxose and D-glucose derivatives, the predominant product possessed the α configuration presumably resulting from an axial addition. An interesting observation was noted in that all reactions proceeded well when silyl groups, ketals, and benzyl protecting groups were used. However, the use of acetate groups facilitated the formation of the illustrated 1,2-*O*-isopropylidene derivative on treatment with dimethylzinc. The mechanism postulated for the formation of this compound is shown in Scheme 2.3.11 and involves displacement of the sulfide by the neighboring acetate forming a stabilized carbocation. Subsequent transfer of a methyl group from dimethylzinc gives the observed product.

Scheme 2.3.11   ***C*-Glycosidations *via***
                        Organozinc Reagents

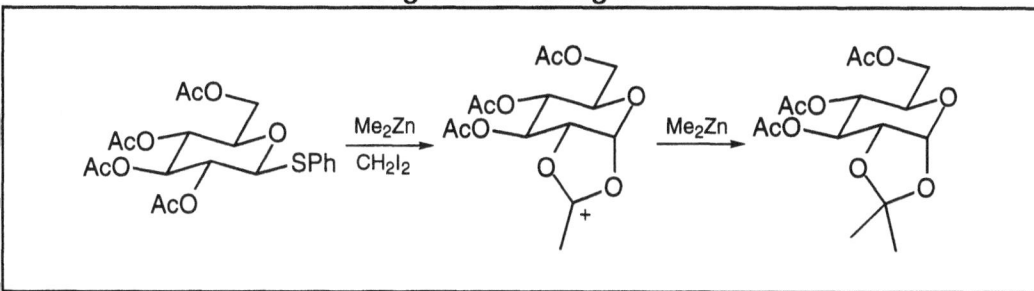

2.3.5   *Use of Modified Sugars and Anionic Nucleophiles*

Scheme 2.3.12   ***C*-Glycosidations *via***
                        Organolithium Reagents

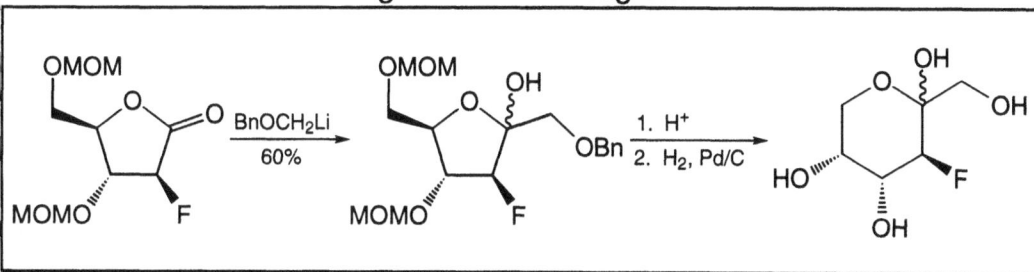

Scheme 2.3.13   ***C*-Glycosidations *via***
                        Organolithium Reagents

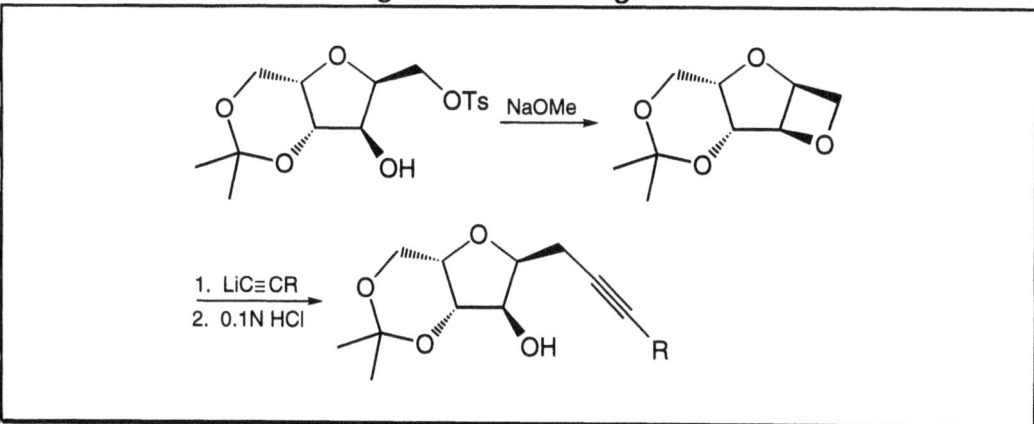

With all the anionic nucleophiles mentioned thus far as reagents useful for the preparation of *C*-glycosides, organolithium compounds have yet to be discussed. While these compounds can be used similarly to other nucleophiles, some of their most interesting chemistry is the result of reactions with modified carbohydrates such as the ribose-derived lactone shown in Scheme 2.3.12. Treatment of this compound with benzyloxymethyllithium gave the alkylated hemiacetal. Furthermore, after removal of the MOM groups, the compound spontaneously rearranged to the illustrated pyranose form.[37] Although the isolated product is a ketose sugar, it bears some properties in common with *C*-glycosides and, as will be discussed later, can be converted to a true *C*-glycoside under reductive conditions.

Scheme 2.3.14   ***C*-Glycosidations *via*
                              Organogrignard Reagents**

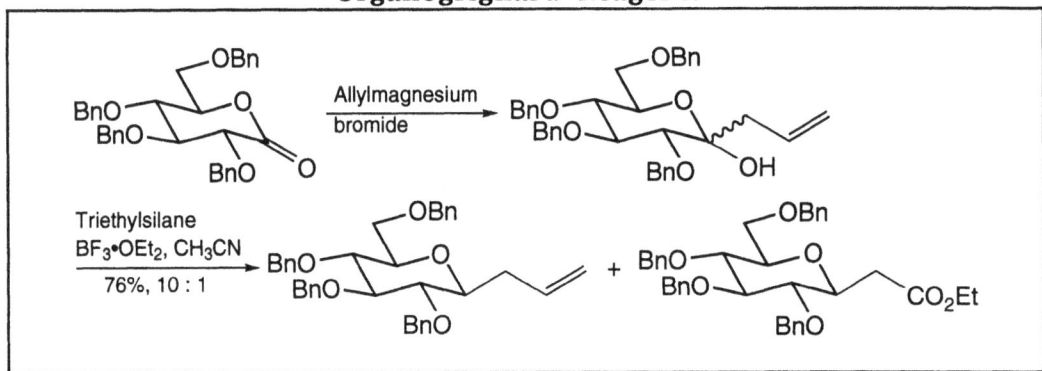

Another excellent example demonstrating the utility of organolithium compounds in the formation of *C*-glycosides from modified carbohydrates was reported by Koll, *et al.*[38] This study, shown in Scheme 2.3.13, involves the addition of a lithium acetylide to a *C*-glycosyl oxetane. The result was the extension of a pre-existing *C*-glycosyl side chain.

A final example regarding the *C*-glycosidation of modified sugars is shown in Scheme 2.3.14 and involves reaction of the illustrated lactone with allylmagnesium bromide. The resulting alcohol was subsequently deoxygenated utilizing triethylsilane in the presence of a Lewis acid thus giving a 76% yield of the illustrated β-anomer.[39] It should be noted that by applying this methodology to the hemiacetals shown in Scheme 2.3.12 true *C*-glycosides can be obtained.

2.3.6   *Lewis Acid Mediated Couplings with Unactivated Olefins*

Anionic nucleophiles provide valuable and versatile routes to the preparation of *C*-glycosides. However, the most extensively used technologies lie within the chemistry of Lewis acid mediated reactions of carbohydrates with unsaturated hydrocarbons and derivatives thereof. In the next series of examples, reactions of sugars and sugar derivatives with olefins, silyl olefins, stannyl olefins, and aluminum olefins are discussed.

Scheme 2.3.15   *C*-Glycosidations with 1-Hexene

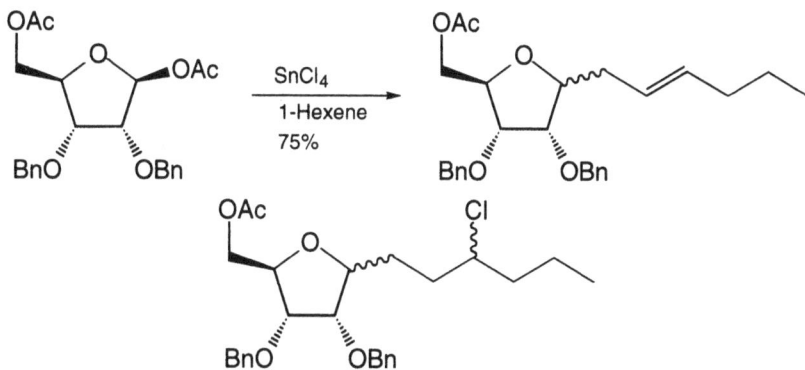

Scheme 2.3.16   *C*-Glycosidations with Olefins

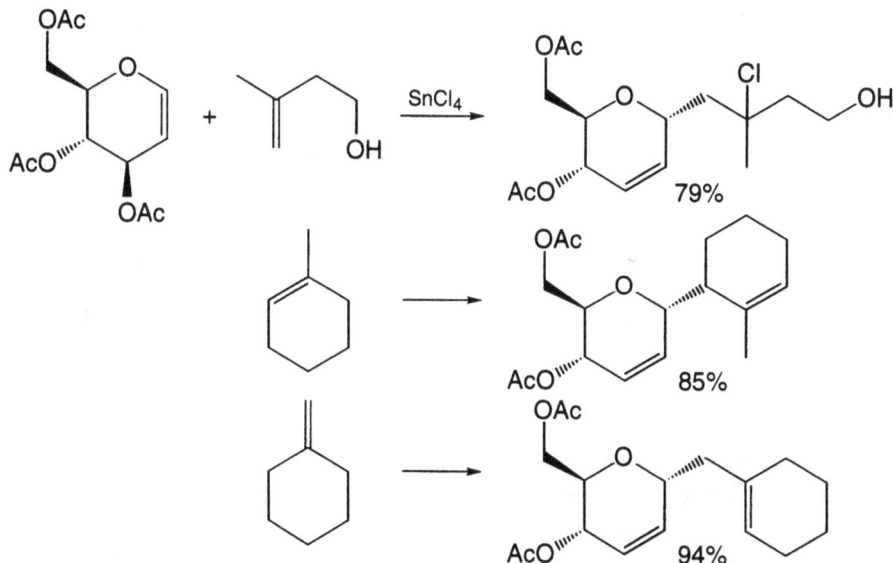

An early example of the reaction of protected furanoses with olefins is shown in Scheme 2.3.15. In this study, Cupps, *et al.*,[40] utilized tin tetrachloride

to effect the formation of a 75% yield of the illustrated C-glycoside. This
reaction showed little anomeric selectivity. Additionally, the chloride, shown,
was identified as a byproduct arrising from the intermediate carbocation.

Scheme 2.3.17   C-Glycosidations with Olefins

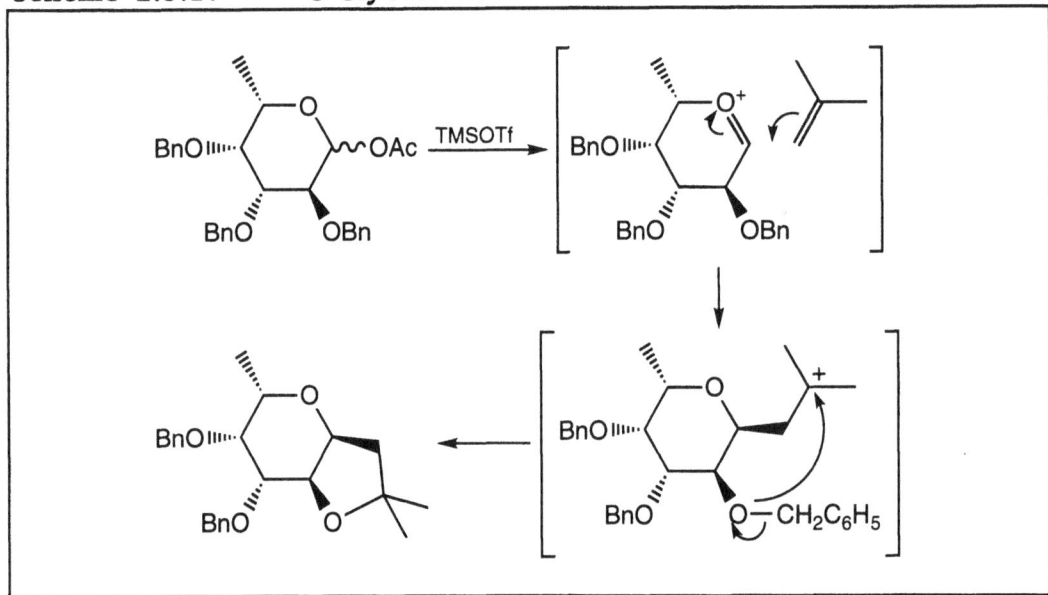

Scheme 2.3.18   C-Glycosidations with Olefins

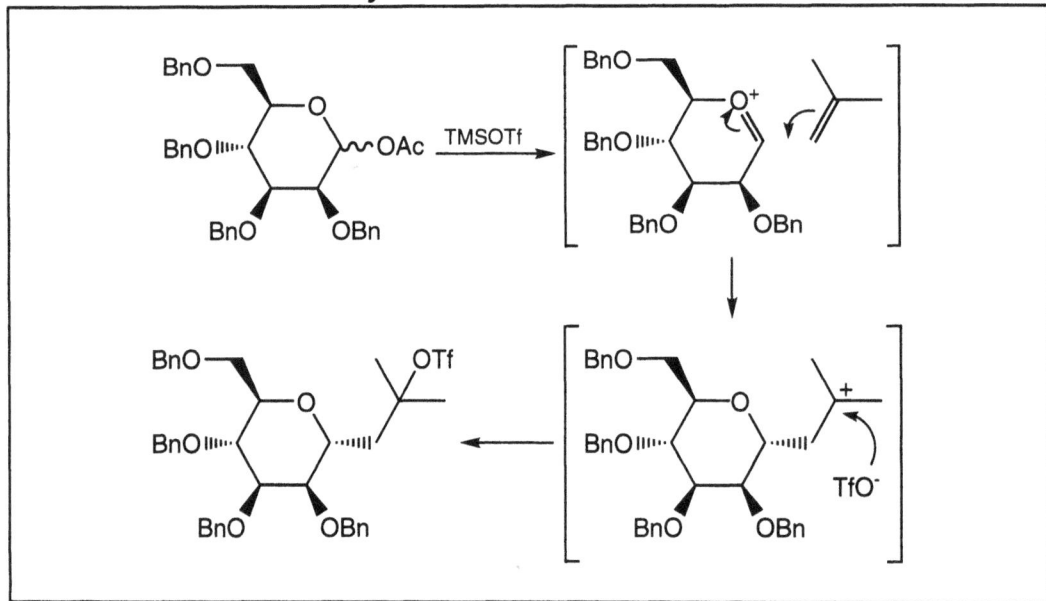

Where the previous reaction involved the use of an unactivated olefin, this was apparently not the source of the lack of stereochemical induction. This is illustrated by work reported by Herscovici, *et al.*,[41,42] shown in Scheme 2.3.16, where several unactivated olefins were successfully used to produce α-*C*-glycosides in good to excellent yields.

Table 2.3.1   **C-Glycosidations with Olefins**

Entry	Sugar	Olefin	Lewis Acid	Product	Yield
A			TMSOTf		74%
B			TMSOTf		76%
C			TMSOTf		21%
D			TMSOTf or BF3·OEt2		71%
E			SnCl4		78%

In a final example demonstrating the utility of unfunctionalized olefins, Levy, *et al.*,[43] reported a novel adaptation in which the formed *C*-glycoside cyclizes to an adjacent oxygen with loss of a benzyl group. As shown in Scheme 2.3.17, 1-*O*-acetyl-2,3,4-tri-*O*-benzyl-L-fucose was treated with isobutylene and an excess of TMSOTf giving the illustrated fused ring *C*-glycoside. This reaction proceeded in >70% yield in both methylene chloride and acetonitrile. Additionally, borontrifluoride etherate could also be used. The proposed mechanism, shown in Scheme 2.3.17, involves the initial elimination of the acetate to form an oxonium ion. This species then undergoes a Prins-type[44] reaction with the olefin giving a tertiary carbocation. Finally, spontaneous cyclization with loss of a benzyl cation gives the observed product. Interestingly, although cyclization relies on the loss of a benzyl group, attempts to induce the formation of acyclic *C*-glycosides failed and no reaction was observed utilizing acetate protecting groups.

In exploring the generality of this reaction, the 1-*O*-acetyl-tetra-*O*-benzyl analogs of D-glucose, D-galactose, and D-mannose, were studied. In all three cases, borontrifluoride etherate was found to be too weak a Lewis acid to effect this reaction and TMSOTf was required. The reactions of these compounds with TMSOTf and isobutylene are summarized in Table 2.3.1. An additional example with L-fucose and methylene cyclohexane is also included therein.

When comparing the data presented in Table 2.3.1, the anomalous results observed in the case of D-mannose (Entry C) can be explained as shown in Scheme 2.3.18. The initial α-*C*-glycosidation gives a tertiary carbocation. The anomeric preference for the addition places the newly formed bond *trans* to the 2-benzyloxy group. The high strain associated with a 6-5 *trans*-fused ring system disfavors attack by the oxygen of the neighboring group allowing the triflate anion to add to the tertiary carbocation.

As *cis*-fused tetrahydrofuran-tetrahydropyran ring systems are found in the halichrondrins,[45] the herbicidins,[46-48] and octosyl acid,[49] this simple one-pot preparation demonstrates unique advantages over previous approaches utilized to target these classes of compounds.[50,51]

2.3.7   *Lewis Acid Mediated Couplings of Glycosides and Silanes*

While the use of unprotected olefins has proven fruitful in the preparation of *C*-glycosides, the most commonly utilized technology involves the use of silyl compounds. Reactions involving the coupling of sugar derivatives with allyl and acetylenic silanes has been initiated with a number of Lewis acids as well as a wide variety of glycosidic activating groups. The first of these to be addressed is the acetate activating group.

An early report demonstrating the generality of Lewis acid mediated *C*-glycosidations with allylsilanes was reported by Giannis, *et al.*,[52] and is shown in Scheme 2.3.19. In this study, allylations were catalyzed by borontrifluoride etherate utilizing either dichloroethane or acetonitrile as solvents. As observed, no stereoselectivity was noted with dichloroethane. However, the more polar acetonitrile, capable of coordinating to intermediate oxonium ions, induced fairly high selectivity. An additional reaction demonstrated the applicability of this technology to disaccharides. In a later report, similar observations were noted in the reaction of penta-*O*-acetyl-D-galactose with allyltrimethylsilane in nitromethane utilizing borontrifluoride etherate as the catalyst (Scheme 2.3.20).[14]

Scheme 2.3.19   *C*-Glycosidations with Allylsilanes

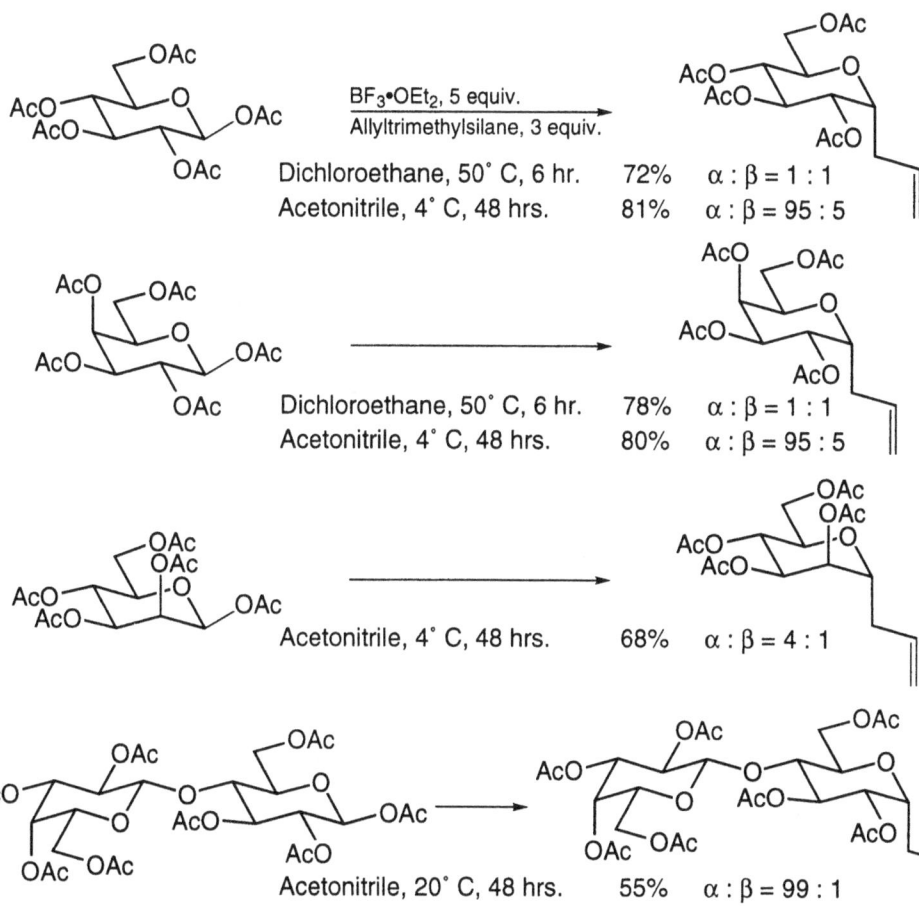

BF$_3$•OEt$_2$, 5 equiv.
Allyltrimethylsilane, 3 equiv.

Dichloroethane, 50° C, 6 hr.   72%   α : β = 1 : 1
Acetonitrile, 4° C, 48 hrs.   81%   α : β = 95 : 5

Dichloroethane, 50° C, 6 hr.   78%   α : β = 1 : 1
Acetonitrile, 4° C, 48 hrs.   80%   α : β = 95 : 5

Acetonitrile, 4° C, 48 hrs.   68%   α : β = 4 : 1

Acetonitrile, 20° C, 48 hrs.   55%   α : β = 99 : 1

Scheme 2.3.20   *C*-Glycosidations with Allylsilanes

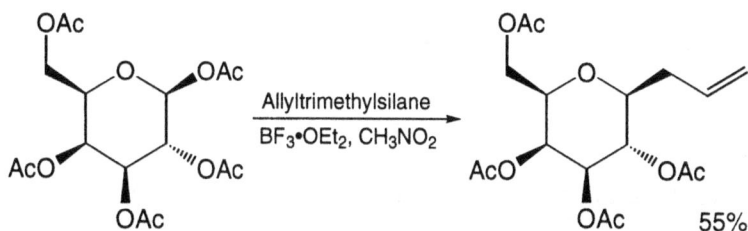

Allyltrimethylsilane
BF$_3$•OEt$_2$, CH$_3$NO$_2$

55%

With respect to the related trifluoroaceoxy glycosides, comparable *C*-glycosidation results were observed. Additionally, as shown in Scheme 2.3.21, the reaction exhibited a similar preference for formation of α anomers.[53]

Scheme 2.3.21   *C*-Glycosidations with Allylsilanes

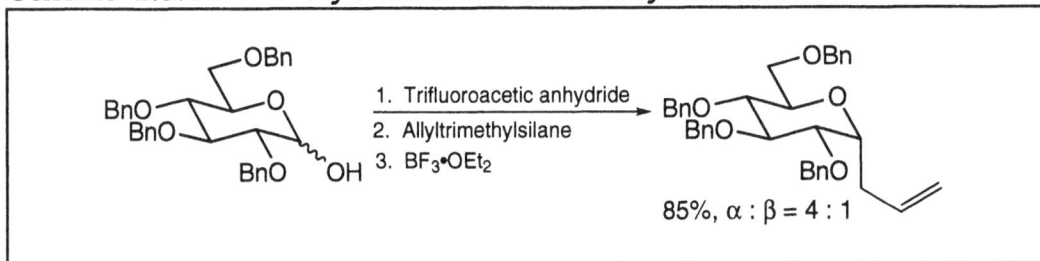

Scheme 2.3.22   *C*-Glycosidations with Allylsilanes

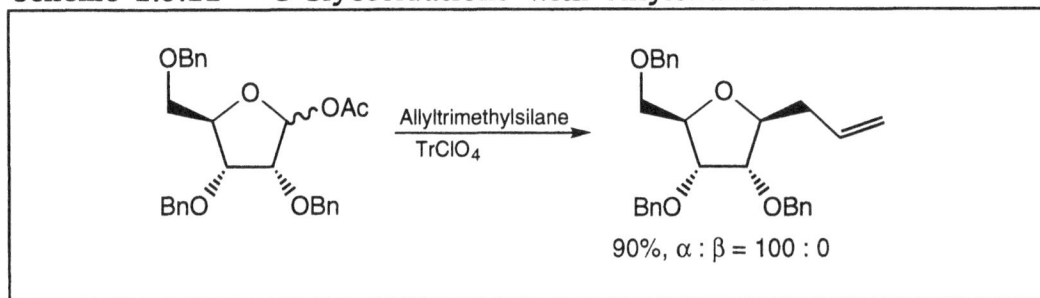

Scheme 2.3.23   *C*-Glycosidations with Allylsilanes

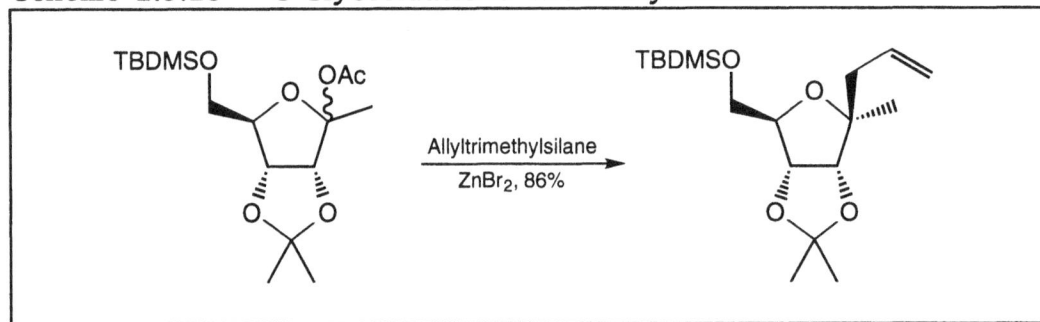

With the usefulness of the chemistry described in the previous three schemes, it is important to note that similar reactions are applicable to furanose sugars. As shown in Scheme 2.3.22, Mukaiyama, *et al.*,[9] effected the illustrated reaction between allyltrimethylsilane and an acetate activated ribose derivative. Utilizing trityl perchlorate as a catalyst, the reaction proceeded in 90% yield with total α selectivity. Concurrently, a similar reaction was reported with a 1-methyl-1-acetoxy furanoside.[54] This reaction, shown in Scheme 2.3.23, not only demonstrates that high stereoselectivity can be obtained with

1-substituted sugars, it also illustrates the compatibility of acetonides and silyl ethers as protecting groups when zinc bromide is used as a Lewis acid.

Scheme 2.3.24   C-Glycosidations with Allylsilanes

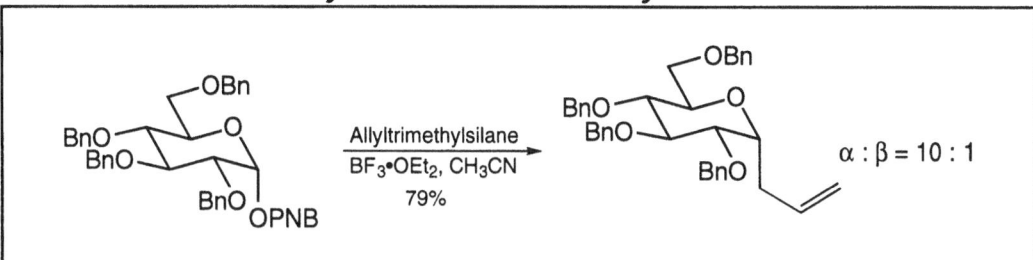

Scheme 2.3.25   C-Glycosidations with Allylsilanes

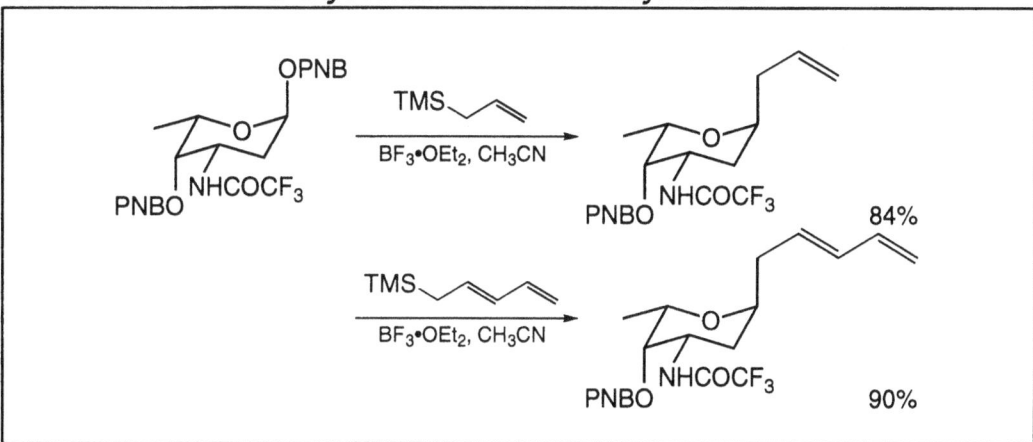

With the current discussion of the acetate activating group used in the preparation of C-glycosides, the chemistry of 1-O-p-nitrobenzoyloxy glycosides deserves mention.

Perhaps one of the earliest examples of the use of 1-O-p-nitrobenzoyloxy (PNB) glycosides as substrates for C-glycosidations was reported by Lewis, et al.[39] In the cited report, the PNB glycoside, shown in Scheme 2.3.24, was reacted with allyltrimethylsilane utilizing borontrifluoride etherate as a catalyst to yield the desired α-C-glycoside. This initial reaction, from its good yield and preference for the α anomer, demonstrated the utility of this activating group applied to pyranose sugars thus setting the stage for further exploration.

Shortly after publication of the previous example, Acton, et al.,[55] reported the reactions shown in Scheme 2.3.25. In this study, the high yielding C-glycosidations were utilized in the preparation of anthracycline C-glycosides via Diels-Alder methodology. In the examples reported, high α selectivity was achieved utilizing borontrifluoride etherate in acetonitrile.

Scheme 2.3.26   *C*-Glycosidations with Allylsilanes

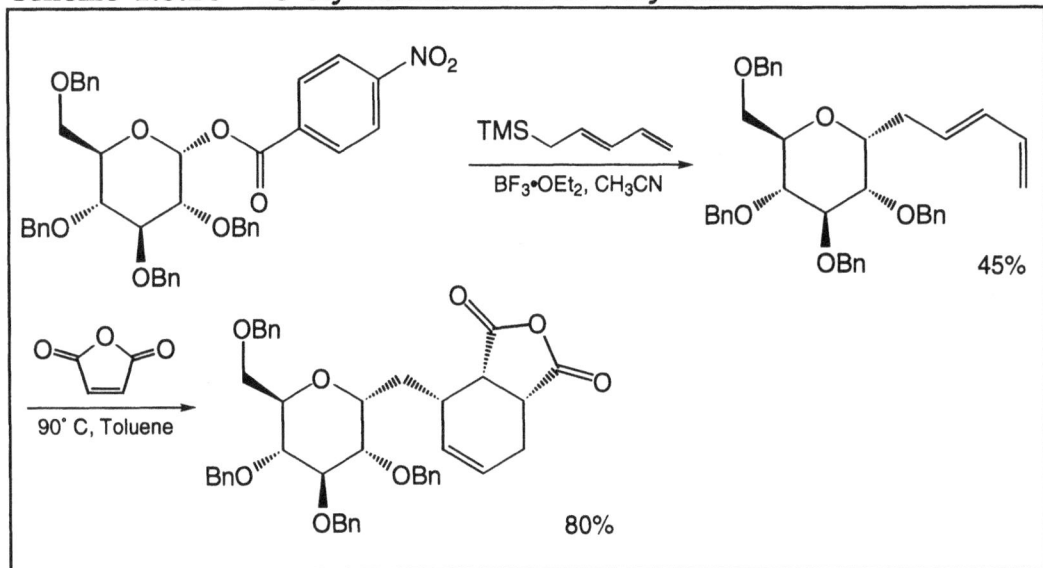

45%

80%

Scheme 2.3.27   *C*-Glycosidations with Allylsilanes

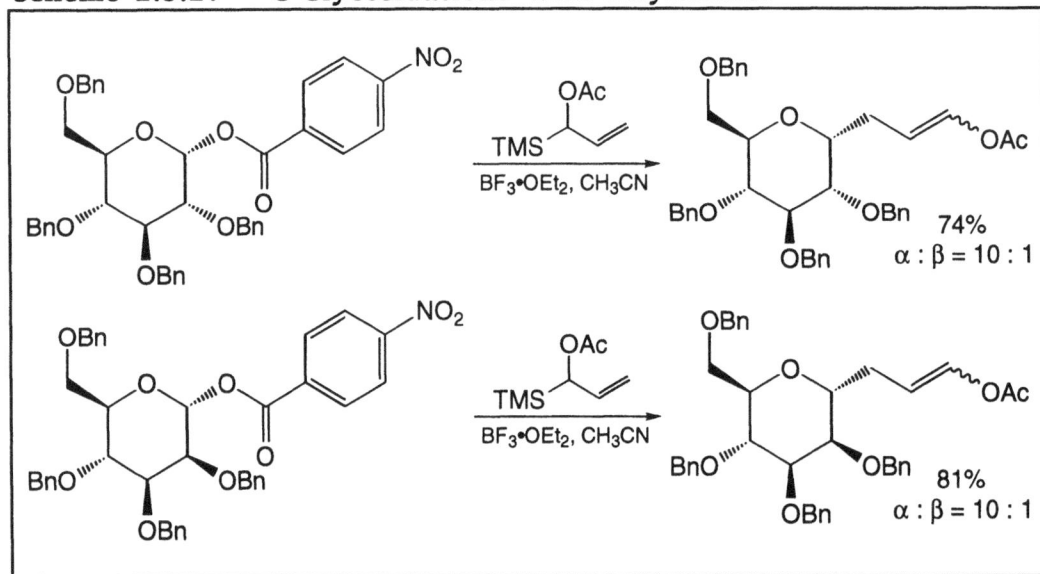

74%
$\alpha : \beta = 10 : 1$

81%
$\alpha : \beta = 10 : 1$

A similar reaction, reported by Martin, *et al.*,[56] is shown in Scheme 2.3.26. However, several differences are apparent in comparing this reaction to the earlier example. Initially, the fact that the previous example is applied to an amino sugar becomes obvious. However, the more important observation is the fact that the amino sugar is also a 2-deoxygenated sugar. The result is an electron deficient sugar compared to the glucose analog shown in scheme 2.3.26. It is, in fact, this observation that may account for the somewhat lower yield reported for this analogous *C*-glycosidation.   As with the previous example, a Diels-Alder reaction was used to further derivatize this *C*-glycoside.

Scheme 2.3.28   **C-Glycosidations with Propargylsilanes**

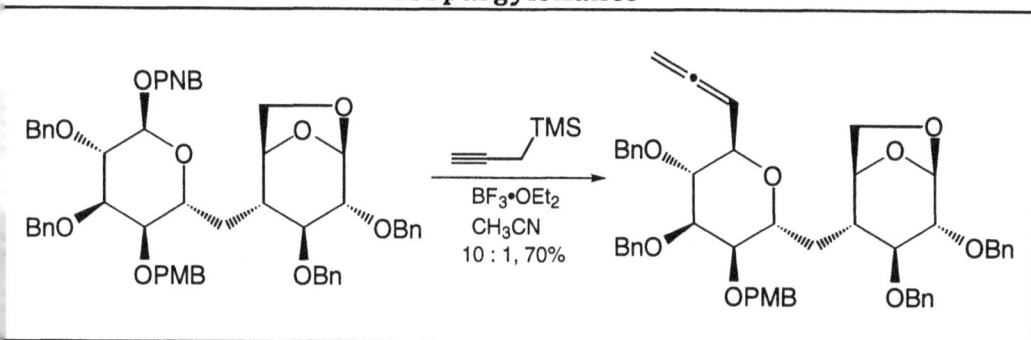

Scheme 2.3.29   C-Glycosidations with Allylsilanes

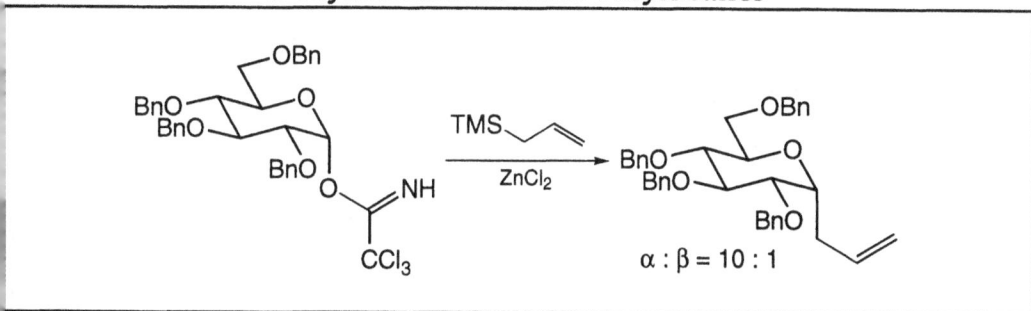

Scheme 2.3.30   C-Glycosidations with Allylsilanes

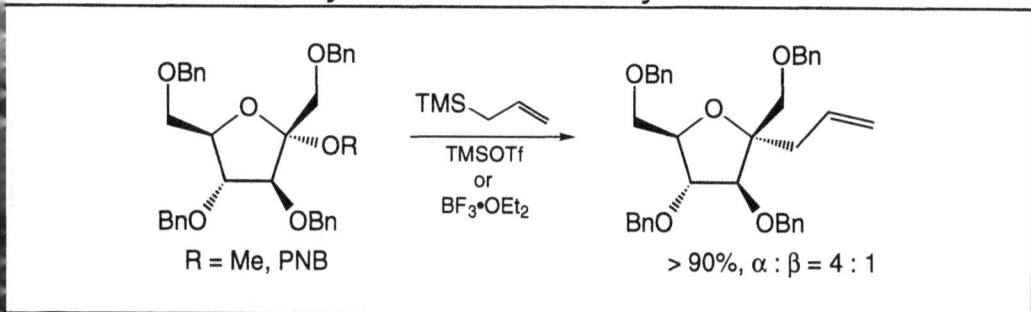

In addition to extended allylsilanes, acetate substituted allylsilanes are useful reagents for *C*-glycosidations. As shown in Scheme 2.3.27, PNB activated glucose   and   mannose   derivatives   were   treated   with   1-acetoxy-allyltrimethylsilane.[57] Utilizing borontrifluoride etherate as the catalyst, high yields of α-*C*-glycosides were obtained. The use of these compounds is apparent as hydrolysis of the acetate groups readily liberates aldehydes available for further elaboration.

Where allyl groups are readily incorporated as *C*-glycosides, propargyl groups are equally useful. As shown in Scheme 2.3.28, Babirad, *et al.*,[58] used propargyltrimethylsilane to produce the desired allene. The yield was reported at 70% with an anomeric ratio of 10 : 1 favoring the α anomer. Interestingly, when identical reaction conditions were applied to the α isomer of the starting *C*-disaccharide, no reaction was observed. These results were comparable to results observed for corresponding reactions with allylsilanes demonstrating a useful versatility in the types of *C*-glycosidations available from silyl compounds.

Scheme 2.3.31   *C*-Glycosidations with Allylsilanes

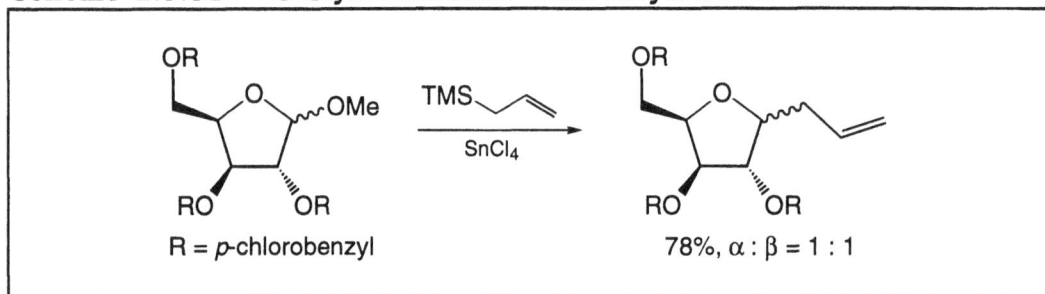

Scheme 2.3.32   *C*-Glycosidations with Allylsilanes

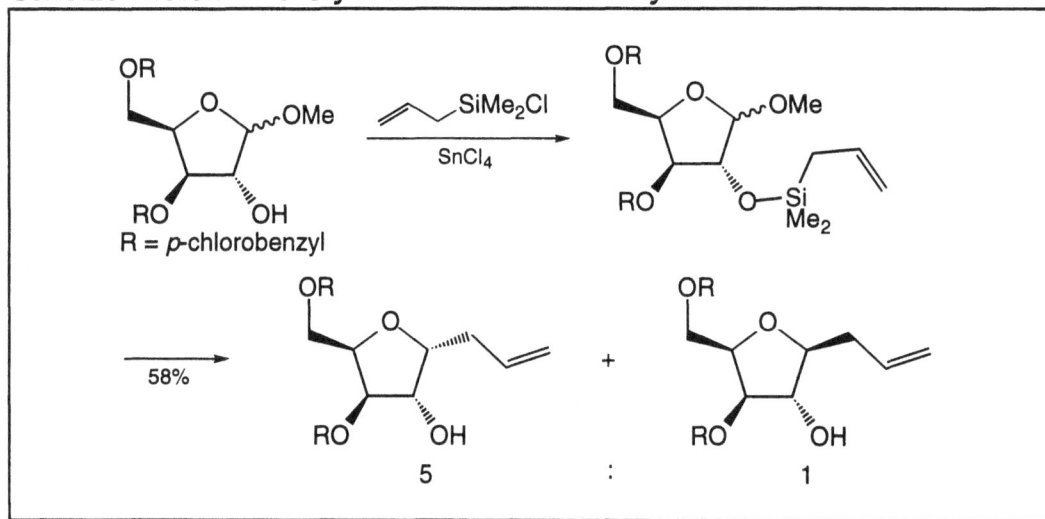

Any discussion involving the preparation of *C*-glycosides from acetate and PNB activated sugars would not be complete without mention of the electronically similar trichloroacetimidate activating group. As mentioned in section 2.2, trichloroacetimidates are easily prepared on reaction of hemiacetals with trichloroacetonitrile. The resulting activating group is subsequently displaced by silyl activated nucleophiles under Lewis acid catalysis. As shown in Scheme 2.3.29, Hoffman, *et al.*,[3] exploited this chemistry in the preparation of *C*-glucosides utilizing allyltrimethylsilane

Scheme 2.3.33   *C*-Glycosidations with Allylsilanes

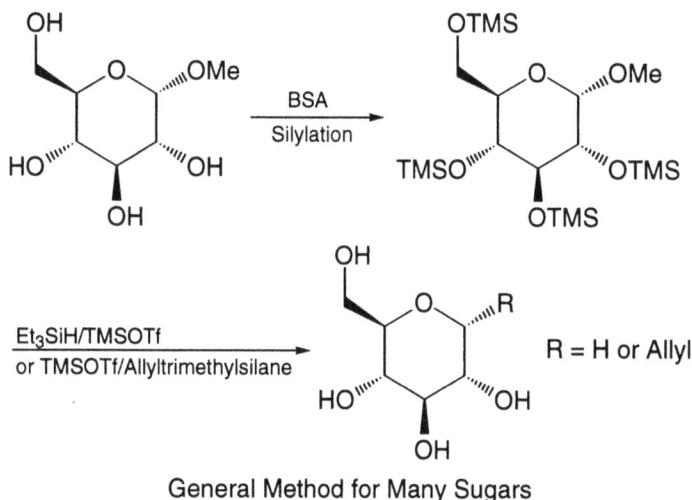

General Method for Many Sugars

Moving from the chemistry of esters as glycosidic activating groups, *O*-methylglycosides have been shown to be extremely useful. As shown in Scheme 2.3.30, Nocotra, *et al.*,[59] utilized both PNB activated and methoxy activated furanosides as substrates for *C*-glycosidations. This study demonstrated essentially identical results for both of these activating groups with yields exceeding 90% and a 4 : 1 preference for the α anomers. Furthermore, no change in the results were observed when alternating between trimethylsilyl triflate and borontrifluoride etherate as Lewis acid catalysts.

Although the above results appear useful for the preparation of *C*-furanosides, these reactions are not always this specific. As shown in Scheme 2.3.31, Martin, *et al.*,[60] attempted a similar reaction with the simpler furanose shown. Although a good yield was achieved for this reaction, the stereochemical outcome presented a 1 : 1 mixture of anomers. However, as shown in Scheme 2.3.32, when intramolecular delivery of the allyl group was effected, the reaction yielded a 5 : 1 mixture favoring the β anomer. Thus, this approach is a potentially useful route to β-*C*-glycosides.

In a demonstration that unprotected sugars may be utilized as substrates for *C*-glycosidations, Bennek, *et al.*,[61] utilized BSA as an *in situ* silylating reagent in the preparation of *C*-pyranosides. Subsequent treatment with either triethylsilane or allyltrimethylsilane in the presence of TMSOTf gave either the deoxy sugar or the allyl glycoside shown in Scheme 2.3.8. This procedure, shown in Scheme 2.3.33, was shown to be a general method applicable to many sugars.

Scheme 2.3.34   *C*-Glycosidations with Allylsilanes

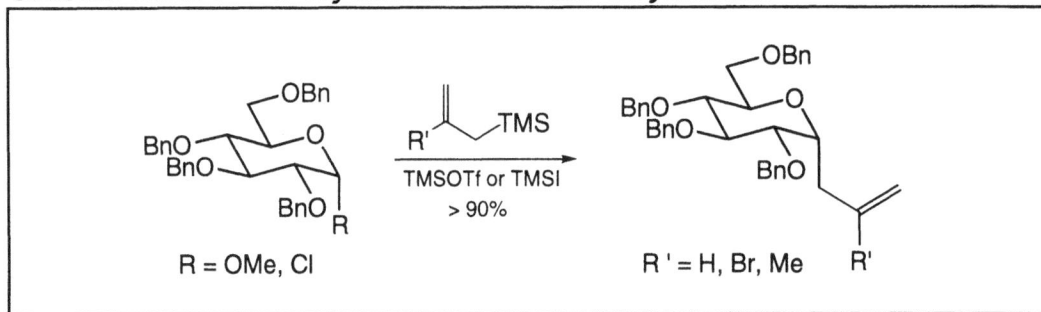

R = OMe, Cl

TMSOTf or TMSI
> 90%

R ' = H, Br, Me

Before changing to a discussion of halogens as glycosidic activating groups, a final example of the use of *O*-methylglycosides deserves mention. This specific example highlights the ability to prepare *C*-glycosides bearing vinyl halides in addition to simple and alkyl substituted olefinic substituents.[62] The cited publication reports applicability to a variety of sugars of which, glucose is shown in Scheme 2.3.34. These examples, like those heretofore mentioned proceed with high α selectivity. Furthermore, as a lead into the discussion of haloglycosides as *C*-glycosidation substrates, chloroglycosides were equally as effective in these reactions.

2.3.8   *Lewis Acid Mediated Couplings of Haloglycosides and Silanes*

Scheme 2.3.35   *C*-Glycosidations with Allylsilanes

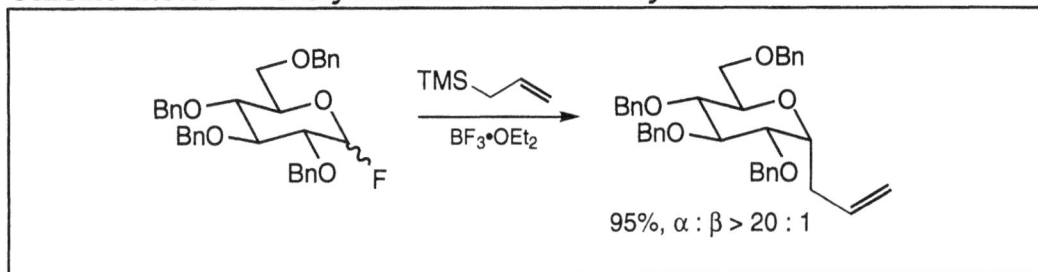

BF₃•OEt₂

95%, α : β > 20 : 1

Haloglycosides, already discussed under cyanation reactions, are useful substrates for *C*-glycosidations utilizing both allylsilanes and silyl enol ethers.

As shown in Scheme 2.3.35, Nicolau, *et al.*,[4] utilize the fluoroglucoside shown and effected the formation of a variety of *C*-glycosides. The example shown proceeded in high yields with excellent stereoselectivity. Similar results, shown in Scheme 2.3.36, were observed by Araki, *et al.*[63]

Scheme 2.3.36   *C*-Glycosidations with Allylsilanes

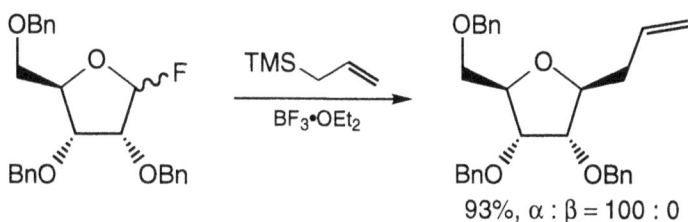

93%, α : β = 100 : 0

2.3.9   Lewis Acid Mediated Couplings of Nitroglycosides and Silanes

Scheme 2.3.37   *C*-Glycosidations with Allylsilanes

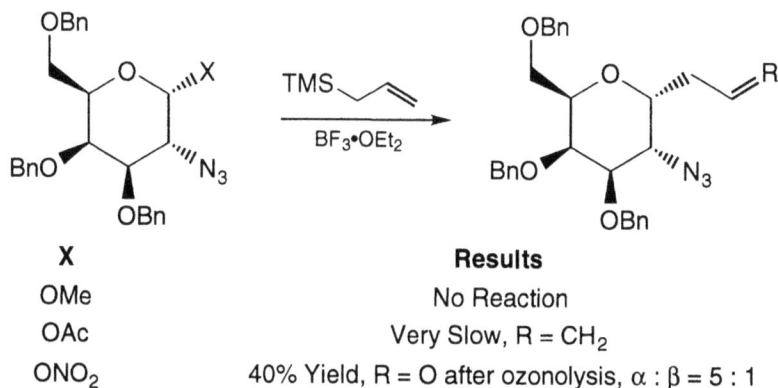

X	Results
OMe	No Reaction
OAc	Very Slow, R = CH_2
ONO_2	40% Yield, R = O after ozonolysis, α : β = 5 : 1

Presently, several classes of anomeric activating groups have been discussed. Additionally many more have been reported. However, no further discussions regarding the formation of alkyl, olefinic, or acetylenic *C*-glycosidations is planned except for the contents of Scheme 2.3.37. This scheme, presenting work reported by Bednarski, *et al.*,[64] is intended to present a final activating group in comparison to some already discussed. As the three groups shown in this report increase in their electronegativities from methoxy to acetoxy to nitro, their reactivities also increase. This is advantageous as applied to the azidosugar shown. Similar results were observed utilizing propargyl trimethylsilane. The take-home lesson is that if *C*-glycosidations fail, changing to more electronegative anomeric activating groups may, in fact, allow the induction of otherwise difficult reactions.

2.3.10   Lewis Acid Mediated Couplings of Glycals and Silanes

Anomeric activating groups provide useful means for the preparation of C-glycosides from naturally occurring sugars. However, glycals have been extremely useful substrates in their complimentarity to sugars. Additionally, they have been used to demonstrate versatility not directly available from other sugar derivatives. In a very nice example of the utility of allylsilanes and silylacetylenes in C-glycoside chemistry, Ichikawa, et al.,[65] demonstrated a preference for the α anomer in all cases. The results, shown in Scheme 2.3.38, included a demonstration that a characteristic nOe can be used to confirm the stereochemical outcome of the reaction with bis-trimethylsilylacetylene.

Scheme 2.3.38   C-Glycosidations with Allylsilanes and Related Compounds

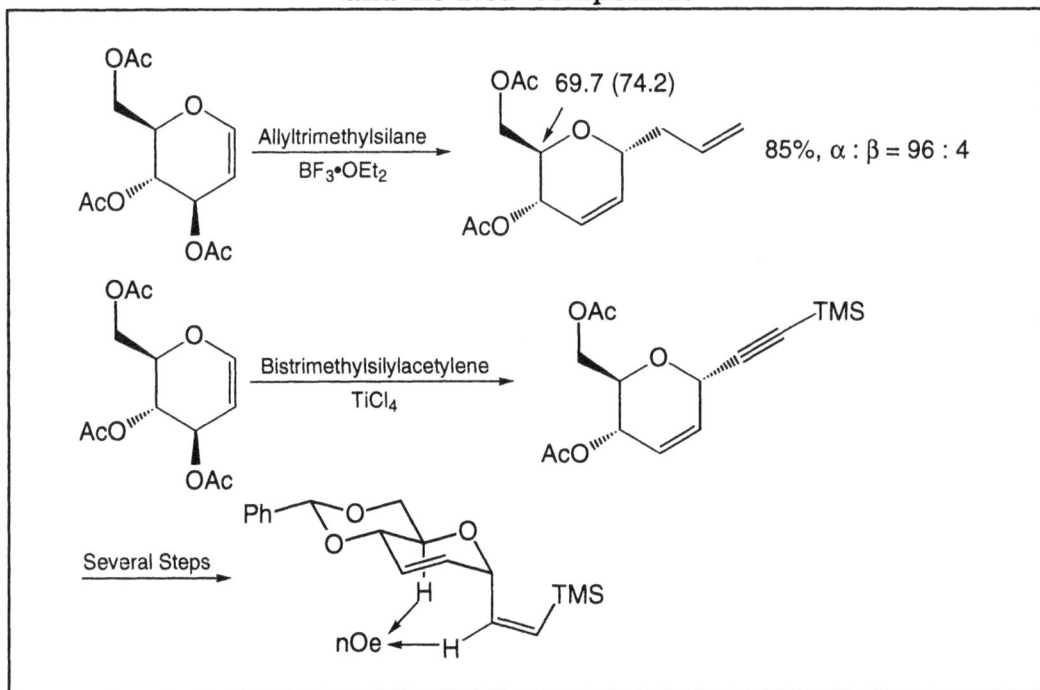

With the advent of glycal chemistry being incorporated into C-glycoside technology came a host of new compounds previously difficult to prepare. As shown in Scheme 2.3.39, this statement is supported by a report by Stutz, et al.,[66] in which a C-linked disaccharide was prepared by the direct coupling of an allylsilane modified sugar with a glycal. While this example is not intended to be representative of the state of *C*-disaccharide chemistry, it does illustrate an extreme regarding the usefulness of glycals.

Scheme 2.3.39   *C*-Glycosidations with Allylsilanes

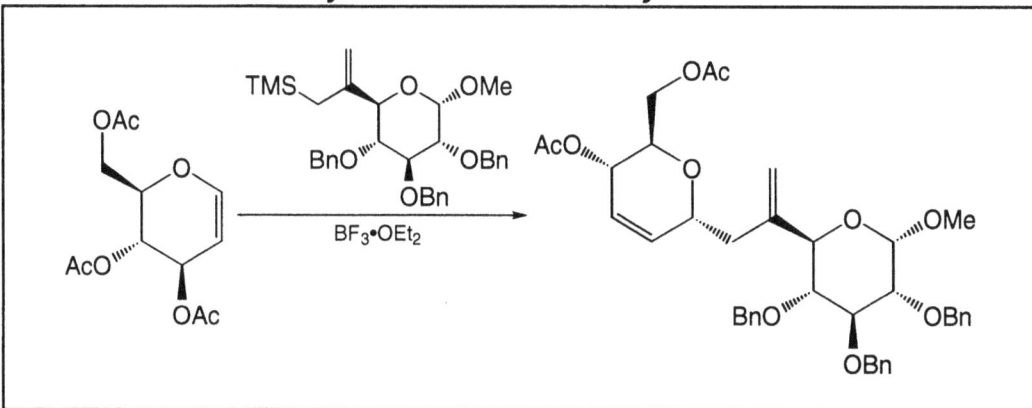

Scheme 2.3.40   *C*-Glycosidations with Allylsilanes

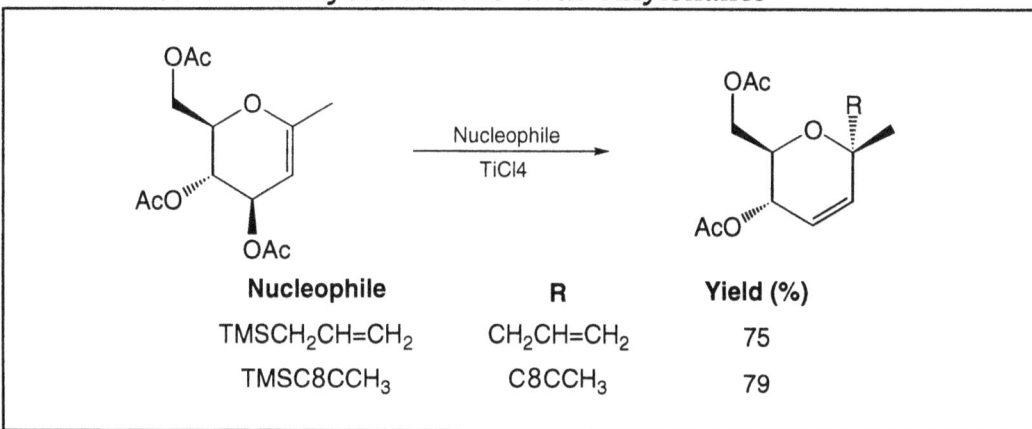

Nucleophile	R	Yield (%)
TMSCH$_2$CH=CH$_2$	CH$_2$CH=CH$_2$	75
TMSC8CCH$_3$	C8CCH$_3$	79

Scheme 2.3.41   *C*-Glycosidations with Allylsilanes

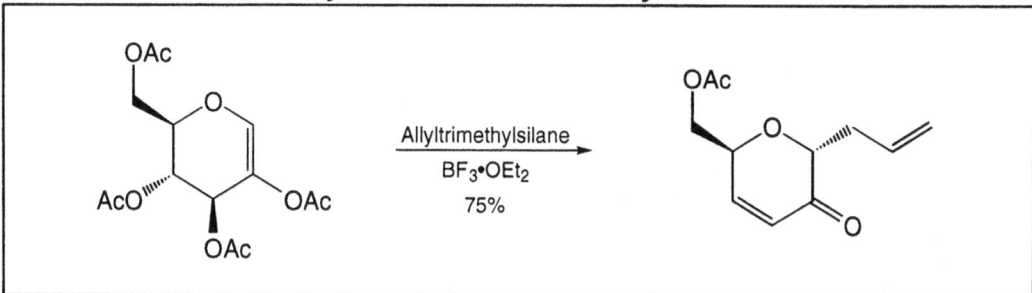

As simple glycals are substrates for *C*-glycosidations, so are 1-substituted glycals. An example of *C*-glycosidations on substituted glycals was reported by Nicolau, *et al.*,[7] and is illustrated in Scheme 2.3.40. In this study, reactions with allyltrimethylsilane and methyl trimethylsilylacetylene were explored. Utilizing titanium tetrachloride as the catalyst, yields in excess of 75% were obtained and the products exhibited the stereochemistry shown.

As with 1-substituted glycals, 2-substituted glycals are known to be useful substrates for *C*-glycosidations. In these cases, the final products are enones. As shown in Scheme 2.3.41, Ferrier, *et al.*,[67] demonstrated the utility of these reactions by effecting the formation of a 75% yield of the product shown. The reaction proceeded utilizing borontrifluoride etherate as a catalyst and displayed a preference for the α anomer. The specific reaction was successfully used in the preparation of trichothecene-related compounds.

Further examples of the utility of 2-acetoxyglycals as *C*-glycosidation substrates were reported by Tsukiyama, *et al.*,[68] and are illustrated in Scheme 2.3.42. This study demonstrated the highly stereospecific high yielding nature of these reactions. Furthermore, the generality of these reactions was demonstrated in their application to a variety of silylated acetylenes.

Scheme 2.3.42 *C*-Glycosidations with Silylacetylenes

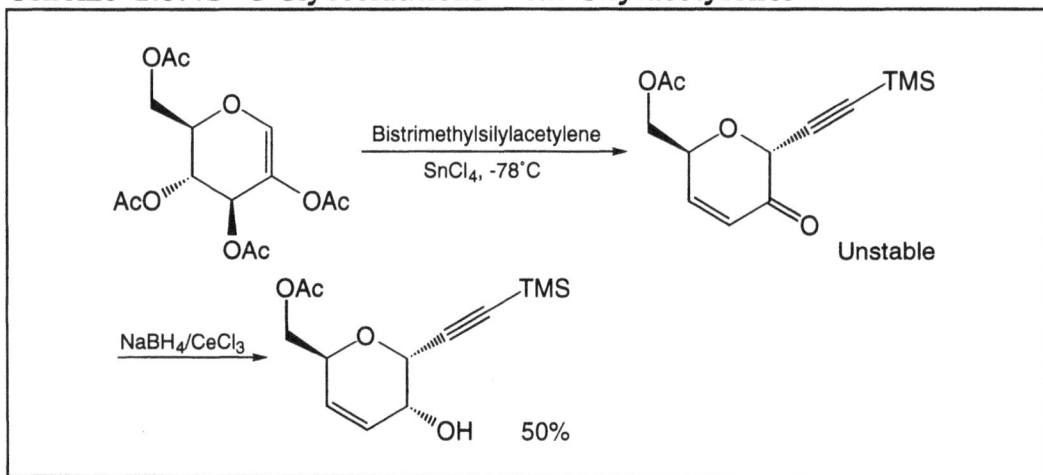

Before leaving the subject of *C*-glycosidations utilizing silylated nucleophiles, the reaction described in Scheme 2.2.7 deserves revisiting. In this report, Smolyadova, *et al.*,[8] formed arylsulfonium compounds on treatment of glycals with *p*-toluene chlorosulfide. These compounds were then treated with cyanating reagents to form *C*-cyanoglycosides. However, as shown in Scheme 2.3.43, allyltrimethylsilane was also reacted with these arylsulfonium compounds allowing the formation of *C*-allylglycosides with high β selectivity. Interestingly, when acetate protecting groups were used, a substantial loss of stereoselectivity was observed with a minor decrease in the yield.

Scheme 2.3.43   Allylations *via* Arylsulfides

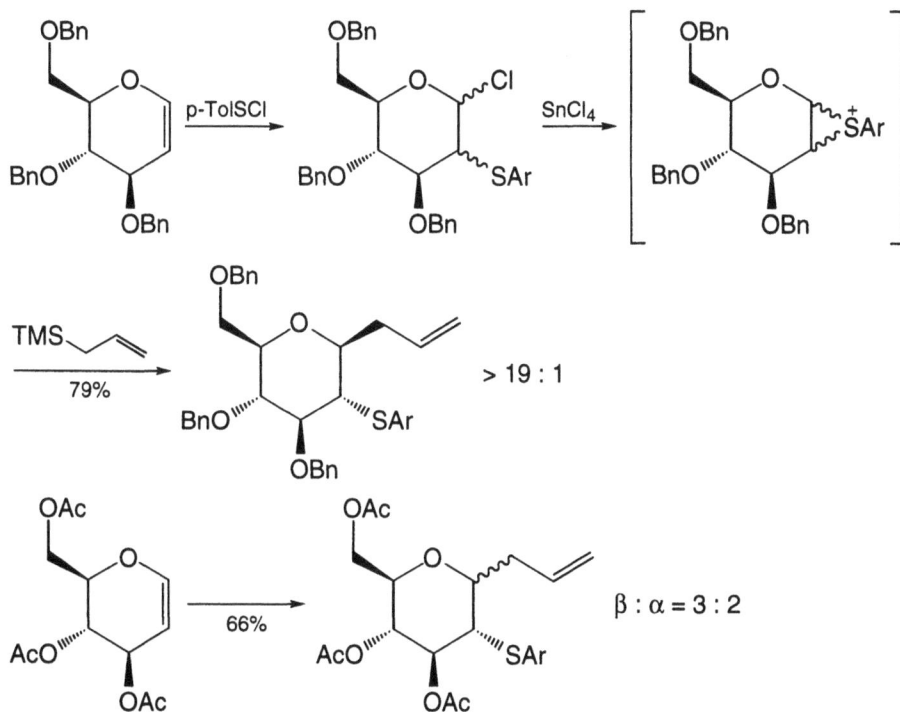

Thus far, the nucleophiles discussed as viable reagents for *C*-glycosidations have centered around silylated compounds. However, no discussion of *C*-glycosides would be complete without addressing the variety of nucleophiles available for the formation of such compounds. Therefore, the remainder of this section will focus on non-silylated reagents and, specifically, tin and aluminum-derived nucleophiles.

2.3.11   *C-Glycosidations with Organotin Reagents*

Allyltin reagents are useful reagents for the preparation of *C*-glycosides. Specifically, their versatility is noted in their ability to produce either α or β *C*-glycosides depending upon the reaction conditions used. As shown in Scheme 2.3.44, Keck, *et al.*,[69] initiated such reactions on phenylsulfide activated pyranose and furanose sugars. Aside from yields exceeding 80% in all cases, specific stereochemical consequences were observed. For example, utilizing pyranose sugars as *C*-glycosidation substrates, photochemical initiated *C*-glycosidations favored the formation of the α anomer while Lewis acid mediated reactions favored the β. Similar results were observed for furanose sugars.   However, an important deviance is the lack of stereospecificity observed under photolytic reaction conditions.

Scheme 2.3.44   *C*-Glycosidation *via*
Allyltin Reagents

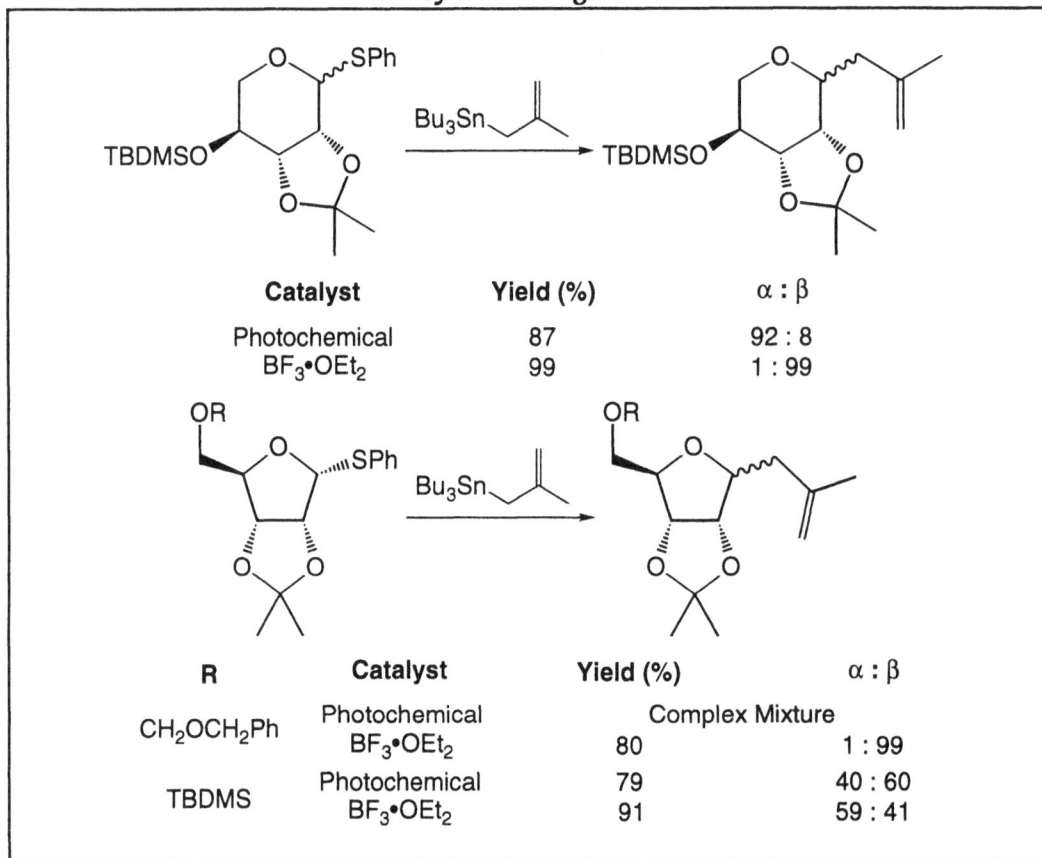

Catalyst	Yield (%)	α : β
Photochemical	87	92 : 8
BF$_3$•OEt$_2$	99	1 : 99

R	Catalyst	Yield (%)	α : β
CH$_2$OCH$_2$Ph	Photochemical	Complex Mixture	
	BF$_3$•OEt$_2$	80	1 : 99
TBDMS	Photochemical	79	40 : 60
	BF$_3$•OEt$_2$	91	59 : 41

Scheme 2.3.45   *C*-Glycosidation *via*
Allyltin Reagents

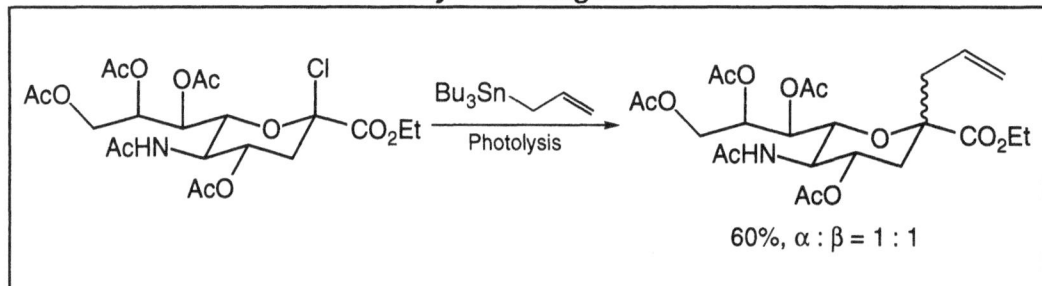

60%, α : β = 1 : 1

Although anomeric selectivity is generally difficult to observe, photolytic reaction conditions have found tremendous utility in the preparation of anomeric mixtures of *C*-glycosides derived from sialic acid and related compounds. As shown in Scheme 2.3.45, Nagy, *et al.*,[70] induced such a reaction to obtain a 60% yield of the desired product mixture. Additional examples of these reactions were reported by Waglund, *et al.*,[71] and are illustrated in Scheme 2.3.46.

Scheme 2.3.46   *C*-Glycosidation *via*
Allyltin Reagents

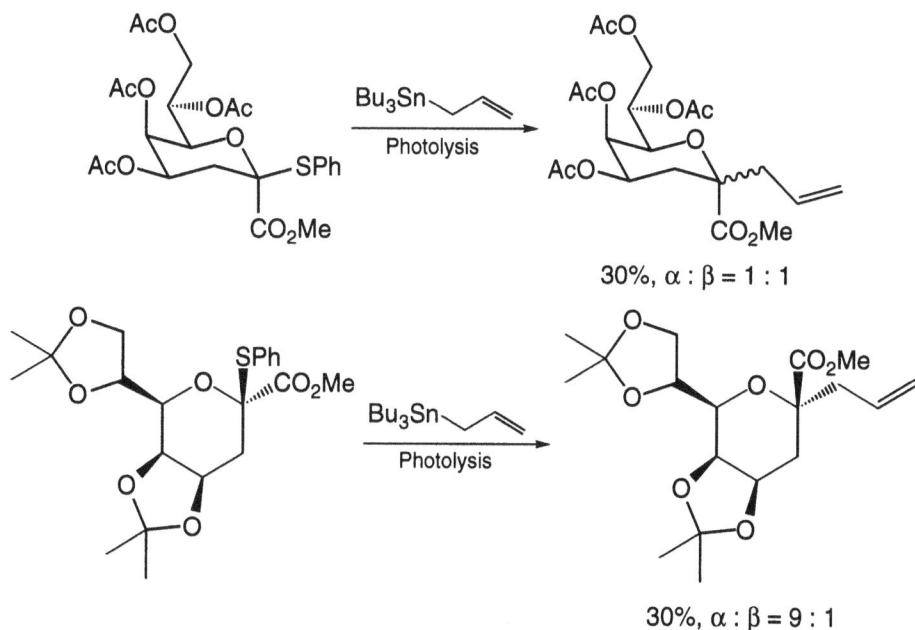

30%, $\alpha : \beta = 1 : 1$

30%, $\alpha : \beta = 9 : 1$

Scheme 2.3.47   *C*-Glycosidation *via*
Acetylenic Tin Reagents

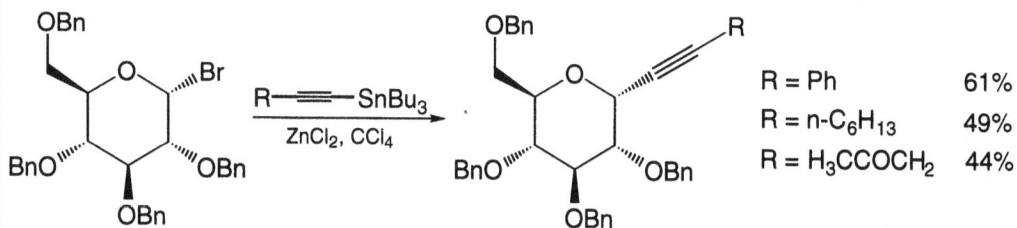

R = Ph	61%
R = n-C$_6$H$_{13}$	49%
R = H$_3$CCOCH$_2$	44%

The chemistry of allyltin compounds, with the exception of photochemical reactions, is analogous to that of allylsilanes and translates to the use of acetylenic tin reagents. As a final illustration of the utility of tin reagents to the formation of C-glycosides, the use of acetylenic tin reagents, shown in Scheme 2.3.47, is addressed. In the cited report, Zhai, et al.,[72] observed the formation of α-C-acetylenic glycosides in good yields. The reactions were catalyzed by zinc chloride and run in carbon tetrachloride. Although the yields were not optimal, the versatility and applicability of this technology is noted in its complimentarity to techniques discussed thus far.

2.3.12   C-Glycosidations with Organoaluminum Reagents

Complimentary to the chemistry of tin nucleophiles utilized in the formation of C-glycosides is the related chemistry surrounding aluminum nucleophiles. The more common examples of this chemistry is applied to haloglycosides as glycosidation substrates. Thus, as shown in Scheme 2.3.48, Tolstikov, et al.,[73] treated 2,3,4,6-tetra-O-benzyl-α-D-glucopyranosyl bromide with a variety of aluminum reagents effecting the formation of C-glycosides in good yields with demonstrated versatility in stereoselectivity. Specifically, no advantageous stereoselectivity was observed utilizing a trimethylsilylacetylene reagent, α selectivity was observed with ethyl and naphthyl reagents, and β selectivity was observed utilizing tri-n-butyl aluminum.

Scheme 2.3.48   C-Glycosidation via
                              Aluminum Reagents

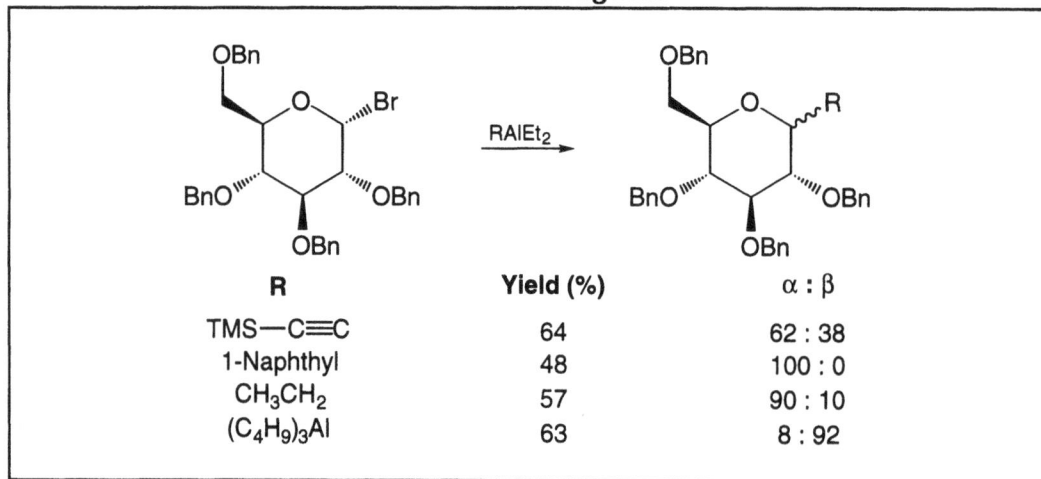

R	Yield (%)	α : β
TMS—C≡C	64	62 : 38
1-Naphthyl	48	100 : 0
CH$_3$CH$_2$	57	90 : 10
(C$_4$H$_9$)$_3$Al	63	8 : 92

Additional examples of the utility of aluminum reagents are taken from a report by Posner, et al.,[74] and are illustrated in Scheme 2.3.49. As shown, utilizing both saturated and unsaturated nucleophiles, good to high yields were obtained on reaction with fluorofuranosides. Additionally, all reactions demonstrated a preference for α selectivity. However, similar reactions, applied to the fluoropyranoside, shown in Scheme 2.3.50, produced high yields with substantially lower anomeric selectivity. Thus, combined with the examples of Scheme 2.3.48, these results clearly demonstrate the utility of organoaluminum reagents for the preparation of *C*-glycosides.

Scheme 2.3.49   *C*-Glycosidation *via*
                                  Aluminum Reagents

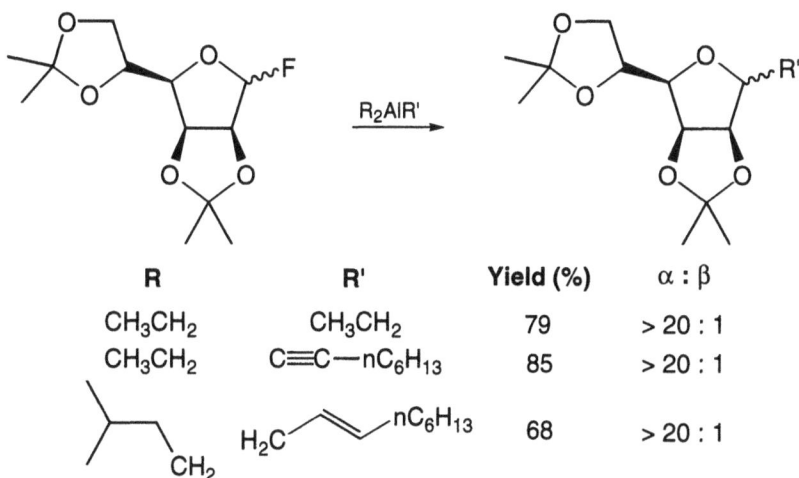

R	R'	Yield (%)	α : β
CH_3CH_2	CH_3CH_2	79	> 20 : 1
CH_3CH_2	$C\equiv C-nC_6H_{13}$	85	> 20 : 1
(isobutyl group) CH_2	H_2C⌒⌒nC_6H_{13}	68	> 20 : 1

Scheme 2.3.50   *C*-Glycosidation *via*
                                  Aluminum Reagents

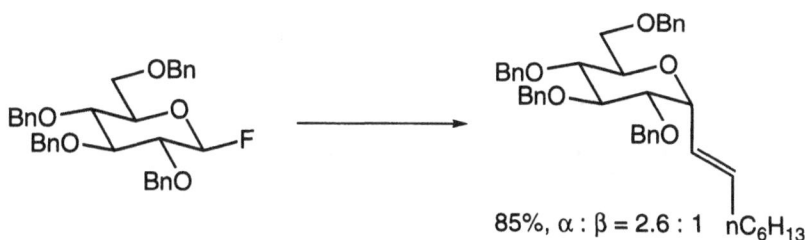

85%, α : β = 2.6 : 1 nC_6H_{13}

2.4   Arylation Reactions

In comparison to the reactions described in the previous section, arylation reactions, related to allylation reactions and illustrated in Figure 2.4.1, can be more convenient and versatile. For example, in most cases, Friedel-Crafts methodology applies. Additionally, not only is the conversion from *O* to *C*-glycosides feasible, but both α and β anomers are accessible.

Before exploring the various preparations of *C*-arylglycosides, it should be noted that an added feature of these compounds is the ease with which stereochemistry can be assigned. Specifically, a report by Bellosta, *et al.*,[75] showed that *C*-arylglucosides and *C*-arylmannosides exhibited characteristic differences in their chemical shifts and C-H coupling constants. As shown in Figure 2.4.2, the C_1 carbon of the α anomers consistently appeared down field from the β anomers. Additionally, the coupling constants of the α anomers were approximately 10 Hz smaller than those for the β anomers.

Figure 2.4.1   Example of Arylation Reactions

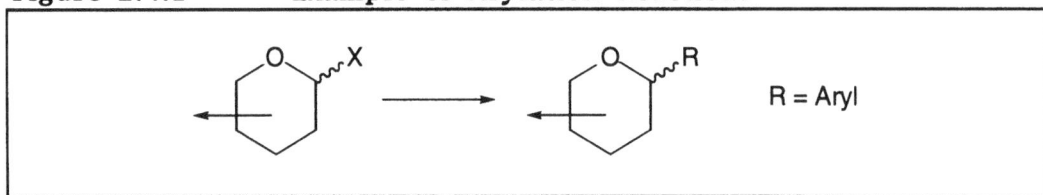

Figure 2.4.2   *C*-Arylglycoside Chemical Shifts and Coupling Constants

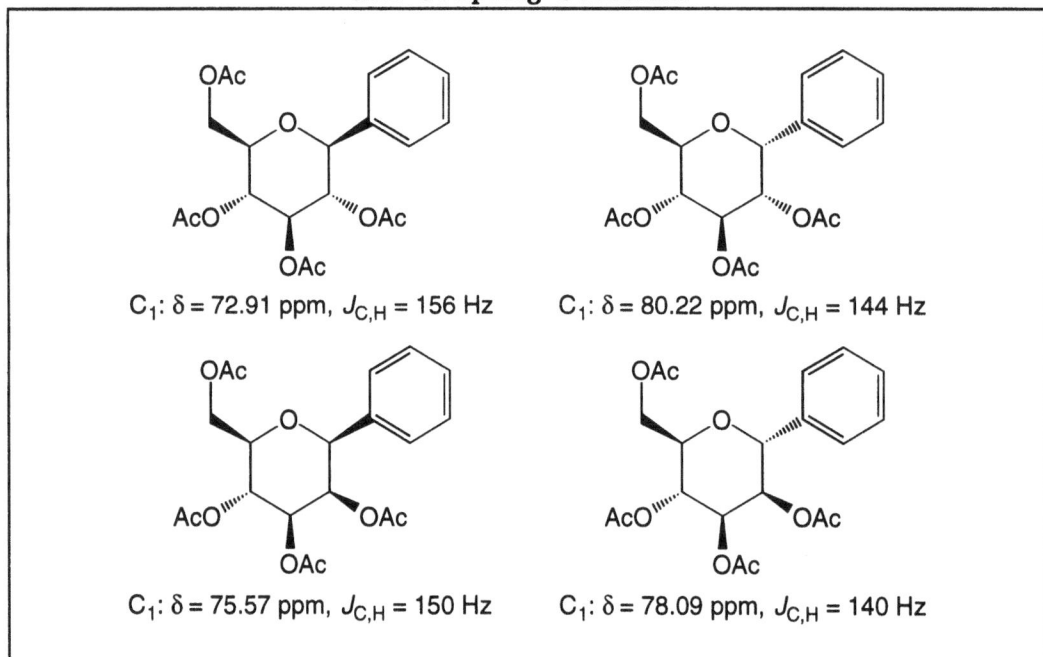

C_1: δ = 72.91 ppm, $J_{C,H}$ = 156 Hz

C_1: δ = 80.22 ppm, $J_{C,H}$ = 144 Hz

C_1: δ = 75.57 ppm, $J_{C,H}$ = 150 Hz

C_1: δ = 78.09 ppm, $J_{C,H}$ = 140 Hz

2.4.1   Reactions with Metallated Aryl Compounds

In beginning a discussion of the chemistry surrounding the preparation of *C*-arylglycosides, let us recall the work illustrated in Scheme 2.3.48 where Tolstikov, *et al.*,[73] treated 2,3,4,6-tetra-*O*-benzyl-α-D-glucopyranosyl bromide with a variety of aluminum reagents effecting the formation of *C*-glycosides in good yields with demonstrated versatility in stereoselectivity. One particular example from this report involved the use of 1-naphthyl-diethylaluminum. This particular reaction, in spite of its 48% yield, exhibited total α selectivity. Observations of the ability of arylaluminum species forming *C*-glycosides have also been noted with heterocyclic aromatic species. As shown in Scheme 2.4.1, Macdonald, *et al.*,[76] utilized furanyl-diethylaluminum in the formation of *C*-furanylglycosides. The remarkable observation noted was the ability of this reagent to effect reaction with retention of the anomeric configuration of the starting sugar. Similar results were also observed with the corresponding pyrrole compound.

Scheme 2.4.1 *C*-Glycosidation *via* Aluminum Reagents

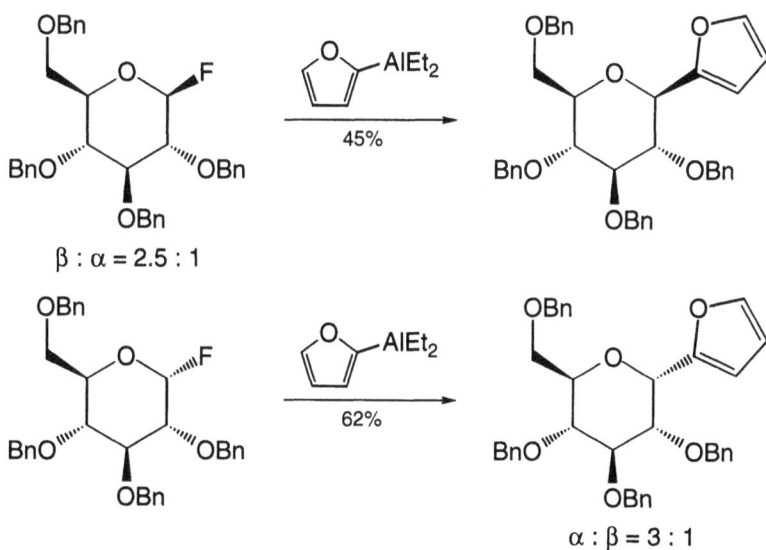

Aryltin reagents have been used similarly to arylaluminum compounds for the preparation of *C*-glycosides. Particularly applied to reactions with furanose glycals, Daves, *et al.*,[77,78] carried our the reactions shown in Schemes 2.4.2 and 2.4.3. Utilizing palladium acetate to mediate the couplings between glycals and aryltributyltin compounds, β anomeric selectivity was observed in poor to good yields.

Scheme 2.4.2   *C*-Glycosidation *via* Tin Reagents

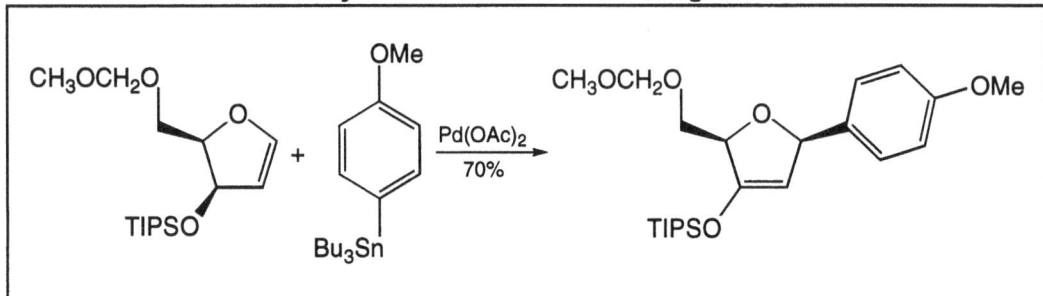

Scheme 2.4.3   *C*-Glycosidation *via* Tin Reagents

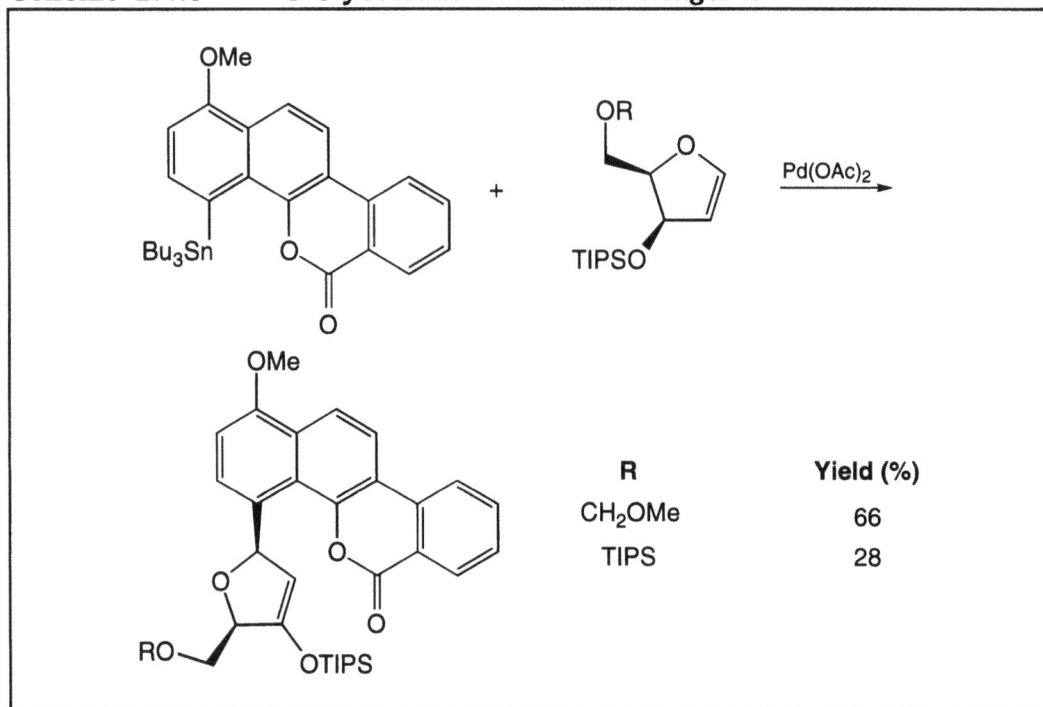

R	Yield (%)
CH$_2$OMe	66
TIPS	28

2.4.2   Electrophilic Aromatic Substitutions with Acetoxy, Trifluoroacetoxy, and PNB Glycosides

Although the use of aluminum and tin reagents have provided fruitful approaches to the preparation of *C*-arylglycosides, the majority of this chemistry centers around reactions utilizing Lewis acid catalysis. Of particular importance are reactions proceeding *via* electrophilic aromatic substitution. This type of chemistry has been widely applied to a sugars bearing a variety of activating groups. One particular example, shown in Scheme 2.4.4, involves the coupling of an acetate activated ribofuranose with 1-methylnaphthalene. This reaction, reported by Hamamichi, *et al.*,[79] proceeded in 79% with the formation of the α anomer comprising only 8.3% of the product mixture.

Scheme 2.4.4   *C*-Glycosidation *via* Electrophilic
Aromatic Substitutions

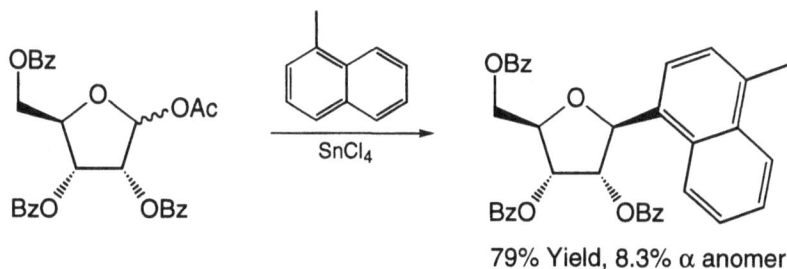

79% Yield, 8.3% α anomer

Scheme 2.4.5   *C*-Glycosidation *via* Electrophilic
Aromatic Substitutions

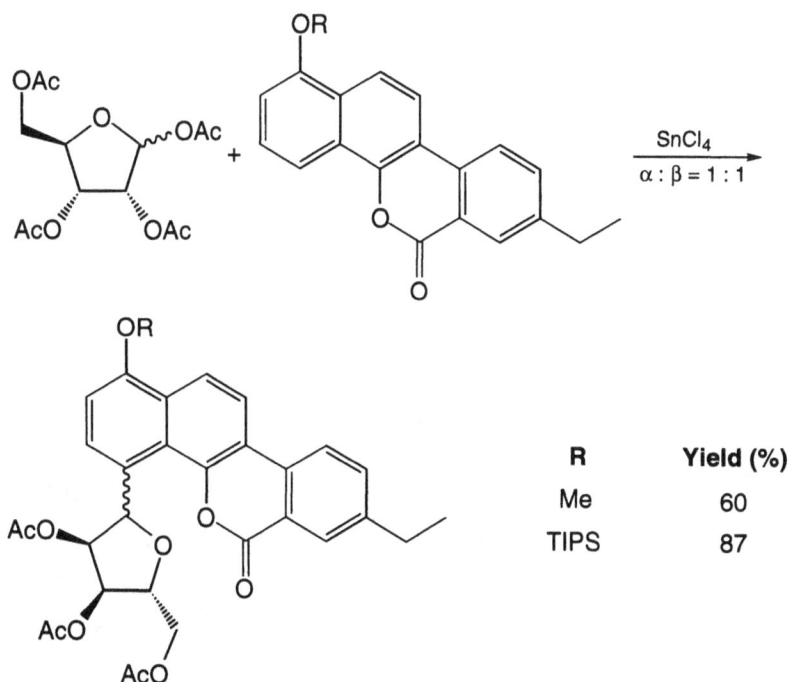

R	Yield (%)
Me	60
TIPS	87

Additional examples utilizing acetate activating groups were reported by Daves, *et al.*,[80,81] and are related to the reaction shown in Scheme 2.4.3. The cited reactions, illustrated in Scheme 2.4.5 proceeded in good yields with low anomeric specificity. However, with the ease of separation of the anomers, functional yields could be obtained on exposing the undesired isomer to tin chloride thus forming a 1 : 1 anomeric mixture ready for separation. The compounds shown in this example were used in the development of synthetic strategies to the preparation of the antibiotics gilvocarcin, ravidomycin, and chrysomycin shown in Figure 2.4.3.

Figure 2.4.3   *C*-Glycosidic Antibiotics

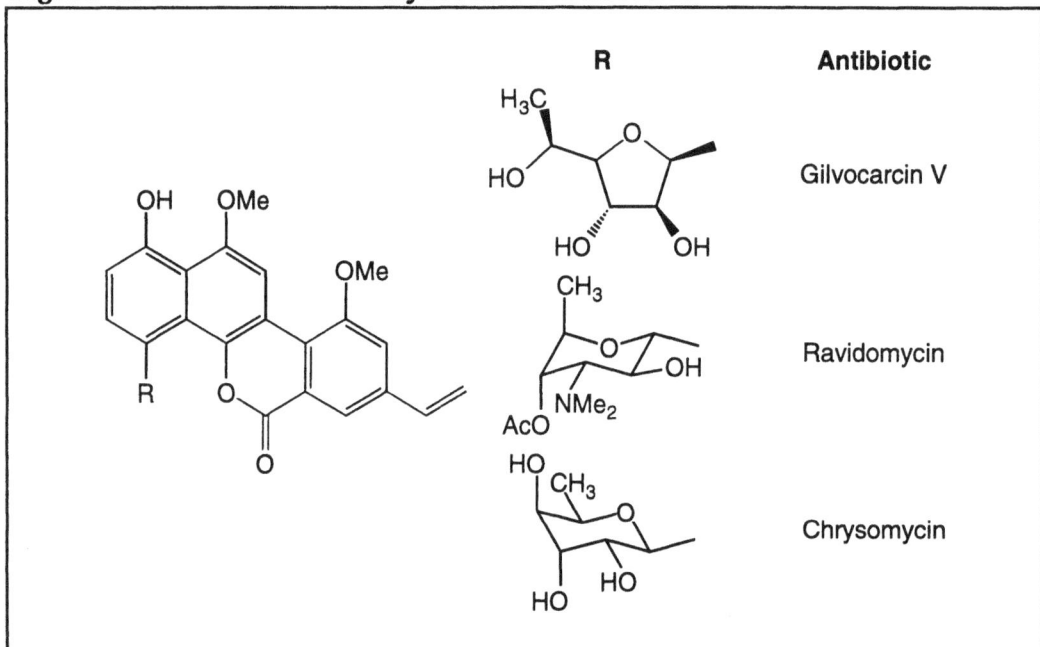

Scheme 2.4.6   *C*-Glycosidation *via* Electrophilic
Aromatic Substitutions

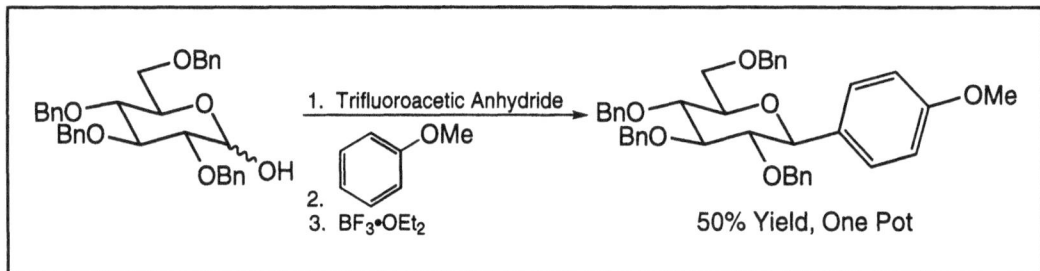

Related to the acetate activating group is the more labile trifluoroacetate group. In a reaction between methylphenol and 1-trifluoroacetoxy-2,3,4,6-tetra-*O*-benzyl-D-glucopyranose, shown in Scheme 2.4.6, Allevi, *et al.*,[53] observed the formation of a 50% yield of the β coupled product. This reaction actually proceeded through a one pot process in which the trifluoroacetate was formed followed by introduction of the nucleophile and addition of the Lewis acid. In spite of the relatively low yield, the ease of this chemistry and selectivity for the β anomer illustrates its usefulness.

**Scheme 2.4.7   *C*-Glycosidation *via* Electrophilic
                                    Aromatic Substitutions**

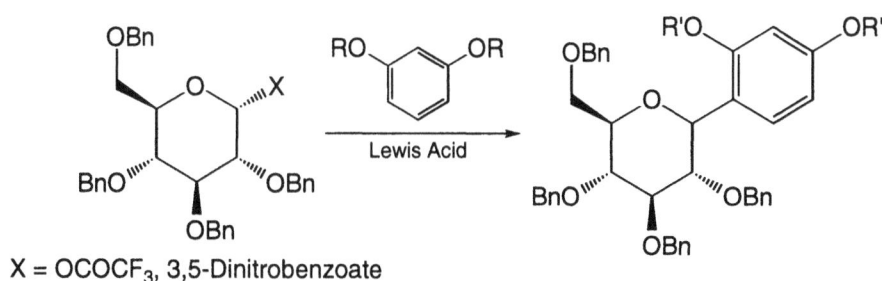

X = OCOCF₃, 3,5-Dinitrobenzoate

Lewis Acid	R	R'	Yield (%)	α : β
BF₃•OEt₂, 5 min.	Me	Me	81	0 : 100
BF₃•OEt₂, 5 min.	TMS	H	95	100 : 0
ZnBr₂, 15 min.	TMS	H	67	0 : 100

**Scheme 2.4.8   *C*-Glycosidation *via* Electrophilic
                                    Aromatic Substitutions**

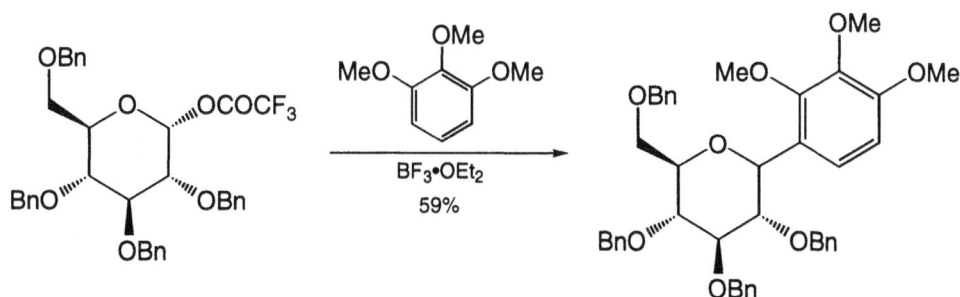

In a subsequent study regarding the α and β selectivity of electrophilic aromatic substitutions, Cai, *et al.*,[82,83] demonstrated that, in addition to aromatic ring substituents, selectivity is also related to the Lewis acid used. The study, illustrated in Scheme 2.4.7, demonstrated a complete reversal in selectivity when the Lewis acid was changed from borontrifluoride etherate to zinc chloride.   Unlike the previous example where only the trifluoroacetate activating group was used, the anomeric position on the sugars in these examples were also activated with the 3,5-dinitrobenzoate group.

As a final example utilizing the trifluoroacetate anomeric activating group, Schmidt, *et al.*,[84] coupled 1,2,3-trimethoxybenzene with the glucose derivative shown.   As illustrated in Scheme 2.4.8, this reaction proceeded with high β selectivity giving the product in 59% yield.   This reaction was utilized in the preparation of *C*-glycosidic bergenin analogs.

2.4.3   Electrophilic Aromatic Substitutions on Trichloroacetimidates

Scheme 2.4.9   *C*-Glycosidation *via* Electrophilic Aromatic Substitutions

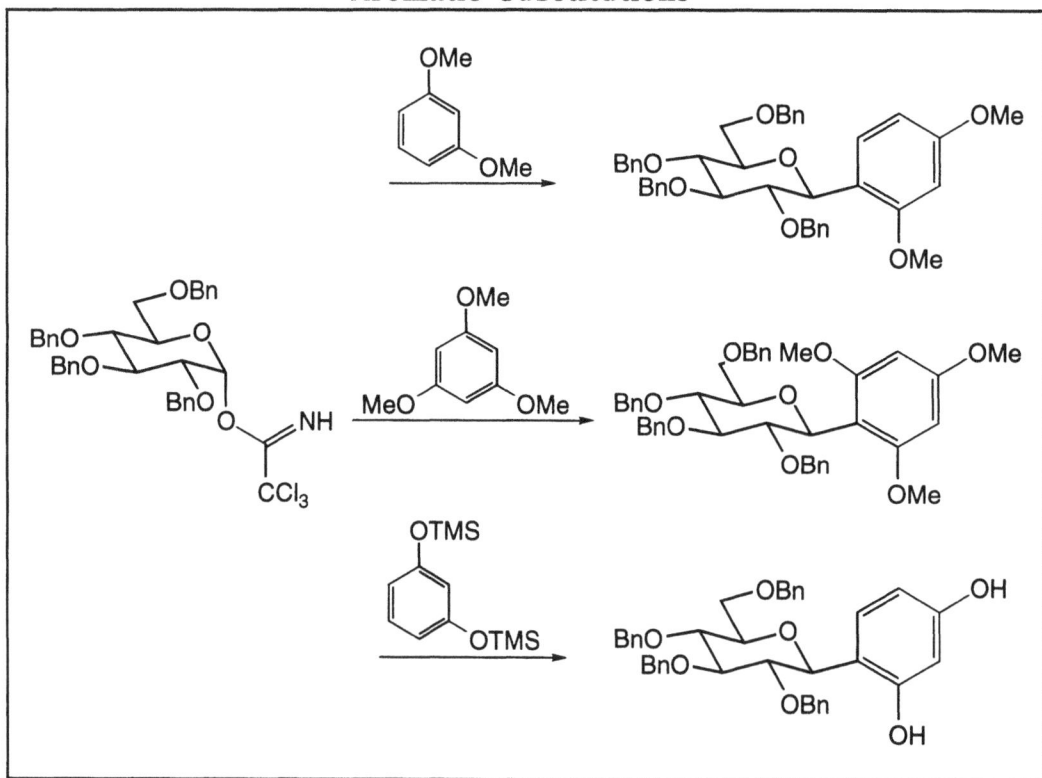

In one of the earliest demonstrations of *C*-arylation chemistry, Schmidt, *et al.*,[85] utilized easily prepared trichloroacetimidates to prepare the products shown in Scheme 2.4.9. In all cases, the β anomer was the preferred product.

An interesting observation is the demonstrated ability to utilize both alkyl and silyl phenolic protecting groups. Furthermore, in an extension of this work to heterocyclic substrates, a variety of furans were utilized.[86] The results from this study, shown in Scheme 2.4.10, exhibited a strong preference for formation of the α anomer as opposed to the observations shown in Scheme 2.4.9.

Scheme 2.4.10   *C*-Glycosidation *via* Electrophilic Aromatic Substitutions

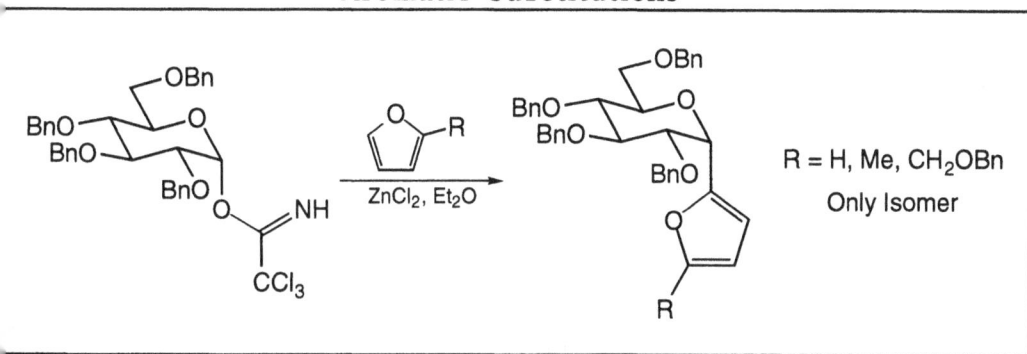

Scheme 2.4.11   *C*-Glycosidation *via* Electrophilic Aromatic Substitutions

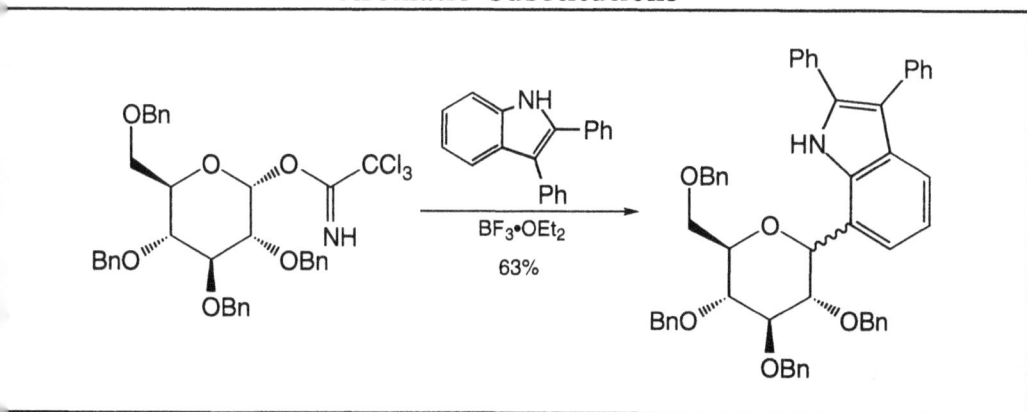

As a final example illustrating the versatility of trichloroacetimidates in the preparation of *C*-glycosides, Schmidt, *et al.*,[87] reported coupling the glucose analog, shown, with 2,3-diphenylindole and related compounds. Although the reaction, shown in Scheme 2.4.11, proceeded with no anomeric selectivity, the 63% yield included none of the *N*-glycosidation product which might have formed and subsequently undergone an *N-C* migration. Interestingly, some *N*-glycosidation was observed when the glucose was activated with an anomeric acetate group.

2.4.4   Electrophilic Aromatic Substitutions on Haloglycosides

As acetate and related activating groups have been useful in the formation of C-arylglycosides, so have halides. In the first of two examples introducing the usefulness of this technology, Matsumoto, et al.,[88] published a report on a zirconicene dichloride-silver perchlorate complex as a promoter of Friedel-Crafts reactions in the preparation of C-glycosides. The reaction studied is shown in Scheme 2.4.12 and some of the results are summarized in Table 2.4.1. In this study, the first three runs proceeded with glycosidation occurring at C_4, exclusively. Runs 4-6 proceeded with glycosidation occurring at C_1, and runs 7-10 followed routes similar to the first three runs. What is important to note is the control over the anomeric configuration of the product. For example where run 4 proceeded with high β selectivity using 5 equivalents of the catalyst, when only 0.2 equivalents were used, the selectivity was reversed giving high α selectivity.

Table 2.4.1   C-Glycosidation *via* Electrophilic
Aromatic Substitutions

Run	Aryl-H	Lewis Acid Equivalents	Reaction Time	Yield (%)	α : β
1	OMe	5.0	20 min	96	0 : 100
2		2.0	45 min	91	1 : 12
3		0.3	2 hr	79	2 : 1
4	MeO	5.0	30 min	77	1 : 13
5		2.0	30 min	58	2 : 1
6		0.2	15 hr	40	21 : 1
7	OMe / OMe	5.0	20 min	59	0 : 100
8		2.0	20 min	64	0 : 100
9		0.2	75 min	50	0 : 100

Scheme 2.4.12   C-Glycosidation *via* Electrophilic
Aromatic Substitutions

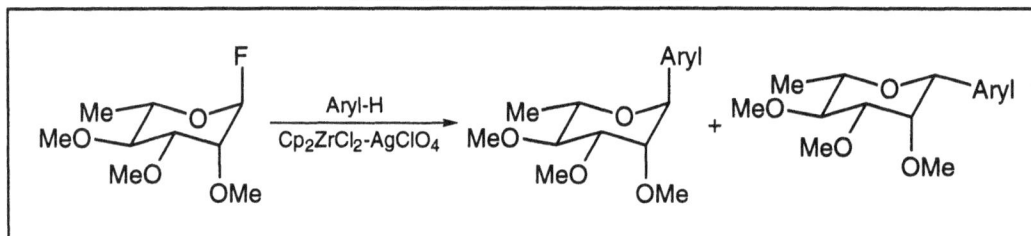

In the second example of the utility of glycosyl halides as substrates for the formation of *C*-arylglycosides, Allevi, *et al.,*[89] demonstrated conditions utilizing silver triflate as the catalyst. This study, summarized in Scheme 2.4.13, proceeded in only 40% yield with total β selectivity. Thus, the developing trend, compared to that in the formation of *C*-allylglycosides is that *C*-arylglycosidations have a tendency to allow the formation of product mixtures highly enriched with β anomers.

Scheme 2.4.13   *C*-Glycosidation *via* Electrophilic Aromatic Substitutions

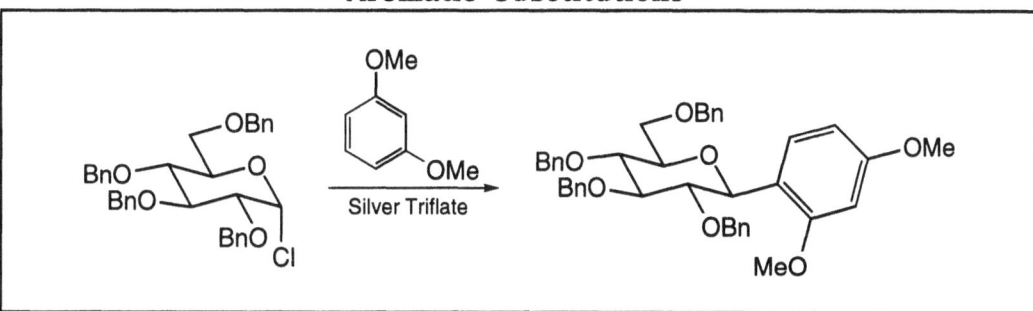

2.4.5   Electrophilic Aromatic Substitutions on Glycals

Scheme 2.4.14   *C*-Glycosidation *via* Electrophilic Aromatic Substitutions

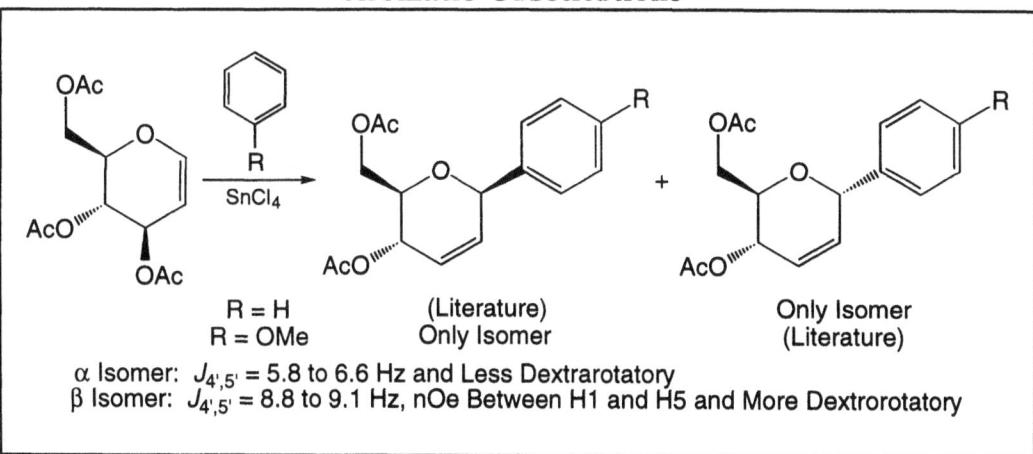

To this point, all examples for the formation of *C*-aryl glycosides utilized Friedel-Crafts type conditions with a propensity for selectively producing β anomers. These reactions were all directed at sugars bearing glycosidic activating groups. However, in 1988, Casiraghi, *et al.,*[90] demonstrated the selective accessibility of both α and β anomers directed by the nature of aromatic ring substitution. This study was applied to tri-*O*-acetylglucal and is illustrated in Scheme 2.4.14. Additionally, the NMR results defining coupling constants and nOe observations characteristic of each anomer are also shown. As stated, "in the D-series, for a given anomeric pair the more dextrorotatory member having $J_{4,5}$ of 8.7-10 Hz and relevant positive nOe between H_1 and H_5 should be assigned as β-D, whilst the other displaying a $J_{4,5}$ value in the range of 3-9 Hz and no nOe between H_1 and H_5 is to be named α-D."

Scheme 2.4.15   *C*-Glycosidation *via* Electrophilic Aromatic Substitutions

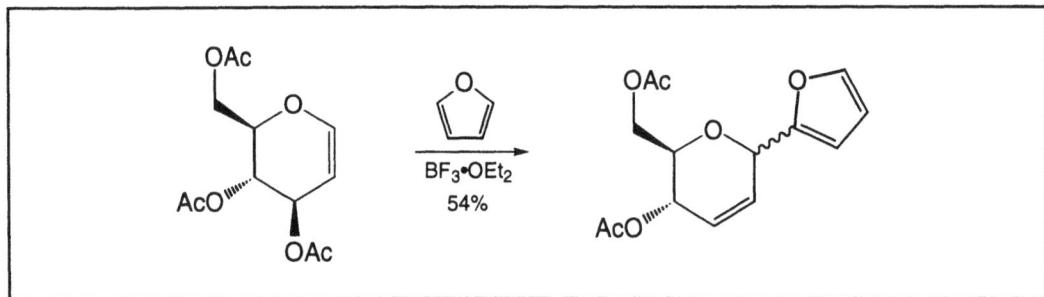

In one additional example of the formation of *C*-arylglycosides from glycals, Ichikawa, *et al.*,[65] illustrated the applicability of this chemistry to heterocyclic aryl species. Specifically, as shown in Scheme 2.4.15, tri-*O*-acetoxy glucal was treated with furan and borontrifluoride etherate. The result was a 54% yield of the desired *C*-furanoglycoside as a 1 : 1 anomeric mixture.

2.4.6   Intramolecular Electrophilic Aromatic Substitutions

Scheme 2.4.16   *C*-Glycosidation *via* Intramolecular Electrophilic Aromatic Substitutions

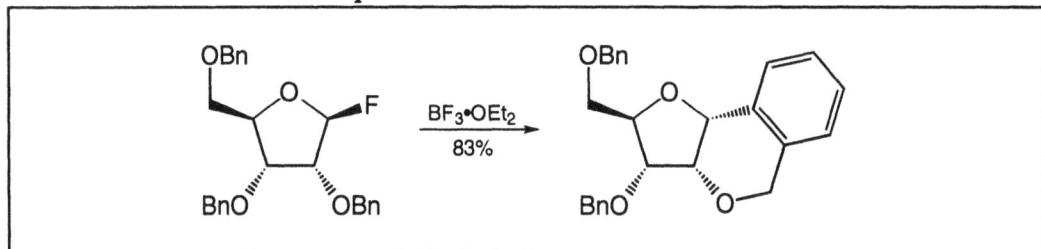

As electrophilic aromatic substitutions have provided access to *C*-arylglycosides *via* intermolecular reactions, similar technology is available for effecting the intramolecular delivery of the aromatic species. For example, utilizing the fluorofuranoside shown in Scheme 2.4.16, Araki, *et al.*,[91] achieved formation of an 83% yield of the bicyclic product on treatment with borontrifluoride etherate.

**Scheme 2.4.17   *C*-Glycosidation *via* Intramolecular
Electrophilic Aromatic Substitutions**

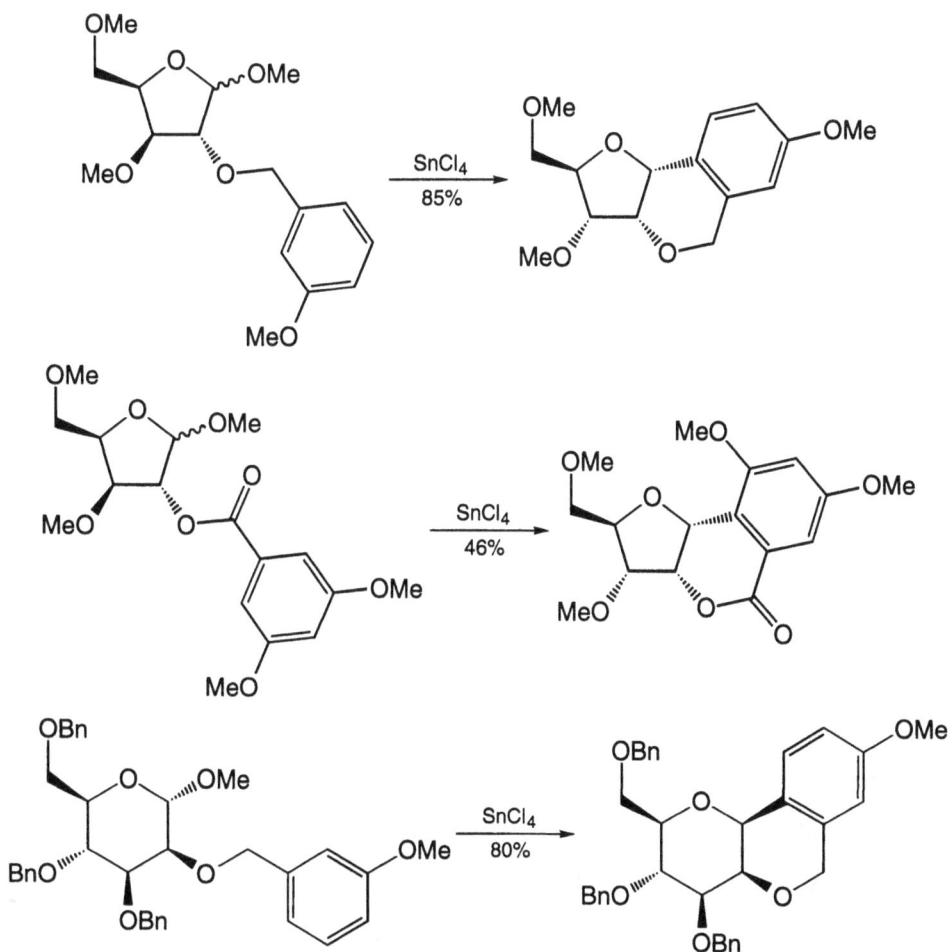

Similar results, observed by Martin, *et al.*,[92] in analogous tin chloride mediated reactions with *O*-methylglycosides. These reactions, shown in Scheme 2.4.17, illustrate the applicability of this chemistry no only to both furanose and pyranose sugars, but also to 2-*O*-benzyl and 2-*O*-benzoylated sugars. In all applicable cases, a detectable amount of the corresponding regioisomer was formed. It is interesting to note that in one case, a double *C*-glycosidation was observed.[93] This reaction, shown in Scheme 2.4.18, proceeded in 63% overall yield with opening of the furanose ring. The resulting product contained a *cis*-fused bicyclic ring system core. In all examples of this technology, the possibilities for further derivitization are apparent through hydrogenation of the benzylic ethers or hydrolysis of the analogous esters.

Scheme 2.4.18   *C*-Glycosidation *via* Intramolecular
Electrophilic Aromatic Substitutions

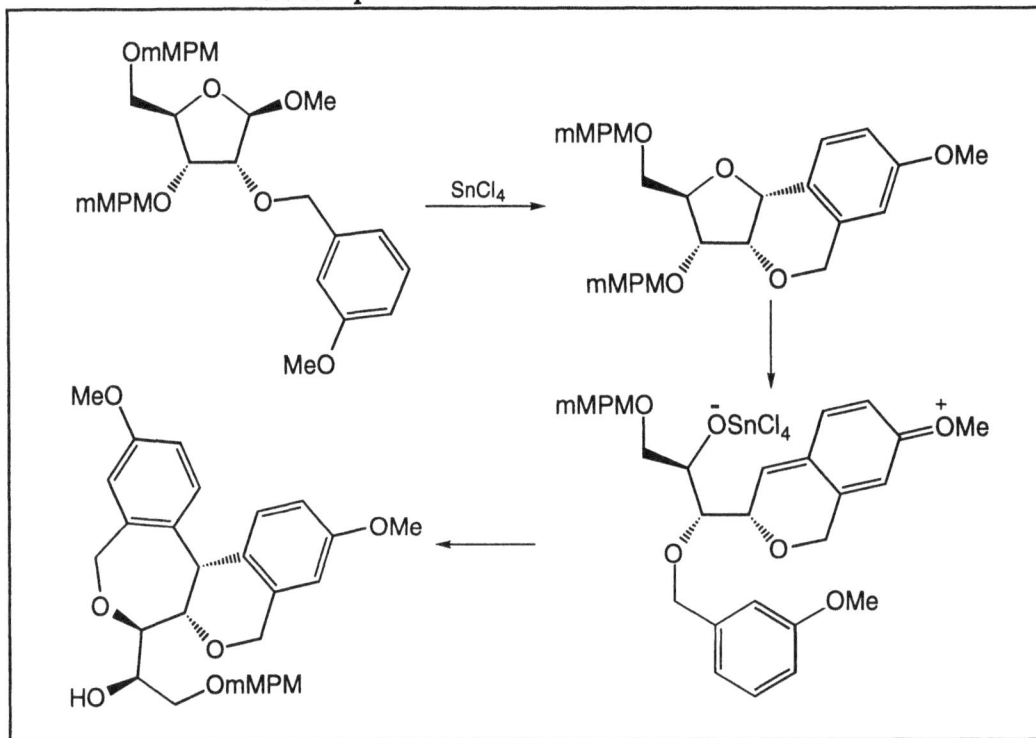

Scheme 2.4.19   *C*-Glycosidation *via* Intramolecular
Electrophilic Aromatic Substitutions

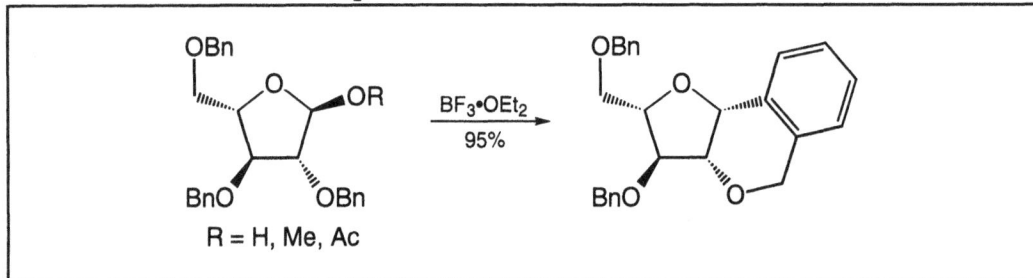

As the methoxy anomeric activating group has been useful in intramolecular *C*-glycosidations, so has the acetoxy group. As shown in Scheme

2.4.19, Anastasia, *et al.*,[94] utilized borontrifluoride etherate to effect this reaction on furanose sugars. In this study, the methoxy group was shown to be equally as useful as the acetoxy group. Furthermore, the free hydroxyl group can be used.

2.4.7   *O-C Migrations*

Scheme 2.4.20   *C*-Glycosidation *via O-C* Migrations

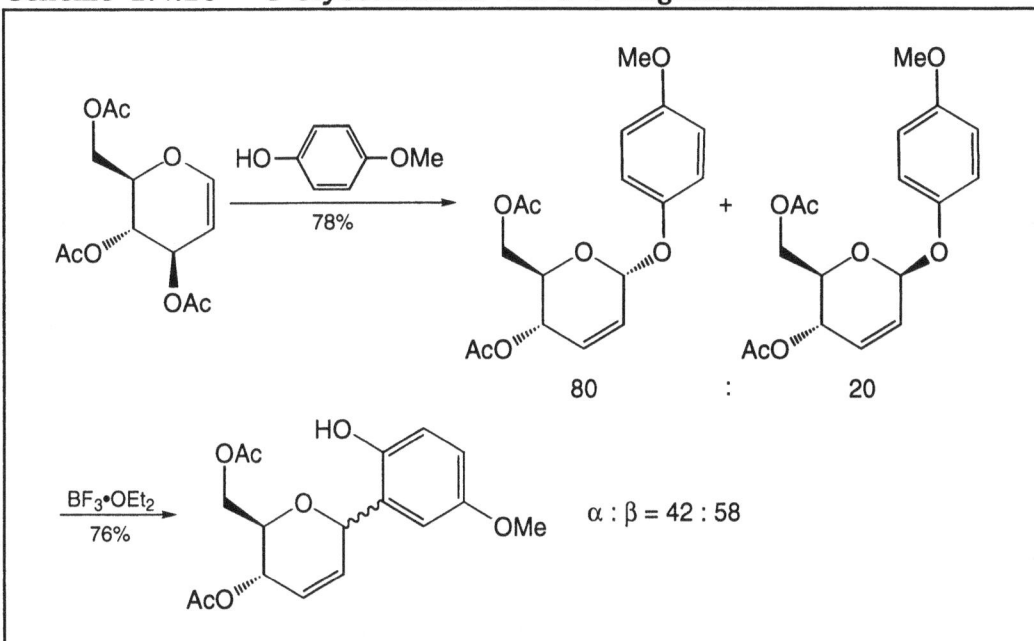

Scheme 2.4.21   *C*-Glycosidation *via O-C* Migrations

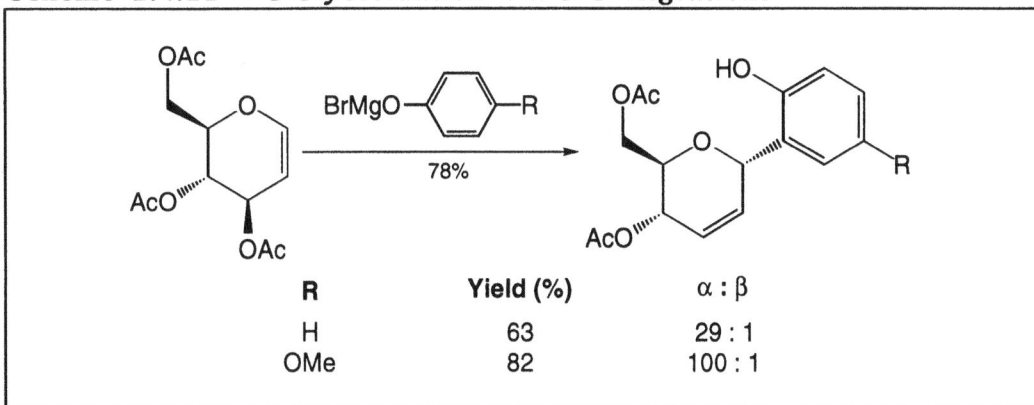

R	Yield (%)	α : β
H	63	29 : 1
OMe	82	100 : 1

Scheme 2.4.22   *C*-Glycosidation *via O-C* Migrations

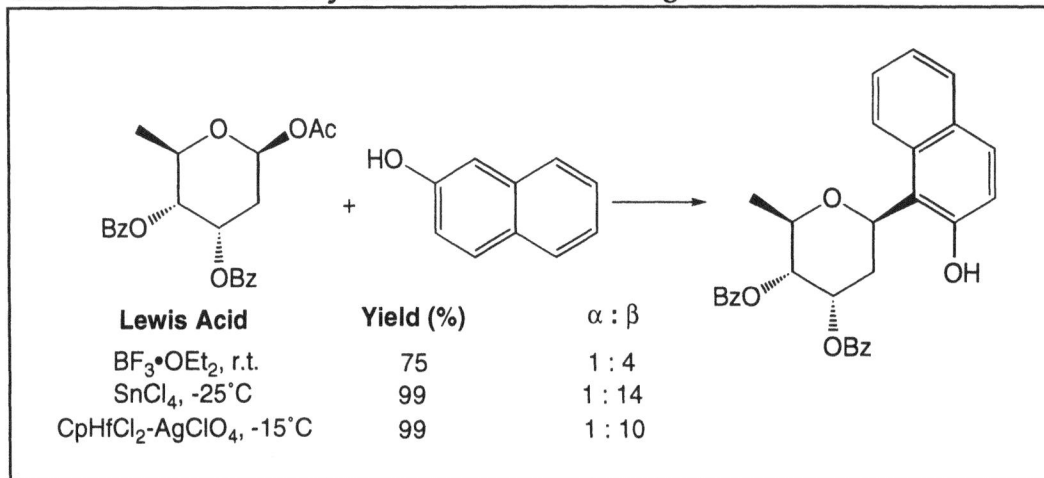

Lewis Acid	Yield (%)	α : β
BF$_3$•OEt$_2$, r.t.	75	1 : 4
SnCl$_4$, -25°C	99	1 : 14
CpHfCl$_2$-AgClO$_4$, -15°C	99	1 : 10

Moving away from electrophilic aromatic substitutions as a means of forming *C*-arylglycosides, perhaps the most notable reaction is the *O-C* migration. This reaction is proposed to proceed through initial formation of an *O*-glycoside. This *O*-glycoside then rearranges, in the presence of Lewis acids, to a *C*-glycoside. Direct evidence for the *O*-glycoside intermediates was provided by Ramesh, *et al.*[95] As shown in Scheme 2.4.20, *para*-methoxyphenol was coupled with the glycal, shown, to give the *O*-glycoside as a mixture of anomers. Regardless of the stereochemical outcome of the initial *O*-glycosidation, all stereochemistry at the anomeric center was lost on formation of the *C*-glycoside.

Scheme 2.4.23   *C*-Glycosidation *via O-C* Migrations

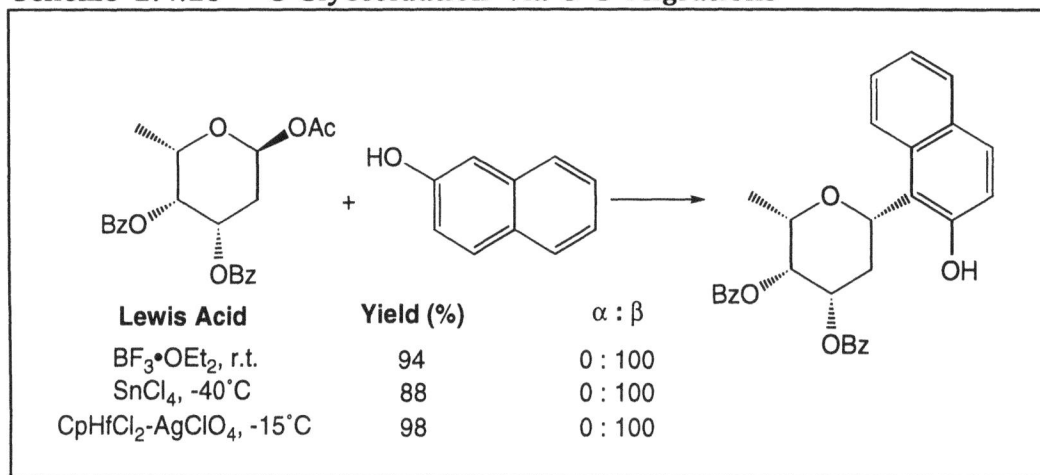

Lewis Acid	Yield (%)	α : β
BF$_3$•OEt$_2$, r.t.	94	0 : 100
SnCl$_4$, -40°C	88	0 : 100
CpHfCl$_2$-AgClO$_4$, -15°C	98	0 : 100

Scheme 2.4.24   *C*-Glycosidation *via* *O-C* Migrations

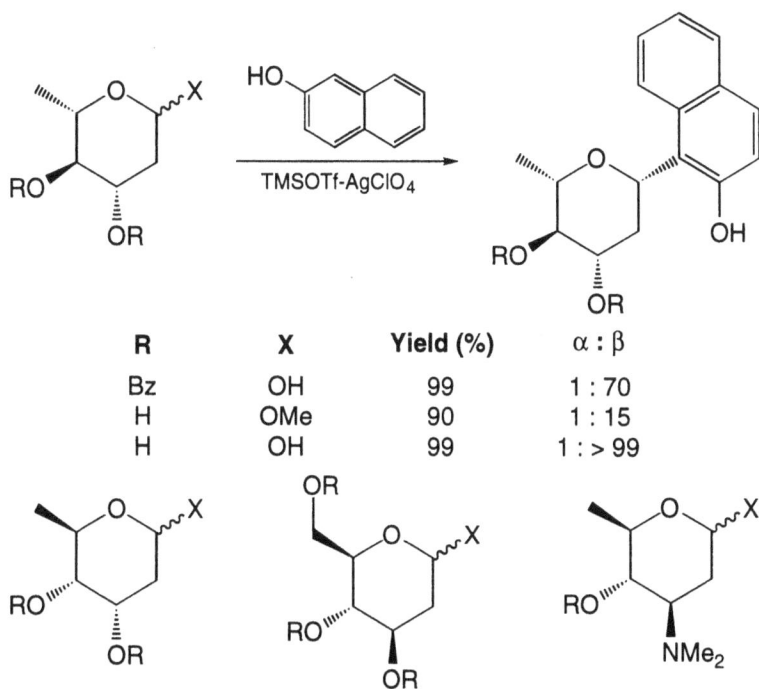

R	X	Yield (%)	α : β
Bz	OH	99	1 : 70
H	OMe	90	1 : 15
H	OH	99	1 : > 99

The study just cited was not the first example of *O-C* migrations. Rather, it was intended to introduce the reaction through a mechanistic insight. Early examples of this reaction were noted using phenolic salts. An excellent example is contained in a report by Casiraghi, *et al.*[96] As shown in Scheme 2.4.21, a glycal derivative was treated with bromomagnesium phenoxide derivatives. In the two examples shown, the yields were quite good and the anomeric preference of this reaction favored the α configuration. Subsequent research further supported these observations.[97] This result is a nice contrast to the β selectivity observed utilizing electrophilic aromatic substitution reaction to effect the formation of *C*-arylglycosides.

As glycals have been useful substrates for *O-C* migrations, so have electron deficient 2-deoxy sugars which undergo more facile *C*-glycosidations. As shown in Schemes 2.4.22 and 2.4.23, Matsumoto, *et al.*,[98] demonstrated that these substrates are easily reacted with 2-naphthol utilizing Lewis acid catalysts. Under these conditions, formation of *C*-naphthylglycosides readily occurred. It should be noted that utilizing borontrifluoride etherate as a catalyst allowed for the isolation of small amounts of *O*-glycosides from the reaction mixtures. The stereoselectivity of these reactions was shown to be influenced by the Lewis acid used as well as by steric effects. These results were confirmed by Toshima, *et al.,*[99,100] and demonstrated to be general to a variety of protected and unprotected sugars. The results, shown in Scheme 2.4.24, also illustrate that the relative reaction conditions are easily applied to unactivated glycosides.

Scheme 2.4.25   *C*-Glycosidation *via O-C* Migrations

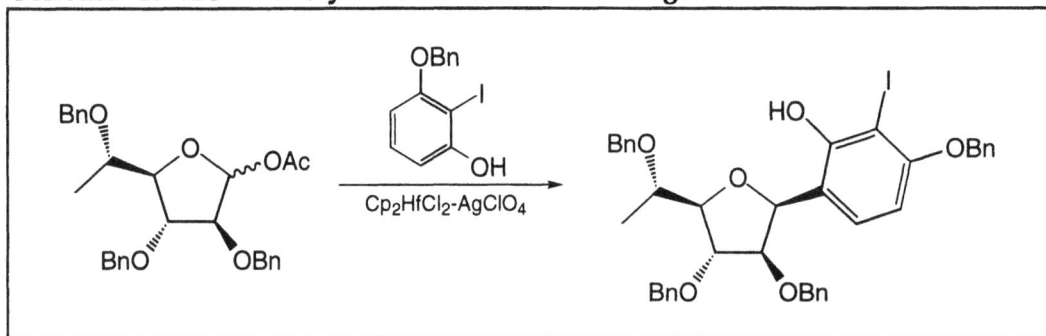

So far, the only examples of *O-C* migrations have applied to pyranose sugars or derivatives thereof. However, it should be noted that such reactions are applicable to furanose sugars. One particular example, reported by Matsumoto, *et al.,*[101] was used in the total synthesis of gilvocarcin M. The specific reaction, shown in Scheme 2.4.25, utilized the acetoxy activated furanoside shown. Treating this compound with the phenol and using the hafnocene dichloride-silver perchlorate complex as the Lewis acid, the desired *C*-arylglycoside was obtained in 87% yield with the α anomer favored in a ratio of 8 : 1.

Scheme 2.4.26   *C*-Glycosidation *via O-C* Migrations

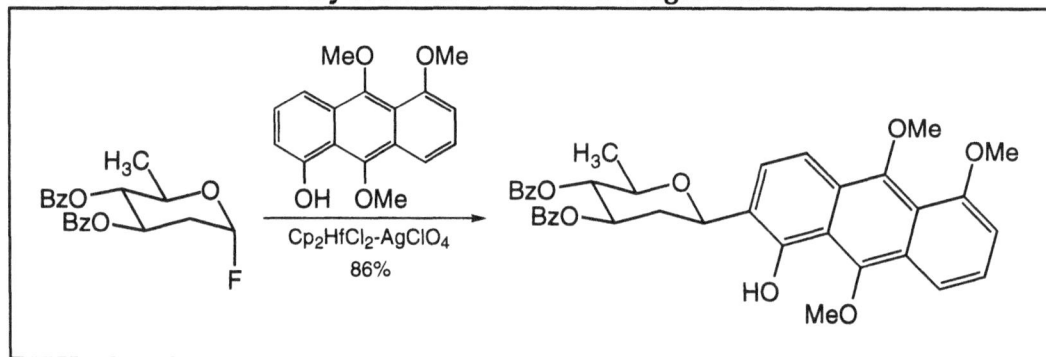

In examining these examples of *O-C* migrations, it should be noted that while the phenolic units in these illustrations are relatively simple, this chemistry has been applied to much more complex systems. An excellent illustration of this statement is contained in a report by Matsumoto, *et al.,*[102]

targeting the total synthesis of the vineomycins. The key reaction, shown in Scheme 2.4.26, involves the application of *O-C* migration chemistry to the anthracene analog shown. The substrate was the 2-deoxyfluoropyranoside shown and the Lewis acid was the hafnocene dichloride-silver perchlorate complex. The desired reaction proceeded in 86% yield with high β selectivity thus illustrating that *O-C* migrations are not necessarily specific for a given anomer. Furthermore, with the versatility contained within arylglycosidations, in general, the preparation of increasingly more complex targets becomes significantly more general.

2.5   Reactions with Enol Ethers, Silylenol Ethers and Enamines

**Figure 2.5.1   Examples of Reactions with
Silylenol Ethers and Related Compounds**

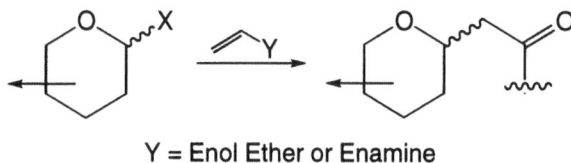

Y = Enol Ether or Enamine

As arylations are complimentary to allylations, condensations of sugars with silylenol ethers and related compounds is directly analogous to condensations with allylsilanes. Consequently, the use of enol ethers, silylenol ethers, and enamines in *C*-glycoside chemistry, illustrated in Figure 2.5.1, provide avenues to additional functionalities on the added groups. The incorporation of these compounds onto carbohydrates is accomplished through relatively general methods. The products are usually of the α configuration. However, the β isomers are accessible through epimerization. In all cases except those involving glycals, benzylated sugars are preferred. Finally, the advantage of being able to run these reactions at low temperature is offset by the need to activate the anomeric position in all cases.

2.5.1   Reactions with Enolates

The need for anomeric activation when utilizing enol ethers and related nucleophiles can be misleading. Examples have been reported illustrating the ability to couple anionic nucleophiles with unactivated sugars. For example, as shown in Scheme 2.5.1, Gonzales, *et al.*,[103] coupled barbiturates with glucosamine to give β-*C*-glycosyl barbiturates. Furthermore, as shown in Scheme 2.5.2, Yamaguchi, *et al.*,[104] combined the illustrated unactivated 2-deoxy sugar with a keto-diester to give the isolated β-*C*-glycoside.

Scheme 2.5.1   *C*-Glycosidations *via* Enolates

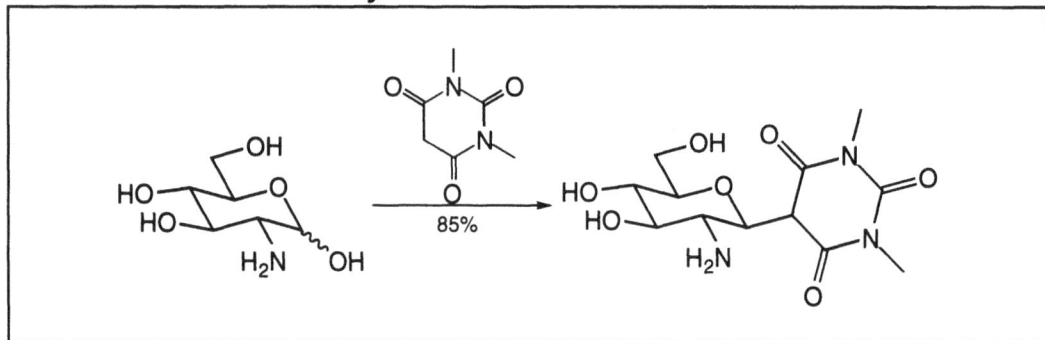

Scheme 2.5.2   *C*-Glycosidations *via* Enolates

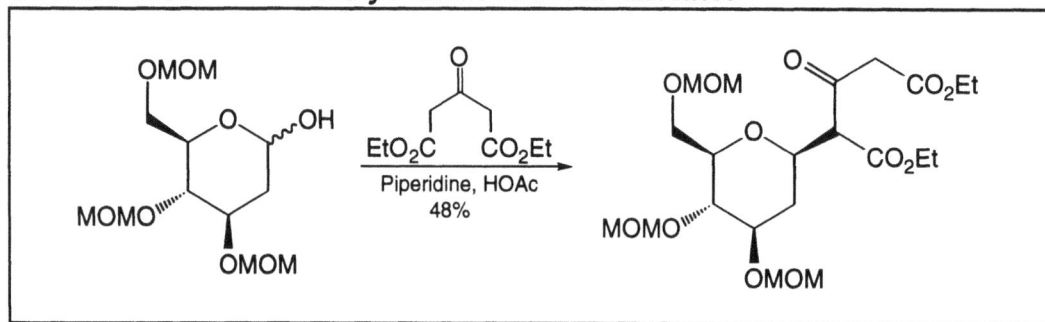

Scheme 2.5.3   Mechanism of *C*-Glycosidations
via Enolates

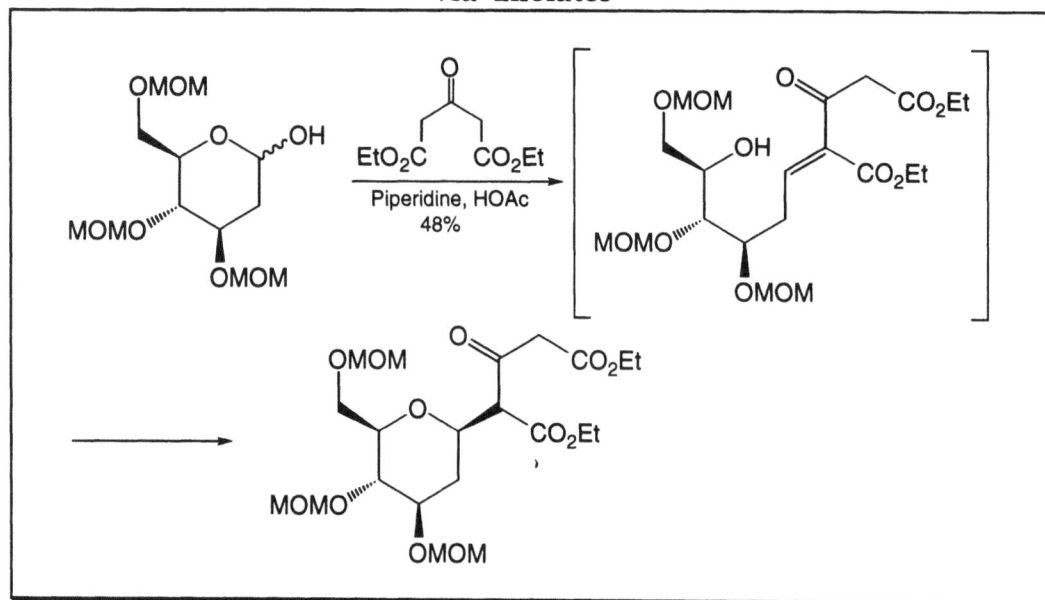

The reactions illustrated in Schemes 2.5.1 and 2.5.2 do not proceed through direct glycosidation pathways. Instead, as shown in Scheme 2.5.3, the nucleophiles adds to the aldehyde of the open sugar. The resulting α,β-unsaturated carbonyl compounds recyclize *via* 1,4-addition of a hydroxyl group allowing axial orientation of the extended unit. The resulting products are the illustrated β-*C*-glycosides.

2.5.2   Reactions with Silylenol Ethers

Scheme 2.5.4   *C*-Glycosidations with Silylenol Ethers

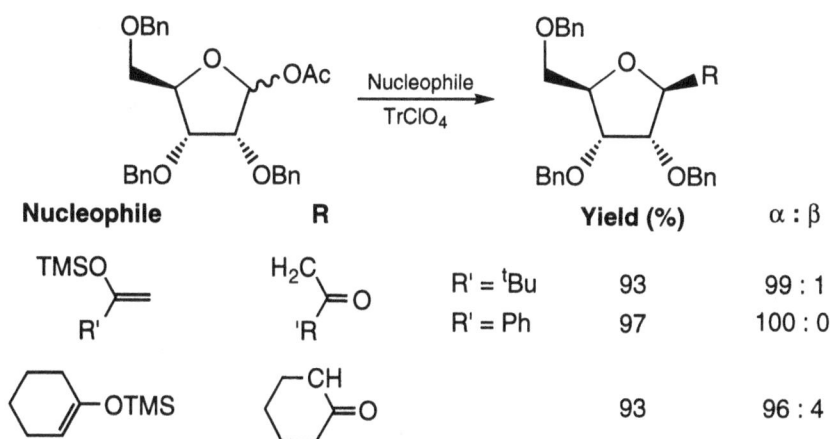

Nucleophile	R		Yield (%)	α : β
		R' = ᵗBu	93	99 : 1
		R' = Ph	97	100 : 0
			93	96 : 4

Scheme 2.5.5   *C*-Glycosidations with Silylenol Ethers

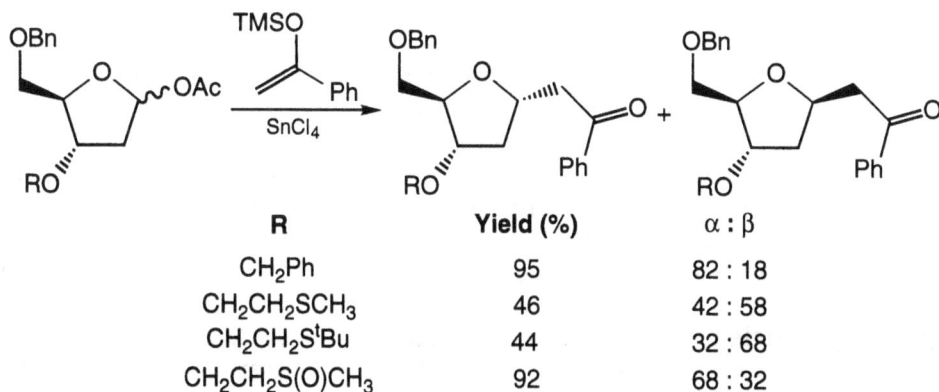

R	Yield (%)	α : β
CH₂Ph	95	82 : 18
CH₂CH₂SCH₃	46	42 : 58
CH₂CH₂SᵗBu	44	32 : 68
CH₂CH₂S(O)CH₃	92	68 : 32

Moving to more conventional glycosidation methods, a variety of anomeric activating groups have been utilized in the direct coupling of silylenol ethers with sugars. As shown in Scheme 2.5.4, Mukaiyama, *et al.*,[9] reported such reactions with a variety of silylenol ethers. Utilizing the acetate activated ribose derivative, shown, the reactions were catalyzed by trityl perchlorate and run in dimethoxyethane. In all cases, the yields exceeded 90% and, unlike the above mentioned couplings with enolates, the α anomer was overwhelmingly the predominant isomer.

Scheme 2.5.6   *C*-Glycosidations with Silylenol Ethers

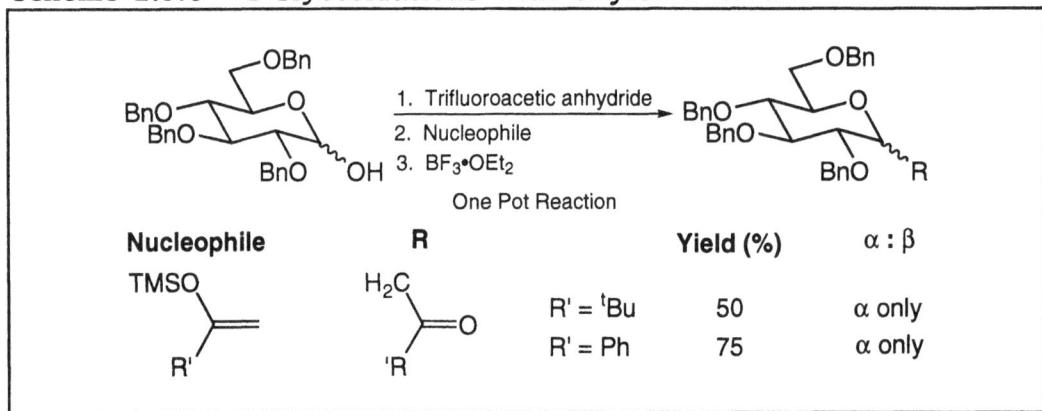

Nucleophile	R		Yield (%)	α : β
TMSO (R')	H₂C (‘R) O	R' = ‘Bu	50	α only
		R' = Ph	75	α only

In a study by Narasaka, *et al.*,[105,106] the variability of the reaction of 2-deoxyribose with silylenol ethers was addressed. As shown in Scheme 2.5.5, the yields, as well as the anomeric ratios, are dependent upon the nature of the protecting group used on the 3-hydroxyl group possibly reflecting steric effects. Consequently, the best results were obtained utilizing benzyl groups.

As with previously discussed *C*-glycosidation methods, the use of trifluoroacetoxyglycosides is complimentary to the use of acetoxyglycosides. In conjunction with couplings to silylenol ethers, Allevi, *et al.*,[53] reported the study summarized in Scheme 2.5.6. The reported results showed that good to high yields are available with high a selectivity. These results are in agreement with those reported for the corresponding acetoxyglycosides.

In comparing acetates and trifluoroacetates, electronically similar trichloroacetimidates must be examined as anomeric activating groups in coupling reactions with silylenol ethers. As mentioned in section 2.2, trichloroacetimidates are easily prepared on reaction of hemiacetals with trichloroacetonitrile. The resulting activating group is subsequently displaced by a variety of nucleophiles under Lewis acid catalysis. As shown in Scheme 2.5.7, Hoffman, *et al.*,[3] exploited this chemistry in the preparation of a variety of *C*-glucosides utilizing silylenol ethers.

Haloglycosides, already discussed with respect to other *C*-glycosidation reactions, are useful substrates for coupling with silylenol ethers. As shown in

Scheme 2.5.8, Nicolau, *et al.*,[4] utilize the fluoroglucoside shown and effected the formation of a variety of *C*-glycosides. The example shown proceeded in high yield. However, the stereoselectivity of the reactions mentioned in this report were relatively low. This does not reflect upon the general utility of this chemistry as will be illustrated in the following example.

Scheme 2.5.7   *C*-Glycosidations with Silylenol Ethers

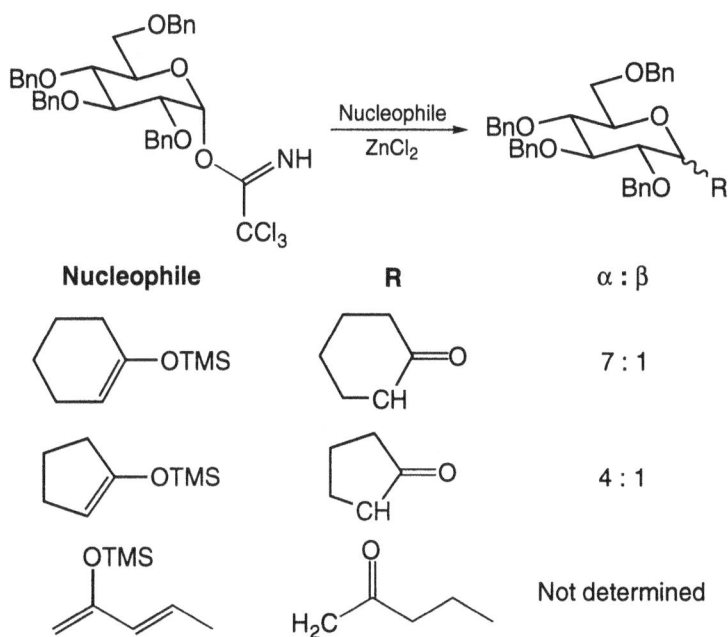

Scheme 2.5.8   *C*-Glycosidations with Silylenol Ethers

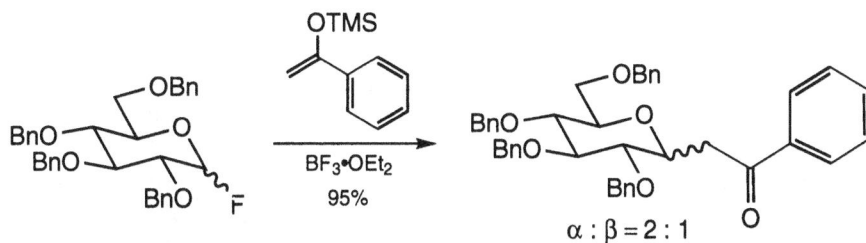

In the application of *C*-glycosidation techniques *via* addition of silylenol ethers to fluoroglycosides, Araki, *et al.*,[63,107] explored their applicability to furanose sugars. As shown in Scheme 2.5.9, although comparable results were obtained using allyltrimethylsilane, substantial improvements over Nicolau's results were observed utilizing the silylenol ether shown.   Thus, fluorofuranosides form C-glycosides on reaction with silylenol ethers in yields exceeding 90% and exhibit high anomeric preferences for the α configuration.

Scheme 2.5.9   C-Glycosidations with Silylenol Ethers

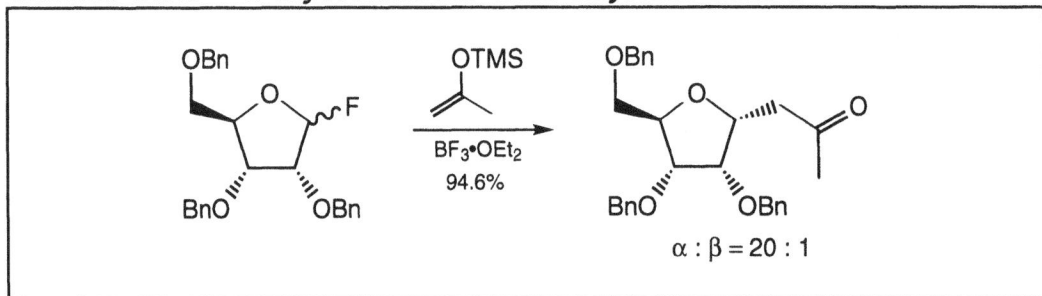

α : β = 20 : 1

Scheme 2.5.10 C-Glycosidations with Silylenol Ethers

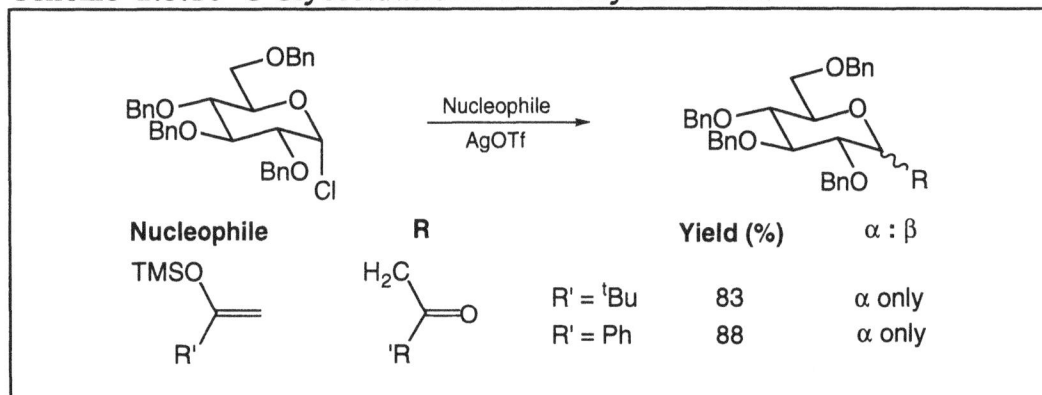

Nucleophile	R		Yield (%)	α : β
		R' = ^{t}Bu	83	α only
		R' = Ph	88	α only

With the ability to make use of fluoroglycosides as C-glycosidation substrates, chloroglycosides have been comparably useful. As shown in Scheme 2.5.10, Allevi, et al.,[89] condensed silylenol ethers with 2,3,4,6-tetra-O-benzyl-α-D-glucopyranosyl chloride. Silver triflate was utilized as the Lewis acid catalyst. In the examples studied, the yields exhibited were consistently greater than 80%. Furthermore, complete α selectivity was observed.

Subsequent research reported by Allevi, et al.,[53,108] demonstrated the feasibility of combining steroidal silylenol ethers with chloroglycosides to give α-C-glycosides. The results of this study are illustrated in Scheme 2.5.11. Furthermore, as these reactions consistently exhibited high α selectivity, conditions for epimerization to the β isomers were demonstrated. As expected, this transformation proceeded easily on treatment with base.

As with arylation reactions, intramolecular delivery of silylenol ethers is a viable approach to the formation of C-glycosides. An interesting example, reported by Craig, *et al.*,[109] demonstrated the formation of a *cis*-fused oxetane from an intramolecular enol ether. The activating group used in this reaction was the phenylsulfide group. Additionally, the formation of only one stereoisomer was observed. The reaction described and its stereoselectivity is rationalized in Scheme 2.5.12.

Scheme 2.5.11 *C*-Glycosidations with Silylenol Ethers

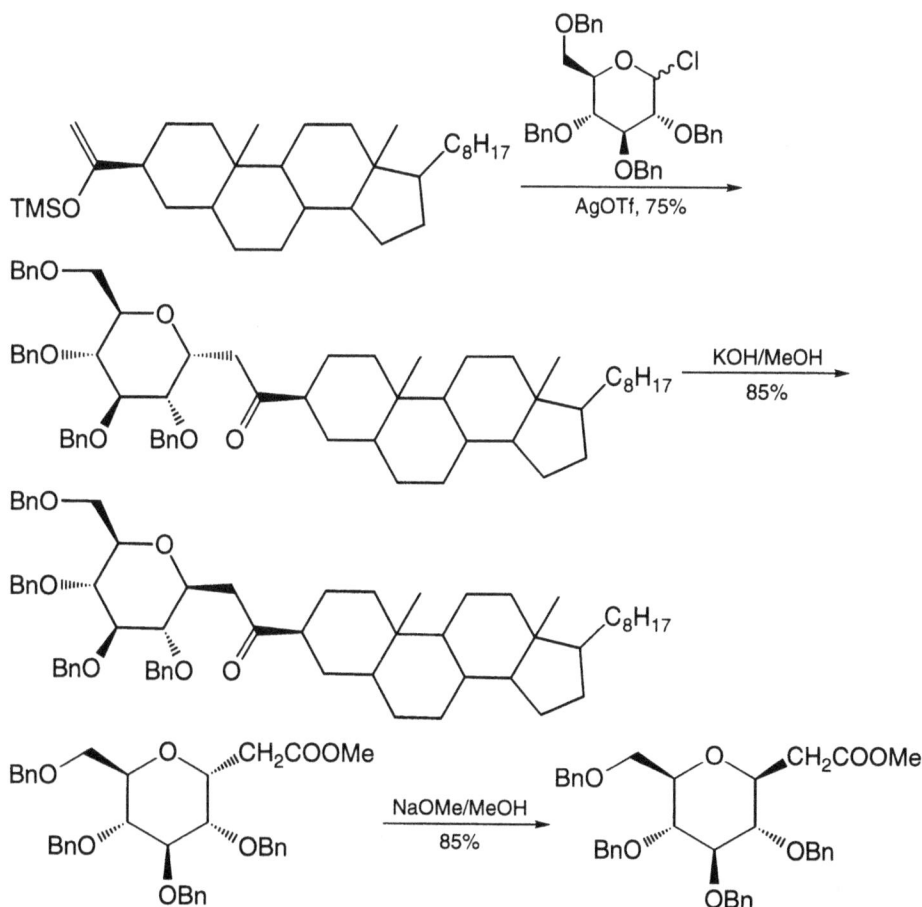

As activated sugars are capable of reacting with silylenol ethers to form *C*-glycosides, so are glycals. In 1981, Fraser-Reid, *et al.*,[110] showed that in addition to forming α-*O*-glycosides from the glycals the same glycals will form α-*C*-glycosides when treated with silylenol ethers. Furthermore, analogous to work previously described, the α anomers were easily converted to the corresponding β isomers on treatment with base. These results are illustrated in Scheme 2.5.13 and summarized in Table 2.5.1.

Scheme 2.5.12 *C*-Glycosidations with Silylenol Ethers

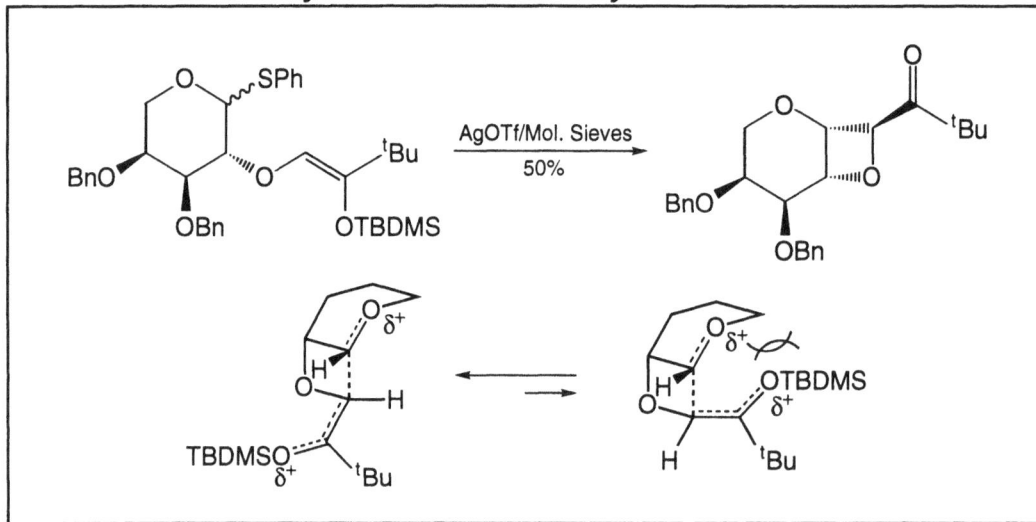

Scheme 2.5.13 *C*-Glycosidations with Silylenol Ethers

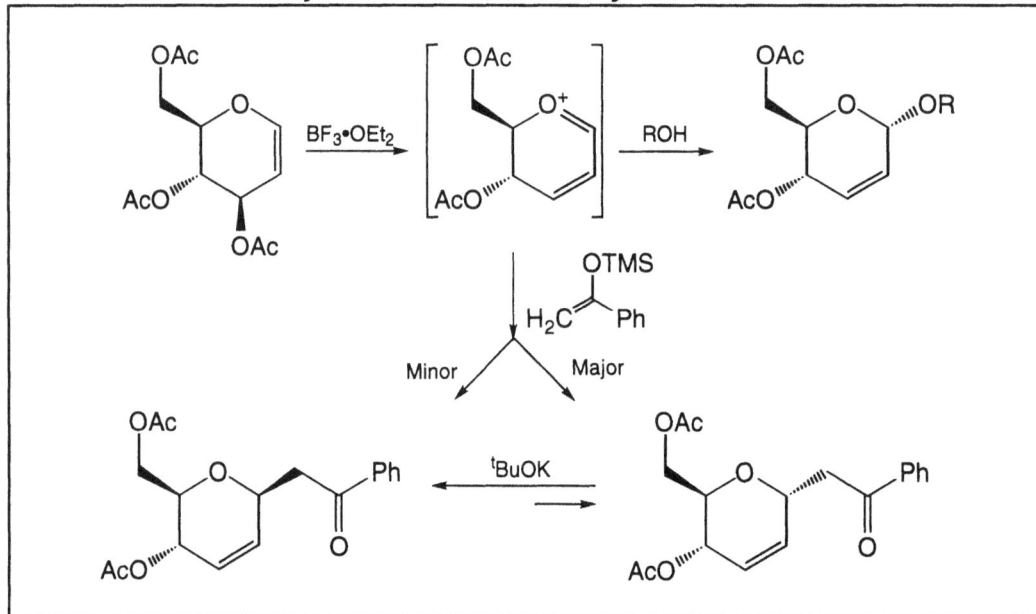

Continuing with glycals, Kunz, *et al.*,[111] not only demonstrated their ability to condense with silylenol ethers, condensations with enamines were also explored. As shown in Scheme 2.5.14, the titanium tetrachloride mediated condensation proceeded in 90% yield with exclusive formation of the β anomer. Alternately, when the enamine was used, exclusive 1,4-addition led to the product shown. As with the silylenol ether, the enamine produced solely the β anomer. However, the isolated yield of 28% was substantially lower.

Table 2.5.1   *C*-Glycosidations with Silylenol Ethers

Solvent	Catalyst	T/°C	t/h	β : α	% Yield
CH_2Cl_2	$BF_3 \cdot OEt_2$	-40 to 0	0.5	4 : 1	99
CH_2Cl_2	$BF_3 \cdot OEt_2$	-78	2.5	- - -	<1
CH_2Cl_2	$AlCl_3$	-40 to 0	1.0	7 : 3	92
CH_2Cl_2	$CF_3SO_3SiMe_3$	23	0.3	7 : 3	75
THF	$BF_3 \cdot OEt_2$	-40 to 23	5.0	- - -	<1
THF	$AlCl_3$	-40 to 0	18	- - -	- - -
THF	$AlCl_3$	23	36	7 : 3	77
CH_3CN	$BF_3 \cdot OEt_2$	-45 to 0	0.75	- - -	<1
CH_3CN	$AlCl_3$	-45 to 10	1.0	4 : 1	97

Scheme 2.5.14   *C*-Glycosidations with Silylenol Ethers and Enamines

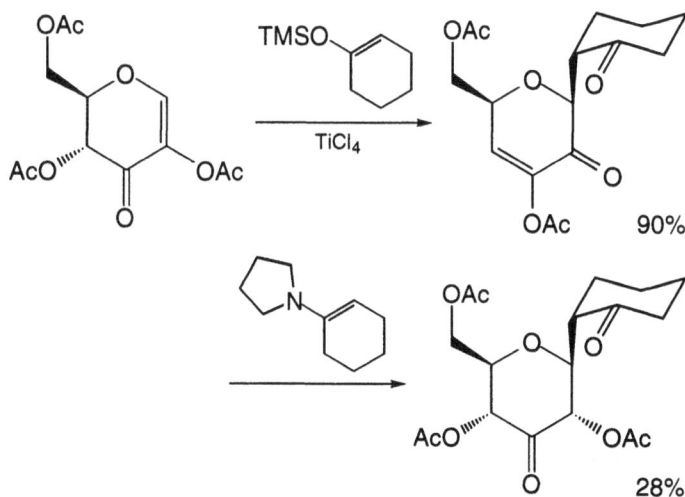

2.5.3   Reactions with Enamines

In a final example of the applicability of enol ether analogs to the synthesis of *C*-glycosides, Allevi, *et al.*,[112] explored a variety of enamines utilizing silver triflate as a catalyst. This study differed from that previously discussed in that the sugar is a glycosyl chloride and the enamines are conjugated to carbonyl functionalities. The results, shown in Scheme 2.5.15, are unlike those in the previous example thus demonstrating a propensity for the formation of α anomers. Furthermore, in all cases, the yields observed were approximately 85%. Thus, the conjugated enamines used in this report appear to have general applicability in condensations with glycosyl halides.

Scheme 2.5.15   *C*-Glycosidations with Enamines

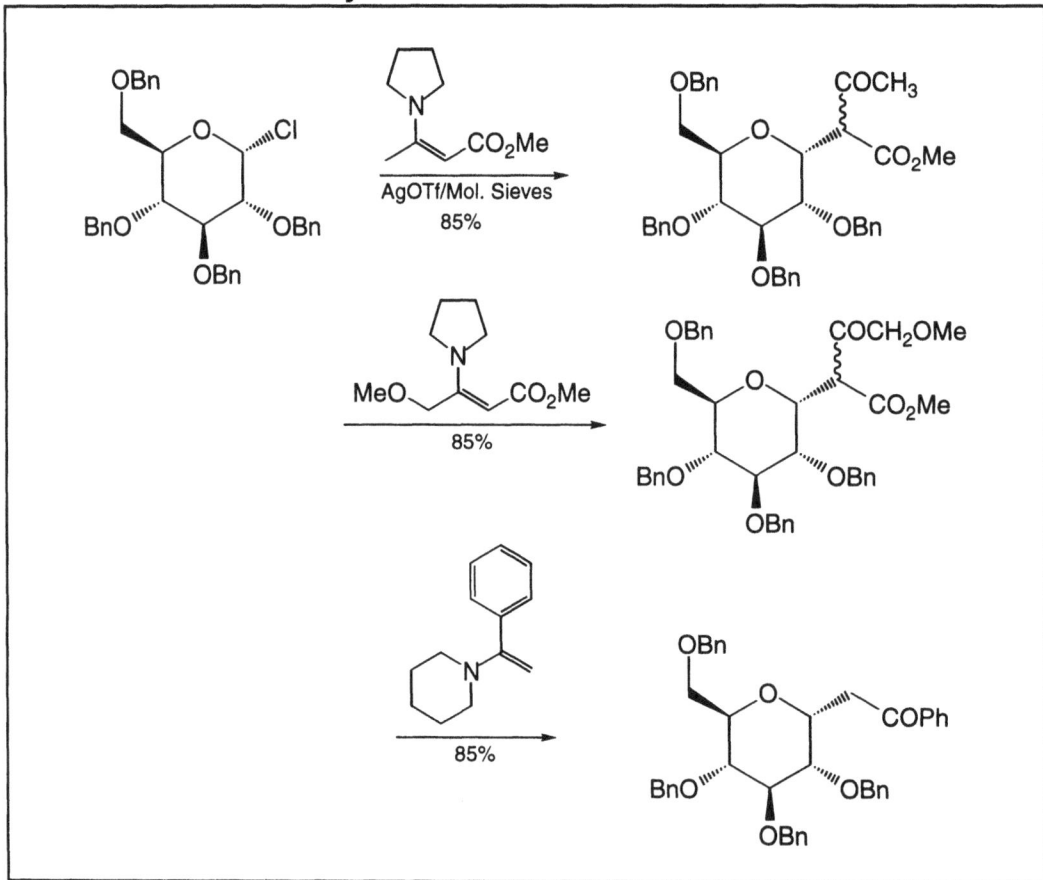

2.6 Nitroalkylation Reactions

The formation of nitroalkyl glycosides, illustrated in Figure 2.6.1, is limited in scope and few examples have been reported. The problems associated with this reaction involve the acidity of nitromethyl groups. Specifically, after formation of the *C*-glycoside, nitromethyl anions are easily formed. Furthermore, when acetate protecting groups are used, nitronates readily form. The few examples illustrated in this section are meant to convey the relatively limited accessibility of these compounds.

Figure 2.6.1   Example of Nitroalkylation Reactions

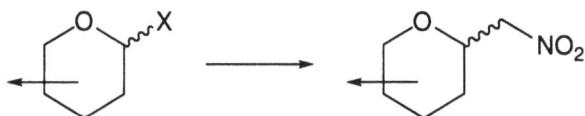

Scheme 2.6.1   Nitroalkylation Reactions

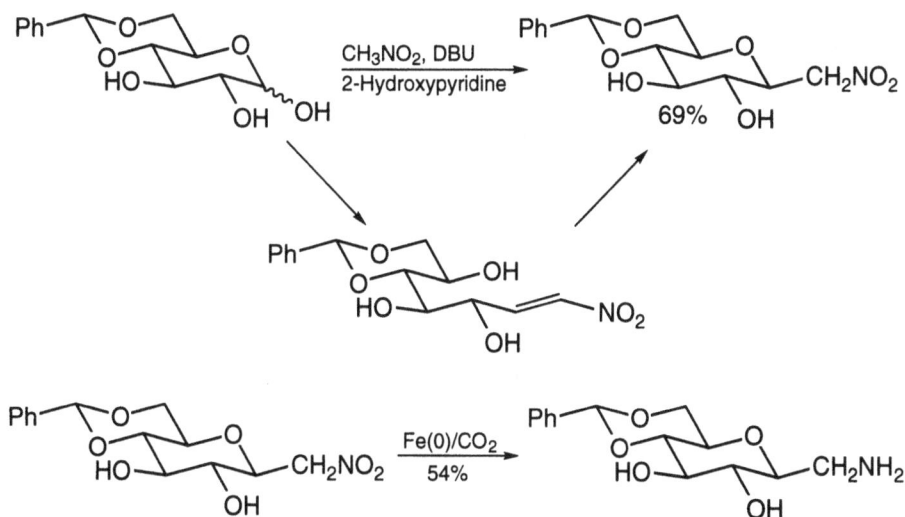

Scheme 2.6.2   Nitroalkylation Reactions

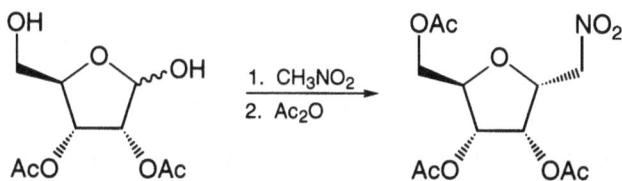

One specific example illustrating the utility of nitroalkylations was reported by Drew, *et al.*,[113] and is shown in Scheme 2.6.1. In this report, a benzylidine protected glucose analog was converted in a 69% yield to a β-*C*-

glycoside. Subsequent reduction of the nitro group gave the corresponding aminomethyl derivative suitable for further functionalization. It should be noted that a potential side reaction arises from nitromethane adding, *via* a 1,4-addition, to the intermediate olefin. Applied to furanose sugars, Koll, *et al.*,[114] reported the analogous reaction shown in Scheme 2.6.2.

Scheme 2.6.3   Synthesis of 1-Deoxy-1-Nitrosugars

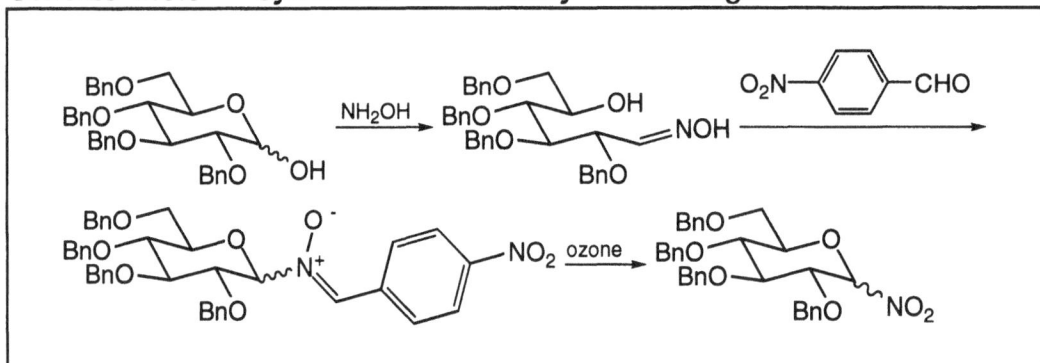

Scheme 2.6.4   Nitroalkylation Reactions

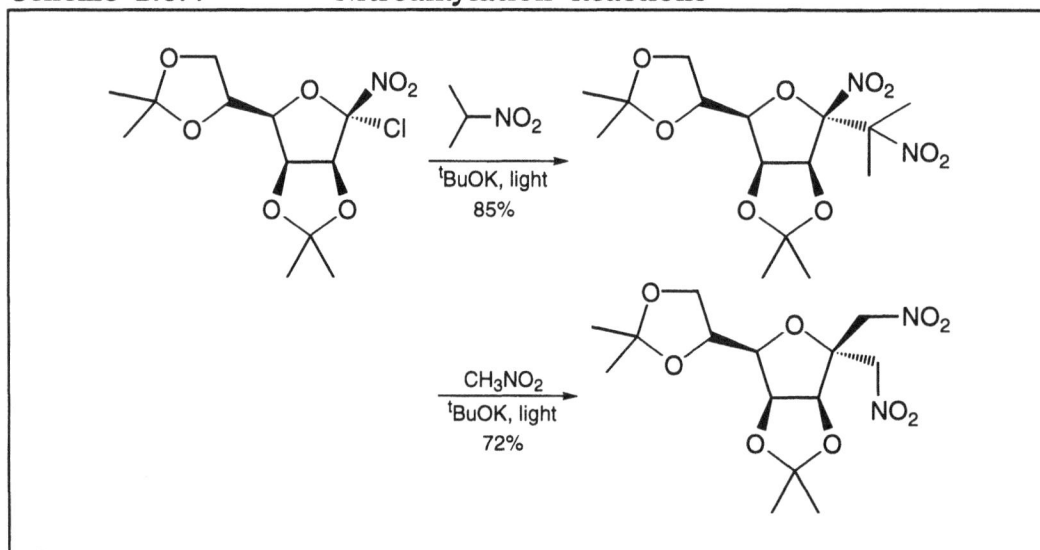

Related to nitroalkylsugars are 1-deoxy-1-nitrosugars. These compounds are relatively easy to prepare *via* oximes. As shown in Scheme 2.6.3, Vasella, *et al.*,[115,116] effected this transformation via addition of an aldehyde to the oxime, shown. Subsequent treatment with ozone gave an anomeric mixture of the desired 1-deoxy-1-nitrosugar in 73% overall yield.

The utility of 1-deoxy-1-nitrosugars was further demonstrated by Vasella, *et al.*,[117] in a study of nitroalkylation reactions. The study, summarized in Scheme 2.6.4, involved the addition of nitromethane and 2-nitropropane. The reactions proceeded in excess of 70% yield. Further studies of the utility of nitrosugars will be discussed later in the formation of *C*-disaccharides.

2.7   Reactions with Allylic Ethers

Figure 2.7.1   Example of Reactions with Allylic Ethers

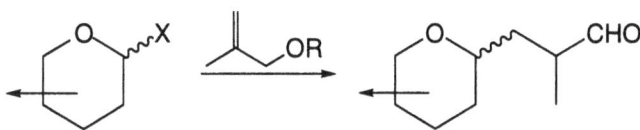

Scheme 2.7.1   Reactions of Glycals with Allylic Ethers

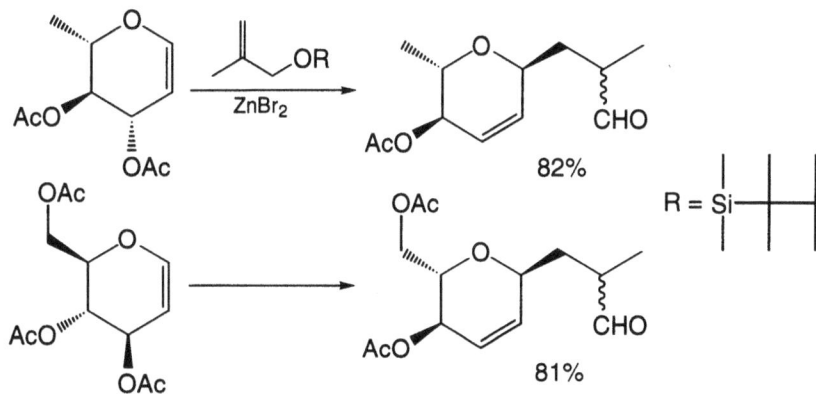

Unlike nitroalkylation reactions, the use of allylic ethers as reagents in *C*-glycoside chemistry has proven to be substantially more useful. These reactions, illustrated in Figure 2.7.1 and alluded to in Scheme 2.3.16,[41,42] show a propensity for α selectivity and are able to provide additional asymmetric centers. These reactions, however, procede in an S_N2'-like fashion and are limited to glycals. Consequently, a wide variety of deoxy sugar derivatives are available.

Although, as shown in Scheme 2.3.16, the chloride was the product observed in the reaction of tri-*O*-acetyl glucal with an allylic alcohol, one can envision the ability of this type of condensation to yield carbonyl compounds if the double bond could be kept intact. This has been accomplished utilizing *O*-

silyl protected allylic alcohols. An early example, shown in Scheme 2.7.1, was reported by Herscovici, *et al.*,[118] and demonstrates applicability to various glycals. As shown, in Scheme 2.7.2, the mechanism of these transformations involves formation of the carbocation followed by the evolution of its equilibrium with the illustrated *O*-glycoside. As the carbocation reacts with the allylic ether, the reaction is driven to provide the desired product.

Scheme 2.7.2   Reactions of Glycals with Allylic Ethers

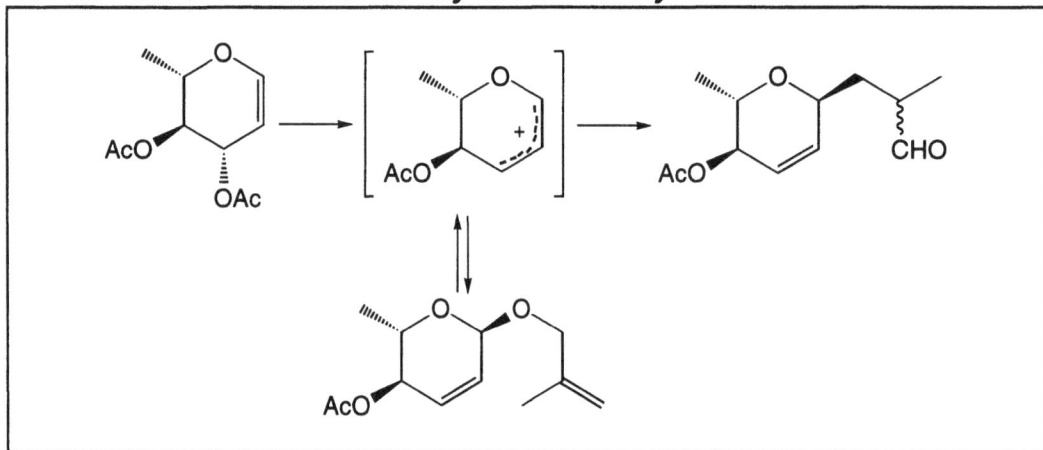

Scheme 2.7.3   Reactions of Glycals with Allylic Ethers

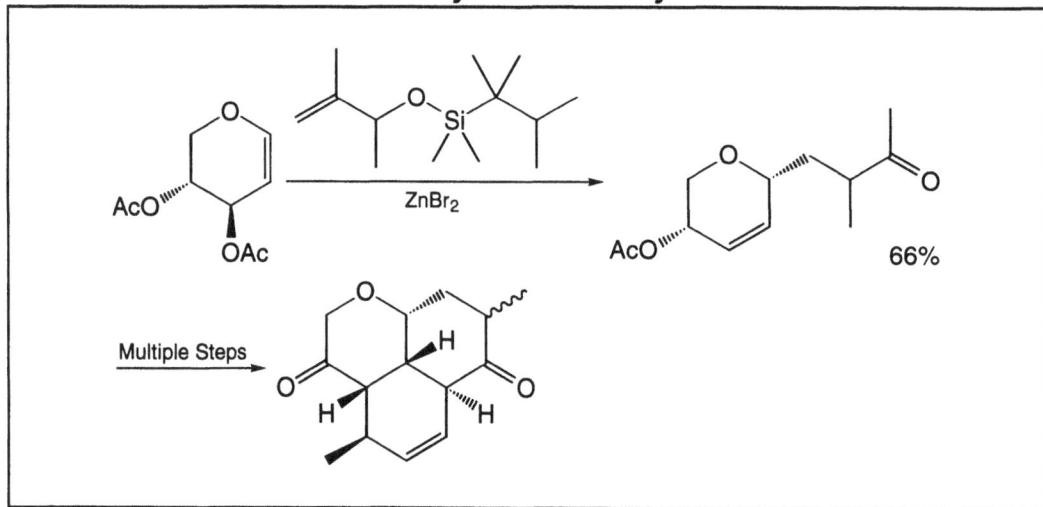

In later reports, Herscovici, *et al.*,[119] demonstrated the applicability of this technology to more complex allylic ethers. The example shown in Scheme 2.7.3 was utilized in an enantiospecific naphthopyran synthesis.

In a final example, summarized in Scheme 2.7.4, Herscovici, *et al.*,[120] utilized several silyl ethers in reactions with various glycals to prepare the products shown. In the same report, a study was carried out involving the use of deuterated allylic ethers. This experiment, shown in Scheme 2.7.5, was designed to prove the involvement of a pinacole-type mechanism in the transformation. The 1,2-deuteride shift supported this hypothesis.

Scheme 2.7.4   Reactions of Glycals with Allylic Ethers

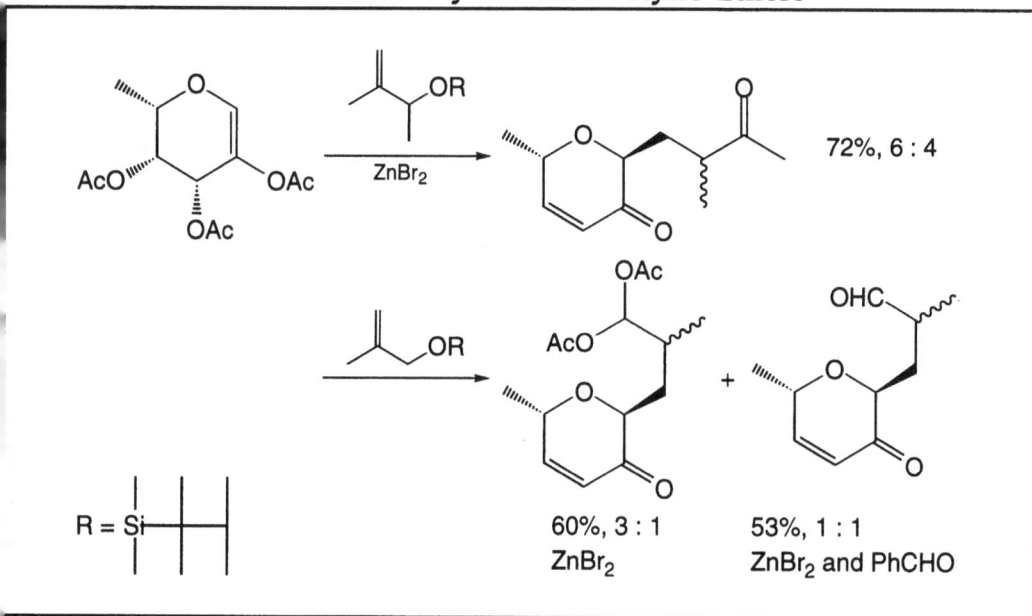

Scheme 2.7.5   Reactions of Glycals with Allylic Ethers

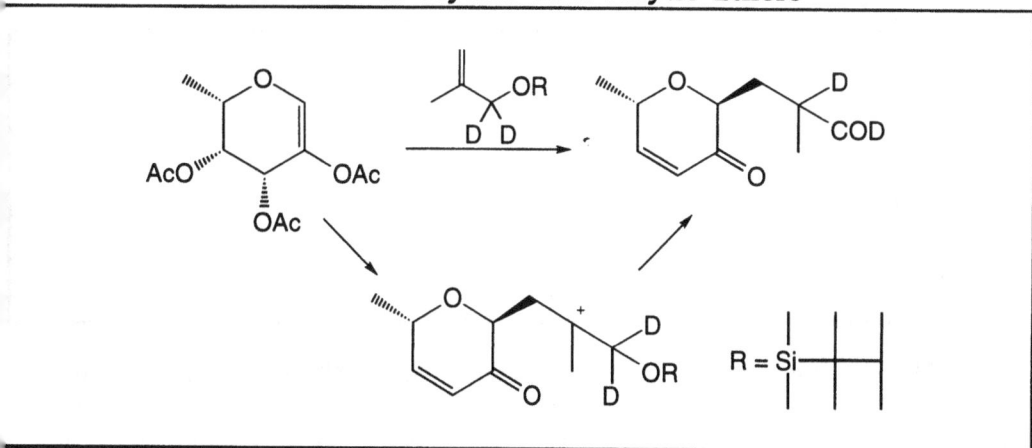

The fact that ketones, aldehydes, and geminal diacetates are readily available from these reactions illustrate their complementarity to reactions with allylsilanes. Specifically, the equivalency of allylic ethers to homoenolates allows for the formation of compounds extended by one carbon unit as compared to the products of couplings with enolate equivalents already discussed.

2.8 Wittig Reactions with Lactols

Where the use of allylic ethers provides convenient routes to the formation of ketones and aldehydes, the Wittig reaction, as illustrated in Figure 2.8.1, may be utilized in the formation of ester functionalized *C*-glycosides. This approach is either α or β selective and works with benzylated, acetylated, and unprotected sugars. A particular advantage is that reaction intermediates can be intercepted and utilized. Furthermore, glycosidic activation is not required for this type of reaction.

Figure 2.8.1   Example of Wittig Reactions with Sugars

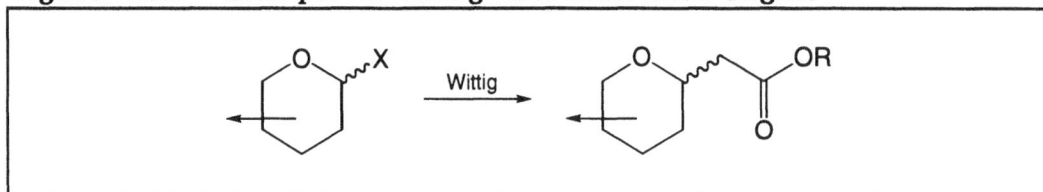

2.8.1   Wittig Reactions

Scheme 2.8.1   Wittig Reactions with Lactols

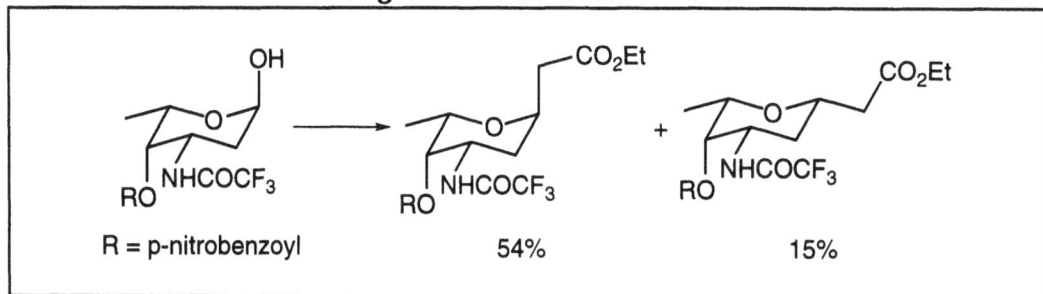

In 1981, Acton, *et al.*,[121] performed the reaction shown in Scheme 2.8.1 to give a 3 : 1 ratio of anomeric esters. Shortly thereafter, Nicotra, *et al.*,[122] ran a similar reaction, shown in Scheme 2.8.2, and studied the ^{13}C chemical shifts of the resulting α and β anomers. This study yielded an easy method for the assignment of the anomeric stereochemistry with the anomeric carbon of the α anomer possessing a chemical shift of approximately 33 ppm and that of the β anomer possessing a chemical shift of approximately 37 ppm.

Scheme 2.8.2   Wittig Reactions with Lactols

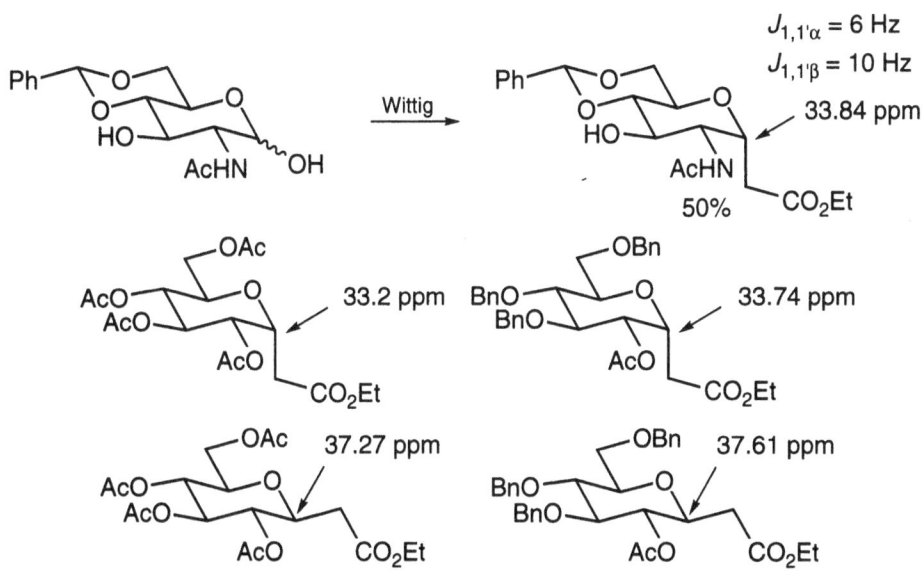

$J_{1,1'\alpha} = 6$ Hz
$J_{1,1'\beta} = 10$ Hz
33.84 ppm
50%

33.2 ppm

33.74 ppm

37.27 ppm

37.61 ppm

Scheme 2.8.3   Mechanism of Wittig Reactions with Lactols

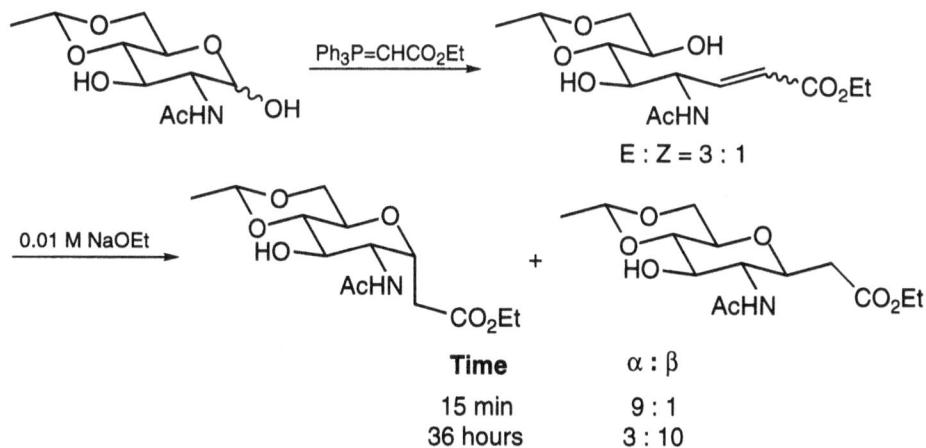

Ph$_3$P=CHCO$_2$Et

E : Z = 3 : 1

0.01 M NaOEt

Time	α : β
15 min	9 : 1
36 hours	3 : 10

The mechanism of Wittig reactions applied to lactols is illustrated in Scheme 2.8.3 in a reaction reported by Giannis, et al.[123] As shown, the Wittig reagent reacts with the aldehyde of the straight-chain form of the sugar. The free 5-hydroxyl group then adds to the resulting α,β-unsaturated ester under basic conditions giving the products shown. Furthermore, the anomeric ratio of the product mixture is dependent upon the time of exposure to base thus demonstrating the ability to convert the α anomer to the β anomer. It is notable that the mechanism of this reaction resembles that of nitroalkylation reactions shown in Scheme 2.6.1. In the cited report, the reaction shown was used in the synthesis of a potential β-D-hexosaminase A inhibitor.

Scheme 2.8.4   Wittig Reactions with Lactols

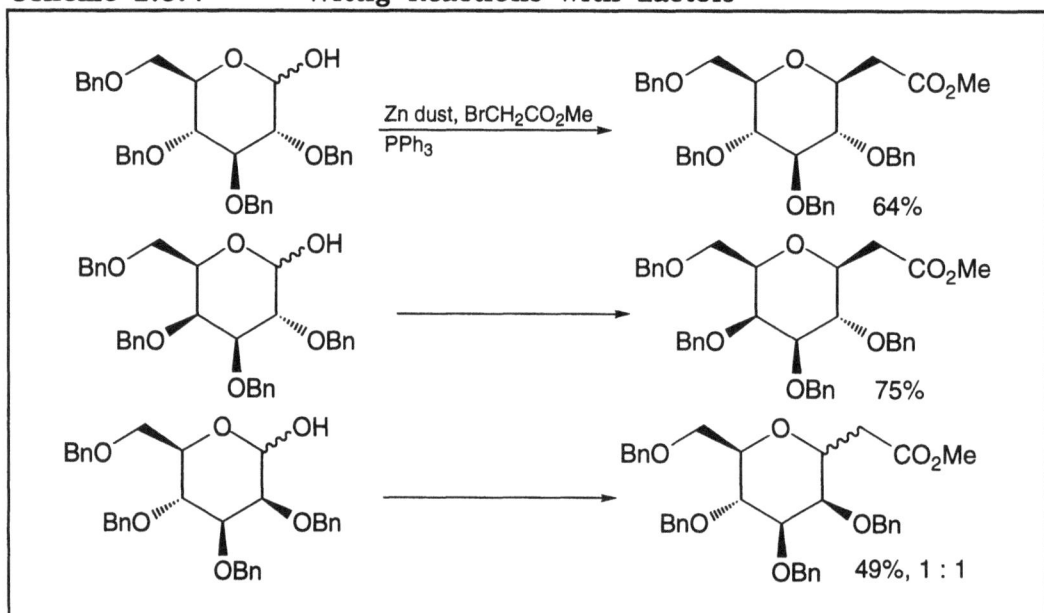

With the utility of Wittig reagents in the formation of C-glycosides, it is important to mention work reported by Dheilly, et al.,[124] in which benzyl protected glucose, galactose, and mannose were treated with methyl bromoacetate, zinc dust, and triphenylphosphine. The procedure described was reported by Shen, et al.,[125] as a one-pot formation and application of Wittig reagents. The original report utilized this methodology in transformations regarding aromatic and aliphatic aldehydes. As shown in Scheme 2.8.4, applying this methodology to protected sugars provides C-glycosidic esters in good yields. In all cases except for mannose, the reaction proceeded with β selectivity. The lack of selectivity for the mannose case may be attributed to steric effects.

Furanose sugars have also been utilized as substrates for Wittig reactions. As shown in Scheme 2.8.5, Fraser-Reid, et al.,[126] applied this methodology to mannose derivatives. The reactions proceeded with both the 5-*O*-trityl protected compound (reacted with carbomethoxymethylene triphenylphosphorane) and the free 5-hydroxyl analog (reacted with acetonyl triphenylphosphorane). In both cases, as mentioned in previous examples, a kinetic and a thermodynamic product was observed. Furthermore, on treatment with 10 mM sodium methoxide, the thermodynamic isomer became more prominent.

Scheme 2.8.5   **Wittig Reactions with Lactols**

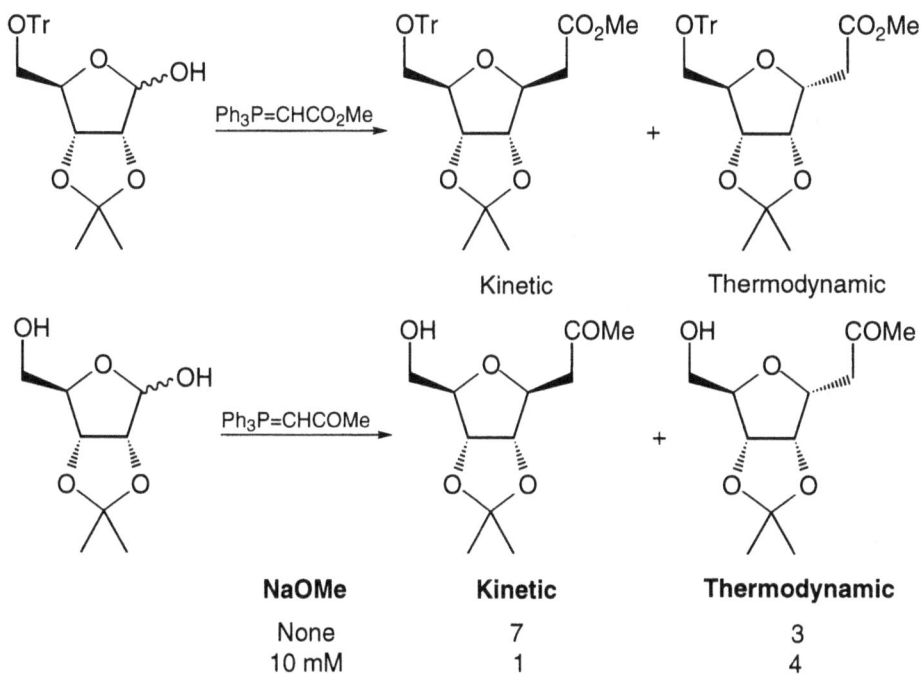

NaOMe	Kinetic	Thermodynamic
None	7	3
10 mM	1	4

Scheme 2.8.6   **Wittig-Type Reactions**
with Sugar Lactones

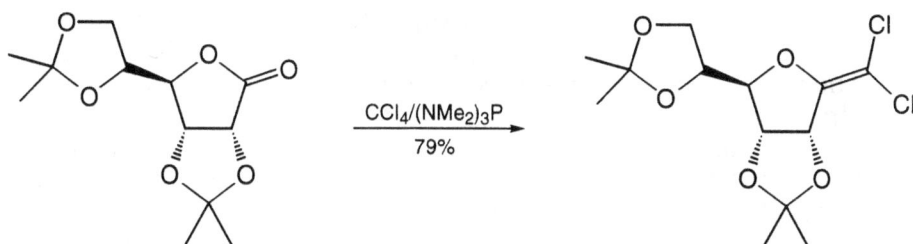

Although outside the scope of this discussion, Wittig-type reactions have been utilized in transformations involving sugar lactones. As shown in Scheme 2.8.6, Chapleur, *et al.*,[127] induced the reaction between the protected sugar, shown, and hexamethylphosphorus triamide-tetrachloromethane to give the desired dichloro olefin in 79% yield. This example is important in the illustration that phosphorus chemistry is capable of providing avenues to a wide variety of desirable functional groups important in the preparation of *C*-glycosides. Further illustrations involving Horner-Emmons methodology are discussed below.

2.8.2   *Horner-Emmons Reactions*

Scheme 2.8.7   **Horner-Emmons Reactions with Lactols**

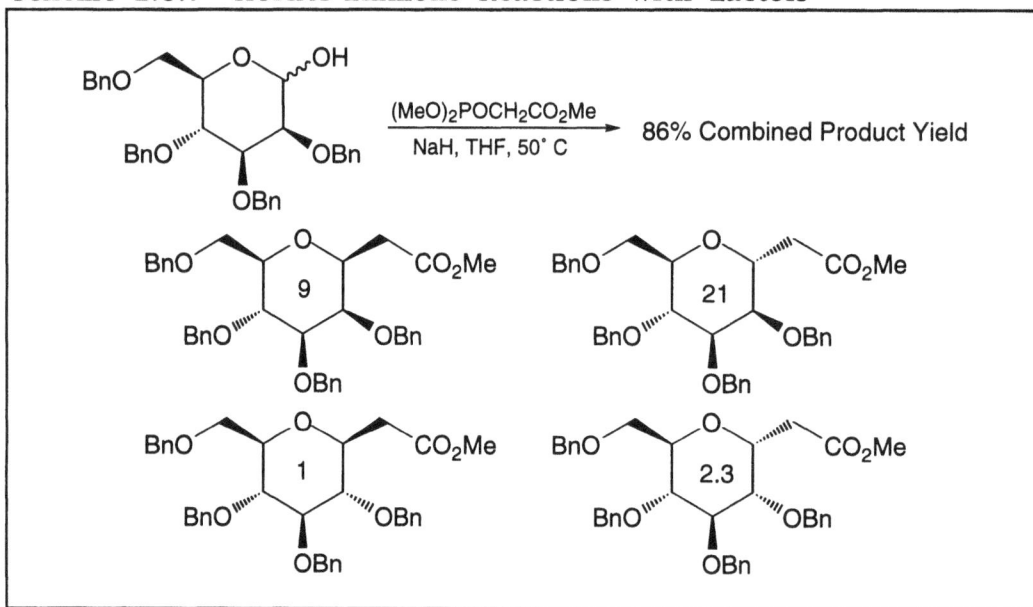

Complimentary to the Wittig reaction is its Horner-Emmons modification. These reactions, unlike the phosphorus ylides utilized in Wittig reactions make use of phosphonate anions as the nucleophilic species. As shown in Scheme 2.8.7, this methodology is applicable to the formation of *C*-glycosides from sugars. The reactions shown, reported by Allevi, *et al.*,[128] involve the coupling of triethylphosphonoacetate with 2,3,4,6-tetra-*O*-benzyl-D-mannopyranose and produce high yields of the desired β-*C*-glycosides bearing ester functionalities. It should be noted, however, that some epimerization of the stereocenter at C_2 was observed. This can be explained by the relative acidity of the proton at C_2 considering the intermediate α,β-unsaturated ester. This is illustrated in the analogous mechanism for the reaction of lactols with phosphorus ylides shown in Scheme 2.8.3. Preferential formation of the α anomer is complimented by its total conversion to the β anomer on treatment with base.

Scheme 2.8.8   **Horner-Emmons Type**
                          Reactions with Lactols

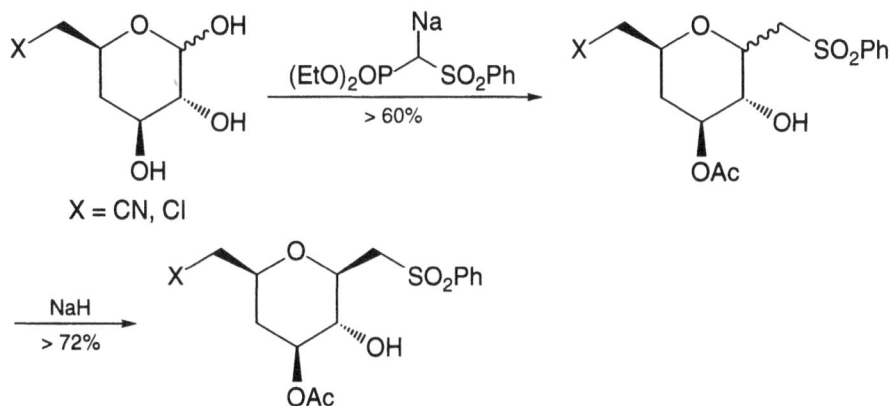

Scheme 2.8.9   **Horner-Emmons Type**
                          Reactions with Lactols

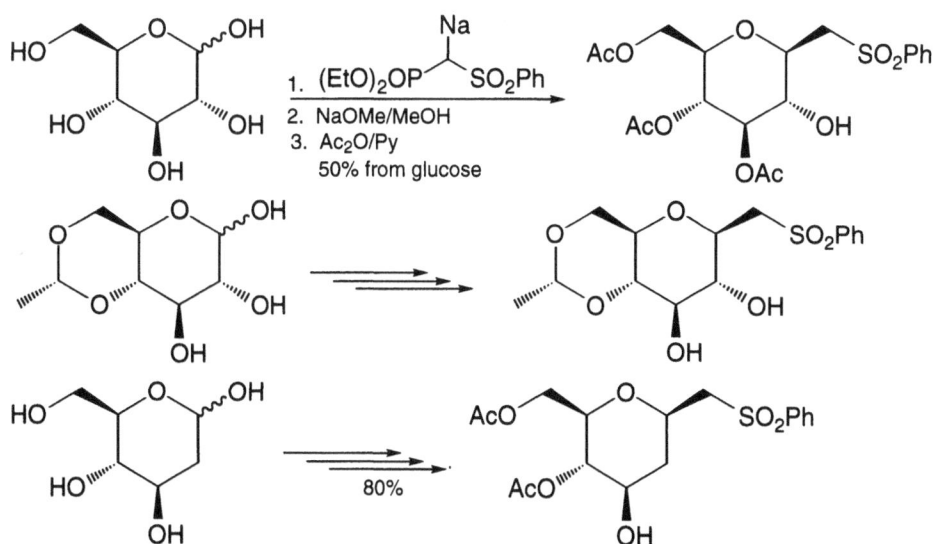

C-Glycosidic esters are not the only functional groups accessible by Wittig and related methodologies. As shown in Scheme 2.8.8, Barnes, *et al.*,[129] effected the formation of analogous compounds utilizing sulphone phosphonates. These reactions, like those with triethylphosphonoacetate formed anomeric mixtures which, on treatment with base, could effectively be converted β anomeric products. Furthermore, subsequent work by Davidson, *et al.*,[130] shown in Scheme 2.8.9, demonstrated the applicability of this technology to free sugars as well as sugars protected with acetal groups.

2.8.3   *Reactions with Sulfur Ylides*

With the utility of phosphonate anions and phosphorus ylides in the preparation of *C*-glycosides, a discussion of the related chemistry surrounding sulfur ylides is warranted. Sulfur ylides, unlike phosphorus ylides, deliver methylene groups to carbonyls thus forming epoxides. As shown in Scheme 2.8.10, Fréchou, *et al.*,[131] exploited this chemistry in the formation of *C*-glycosides. As shown, the sulfur ylide was added to the benzyl protected glucose to give the olefinic epoxide shown. Cyclization could then be effected on treatment with base thus forming the *C*-glycoside with opening of the epoxide. Similar results were observed utilizing the sulfoxide ylide and the acetonide protected furanose sugar shown. It should be noted that spontaneous cyclization occurred only when spontaneous elimination did not.

**Scheme 2.8.10   Reactions of Sulfur Ylides
with Lactols**

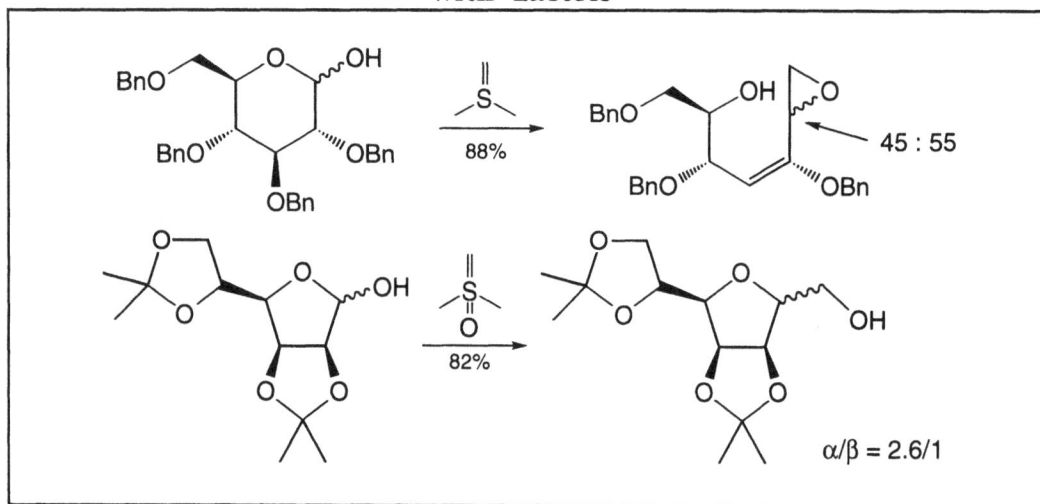

2.9 Nucleophilic Additions to Sugar Lactones Followed by Lactol Reductions

Sugar lactones are useful substrates for the formation of *C*-glycosides and are readily available from the oxidation of lactols. As shown in Figure 2.9.1, the use of this approach involves the addition of a nucleophile to the lactone followed by subsequent reduction of the resulting lactol. The two step process provides *C*-glycosides in high yields and is useful with both benzyl and silyl protected sugar lactones. Furthermore, as the hydride is generally delivered to the axial position, these reactions produce β-*C*-glycosides.

Figure 2.9.1   **Example of *C*-Glycosides**
from Sugar Lactones

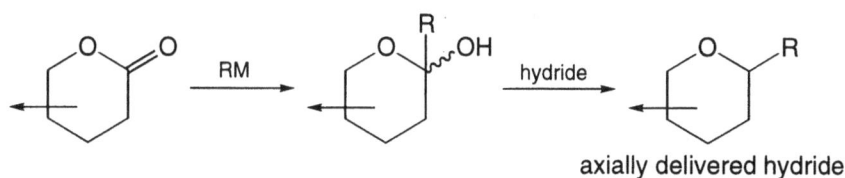

axially delivered hydride

2.9.1   Lewis Acid-Trialkylsilane Reductions

Actual applications of this methodology have already been presented and are now reviewed. In Scheme 2.3.12, the addition of alkyllithium reagents to furanose-derived lactones was addressed. Although reduction of the resulting lactol was not discussed, this example illustrated a convenient conversion of the furanose lactol to a pyranose lactol. In Scheme 2.3.14, the use of allyl Grignard reagents was presented followed by the Lewis acid-triethylsilane reduction of the lactol to the desired *C*-glycosides. In the following paragraphs, expansions upon these examples, along with the introduction of complimentary methods for the formation of *C*-glycosides from sugar lactones, are discussed.

As shown in Scheme 2.9.1, the versatility of products available from the addition of nucleophiles to sugar lactones was addressed by Horton, *et al.*[132] Following the illustrations in the scheme, gluconolactone was silylated and treated with a lithium reagent to give the observed lactol. This lactol was then acetylated and reduced, with raney nickel, to the *C*-methylglycoside. Alternately, the lactol was partially acetylated. Subsequent raney nickel reduction gave 1-methylglucose. In both cases, a byproduct of the acetylations was the open chain glucose, **A**.

Moving away from alkyllithium species, Scheme 2.9.2 illustrates the use of lithium acetylide derivatives as useful nucleophiles for the formation of *C*-glycosides. As described by Lancelin, *et al.*,[133] lithium acetylides were added to sugar lactones giving lactols. The lactols were subsequently reduced to β-*C*-alkynylglycosides with complete stereospecificity. This reaction was subsequently shown to be a general method for a variety of sugars and alkynyllithium reagents.

Scheme 2.9.1   *C*-Glycosides from Sugar Lactones

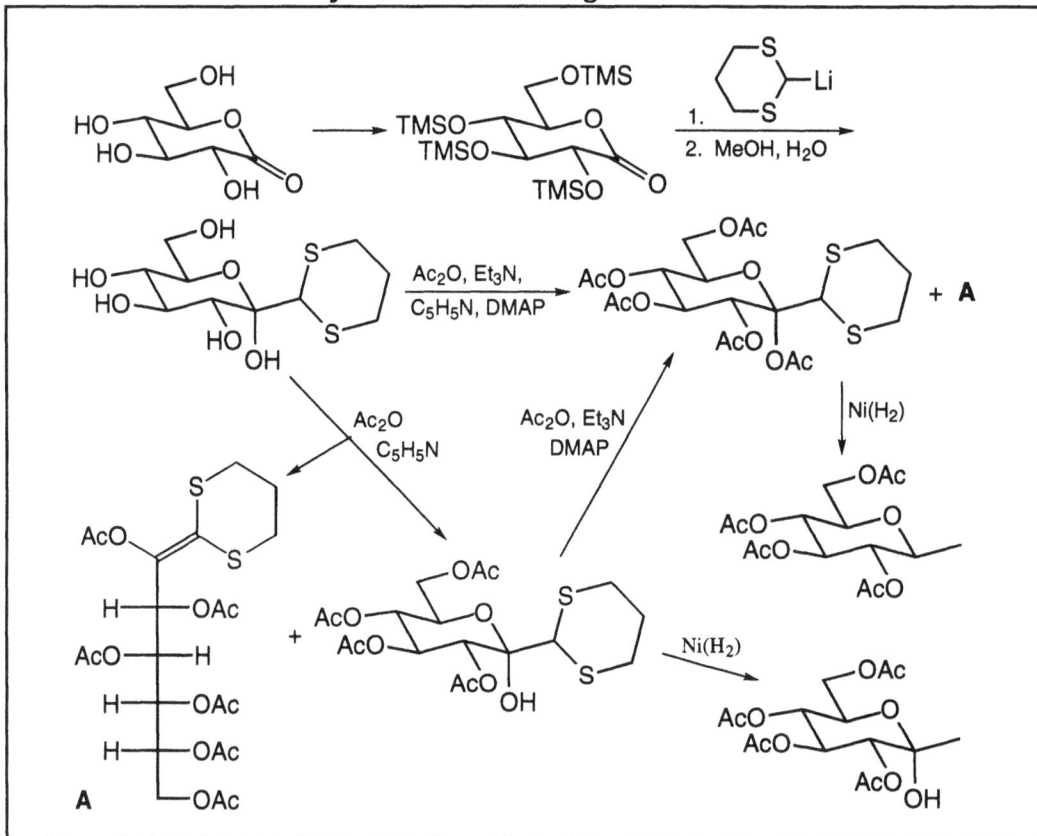

Scheme 2.9.2   *C*-Glycosides from Sugar Lactones

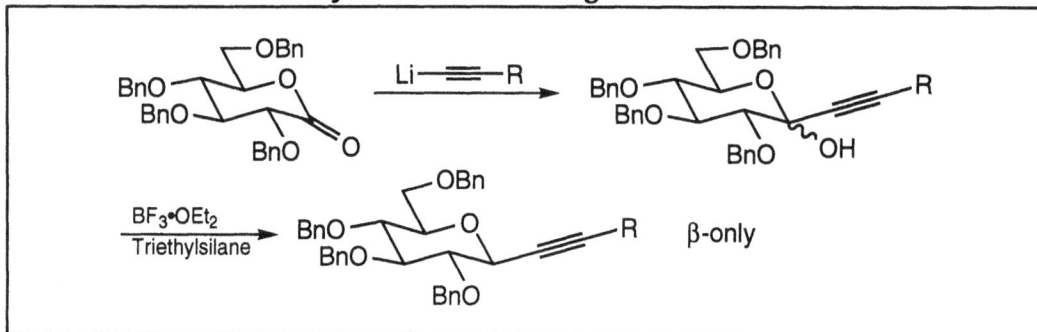

Scheme 2.9.3   *C*-Glycosides from Sugar Lactones

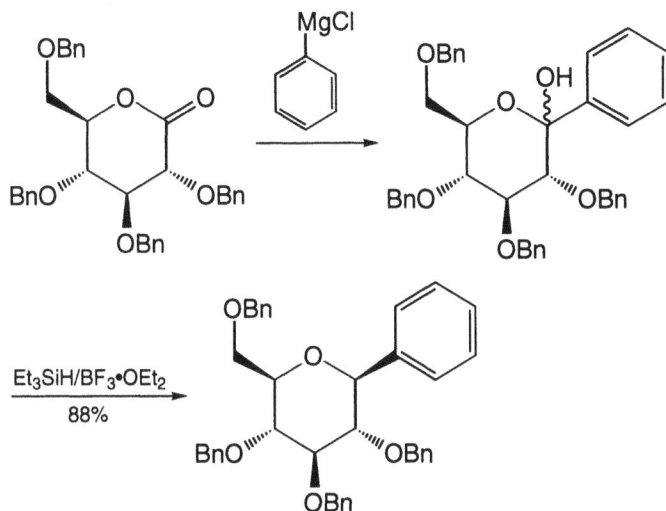

Scheme 2.9.4   *C*-Glycosides from Sugar Lactones

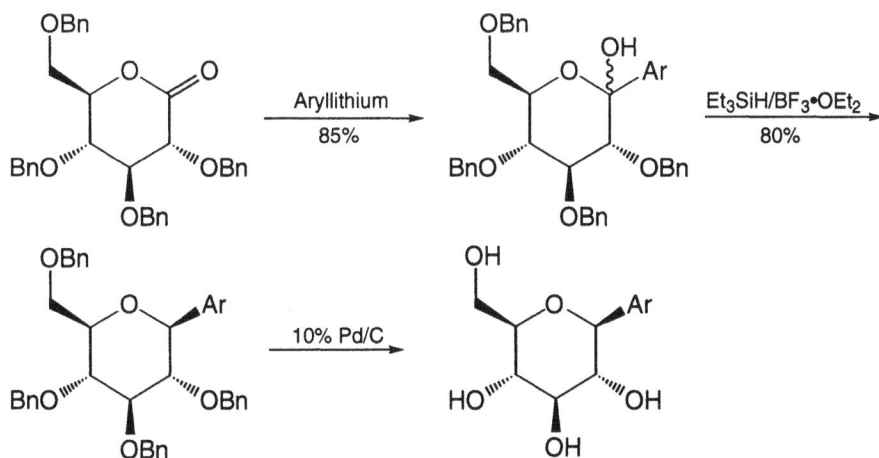

Utilizing aryl groups as nucleophiles, the related reactions, shown in Schemes 2.9.3 and 2.9.4, were carried out by Kraus, *et al.*,[134] and Czernecki, *et al.*,[135] respectively. In both cases, the predominant product possessed the β configuration at the anomeric center. Furthermore, the reactions proceeded in greater than 80% yield. When viewed along with the methods discussed thus far, these reactions affirm the generality of organolithium reagents and Grignard reagents for the preparation of *C*-glycosides from sugar lactones.

Scheme 2.9.5   *C*-Glycosides from Sugar Lactones

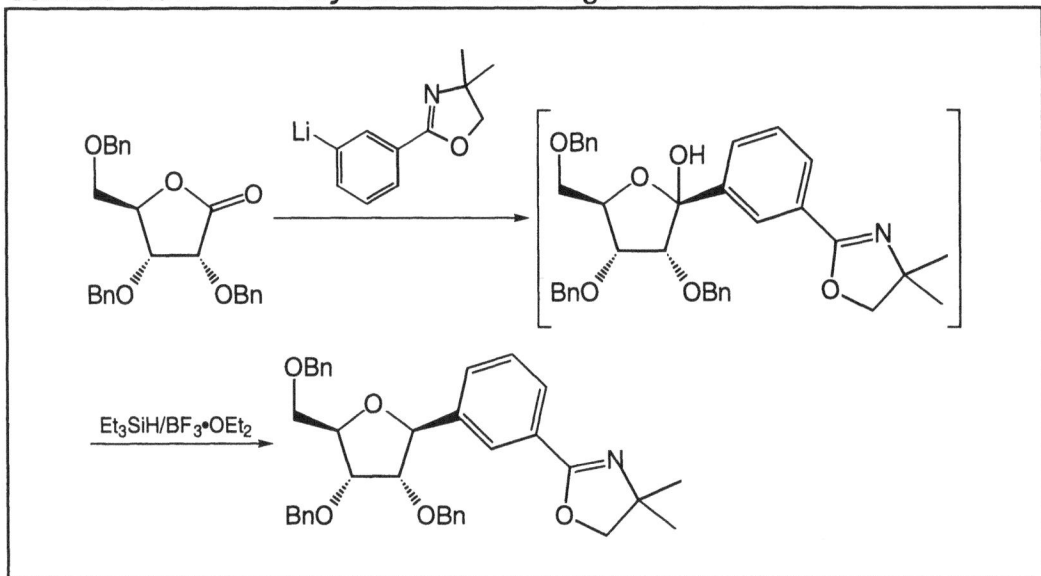

One final example of the use of aryllithium species as *C*-glycosidation agents when applied to sugar lactones is shown in Scheme 2.9.5. In this report, Krohn, *et al.*,[136] applied the technology to furanose lactones. After the reductive deoxygenation, the illustrated product was isolated as a single isomer in 87% yield over the two steps. As with the examples described above where pyranose sugars were used, β selectivity is the general stereochemical outcome for these reactions.

2.9.2   *Other Reductions*

Until now, all methods mentioned for the formation of *C*-glycosides from sugar lactones involve the incorporation of Lewis acid-trialkylsilane methodology for the deoxygenation. There are, however, other methods for accomplishing this reaction. One particular example, shown in Scheme 2.9.6, was reported by Wilcox, *et al.*,[137] and involves the use of sodium cyanoborohydride in the presence of dichloroacetic acid. This reaction proceeded in 68% yield. Furthermore, the use of *p*-toluenesulfonic acid instead of dichloroacetic acid totally blocked the desired transformation. Perhaps the most interesting aspect of this reaction is the fact that the α configuration is made accessible through sugar lactone transformations.

Scheme 2.9.6   **Lactol Reduction with**
Sodium Cyanoborohydride

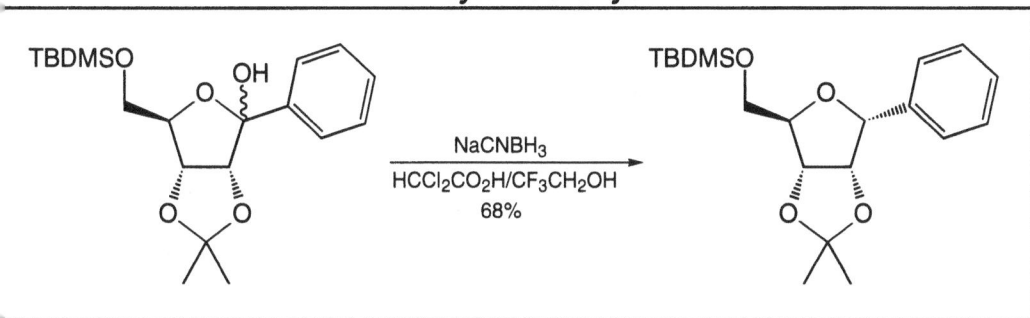

$$\xrightarrow[\text{68\%}]{\begin{array}{c}\text{NaCNBH}_3\\\text{HCCl}_2\text{CO}_2\text{H/CF}_3\text{CH}_2\text{OH}\end{array}}$$

2.9.3   *Sugar-Sugar Couplings*

Before closing this section, the application of nucleophilic additions to
sugar lactones with respect to the joining of two sugar units deserves mention
and will be further elaborated upon in Chapter 8. As shown in Scheme 2.9.7,
Sinay, *et al.*,[138] utilized the lithium anion of a *C*-acetylenic glycoside to effect
formation of an acetylene bridged *C*-disaccharide. Furthermore, as shown in
Scheme 2.9.8, Schmidt, *et al.*,[139] utilized a similar approach involving a
carbohydrate-derived alkyllithium compound.   These examples are not
presented to preempt the topics encompassed within Chapter 8, but to
demonstrate that with the expansion of the chemistry surrounding *C*-glycosides,
the preparation of increasingly more complicated structures becomes possible
through substantially simpler processes.

Scheme 2.9.7   *C*-**Glycosides from Sugar Lactones**

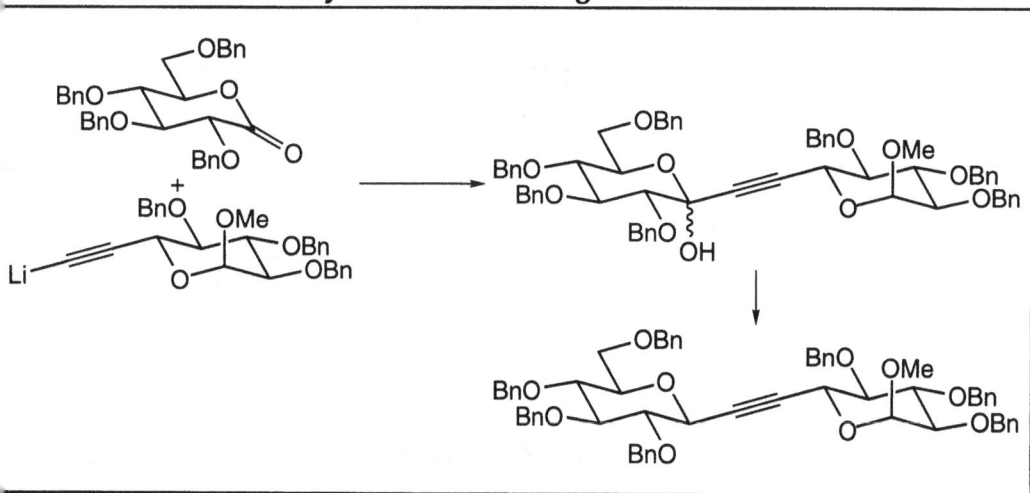

Scheme 2.9.8   *C*-Glycosides from Sugar Lactones

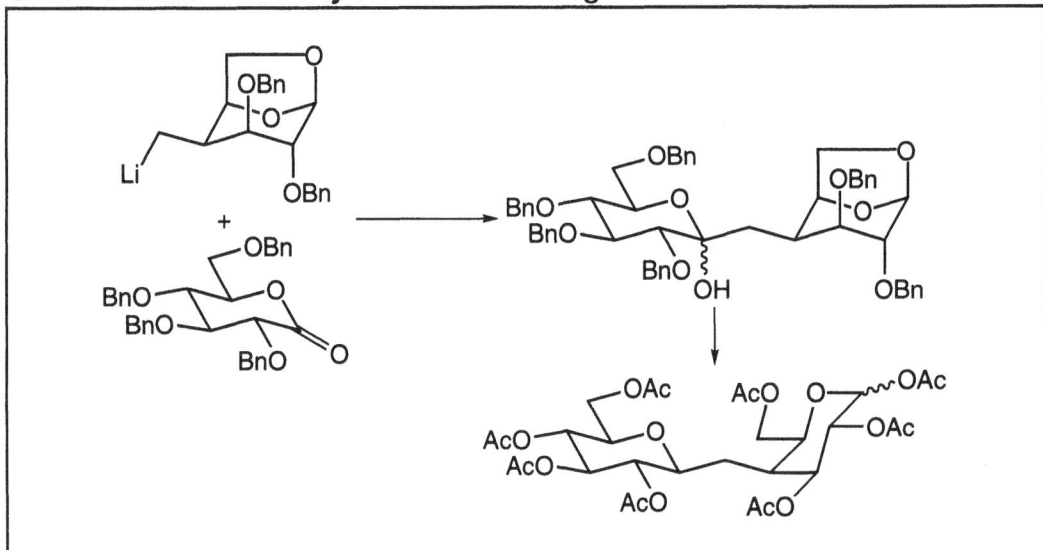

2.10 Nucleophilic Additions to Sugars Containing Enones

Figure 2.10.1   Sugar-Derived Enones or 3-Ketoglycals

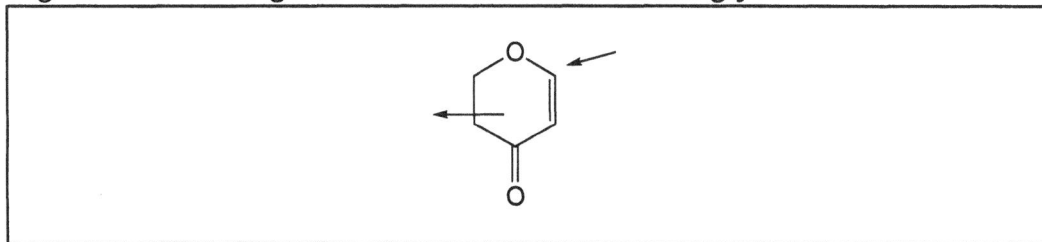

**Scheme 2.10.1   *C*-Glycosides from
Sugar-Derived Enones**

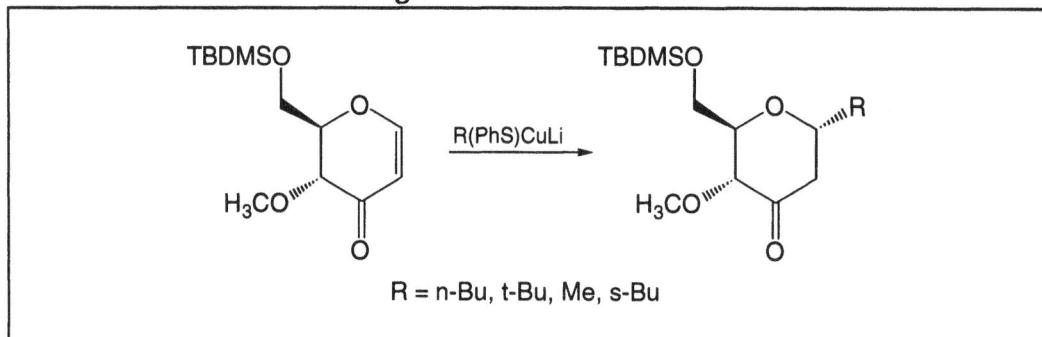

R = n-Bu, t-Bu, Me, s-Bu

Scheme 2.10.2   C-Glycosides from
                                Sugar-Derived Enones

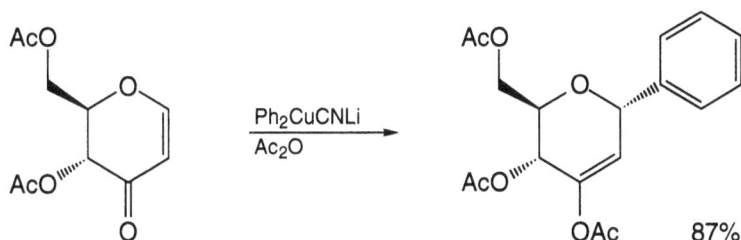

Continuing from sugar lactones, a logical progression is the utilization of sugar-derived enones illustrated in Figure 2.10.1. These enones are readily available from the oxidation of glycals and the stereoselectivity of addition reactions is dictated entirely by stereoelectronic effects. These reactions are particularly useful in that 2-deoxy C-glycosides are the products.

Figure 2.10.2   Stereochemistry of C-Glycosidations
                 Resulting from Cuprates
                 and Sugar-Derived Enones

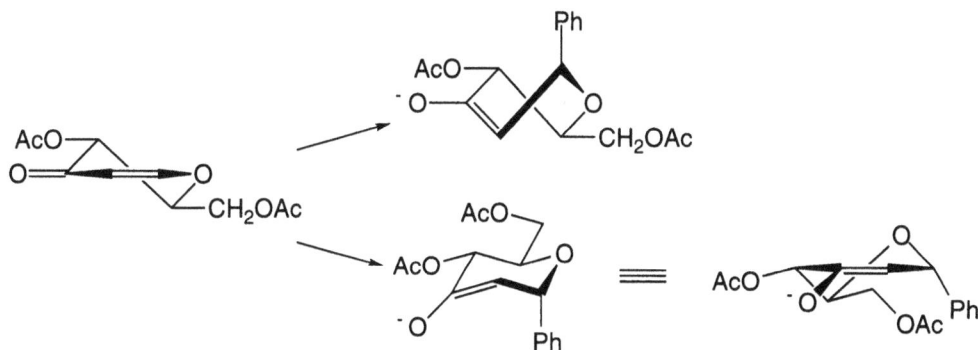

The formation of C-glycosides from enones has already been illustrated in Schemes 2.2.16 and 2.2.17. Recalling these schemes, cyanoglycosides were the products obtained on treating sugar-derived enones with cyanohydrins. However, the incorporation of enones as C-glycosidation substrates was known much earlier. For example, as early as 1983, Goodwin, et al.,[140] utilized various organocuprates to effect the formation of C-glycosides. As shown in Scheme 2.10.1, a variety of alkyl groups, including the bulky tert-butyl group, were efficiently incorporated to the glycosidic center. In this report, the products exhibited the α configuration at the anomeric center.

Scheme 2.10.3   *C*-Glycosides from
                          Sugar-Derived Enones

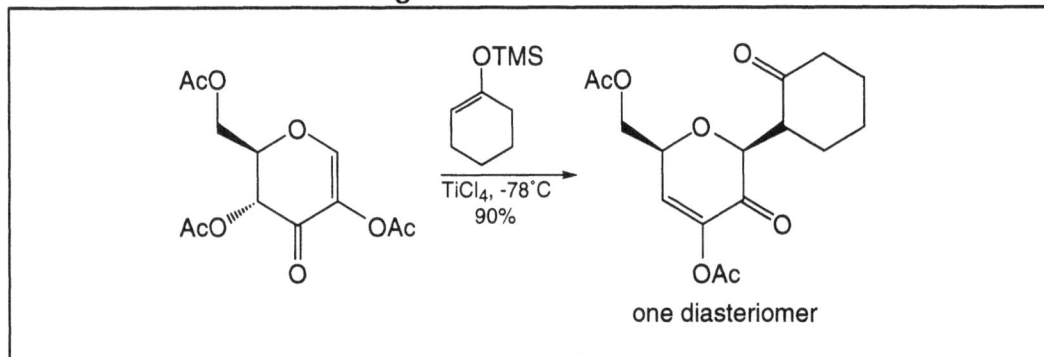

one diasteriomer

A further example of the use of organocuprates in the formation of *C*-glycosides from sugar-derived enones was reported by Bellosta, *et al.*,[141] and is illustrated in Scheme 2.10.2. In addition to the formation of the desired product in 87% yield, a high propensity for the formation of the α anomer was observed. This observation was in good agreement with the previous example. Furthermore, as shown in Figure 2.10.2, this report attempted to rationalize the stereoselectivity through stereoelectronic considerations. Through these considerations, a mechanism proceeding through a twist boat conformation is predicted. The α selectivity, in this case, arises from the pseudoequatorial approach of the nucleophile to the enone. One additional point with respect to the presently discussed example is the ability to utilize acetate protecting groups in the reaction complimentary to the example shown in Scheme 2.10.1 where silyl groups were used.

Scheme 2.10.4   *C*-Glycosides from
                          Sugar-Derived Enones

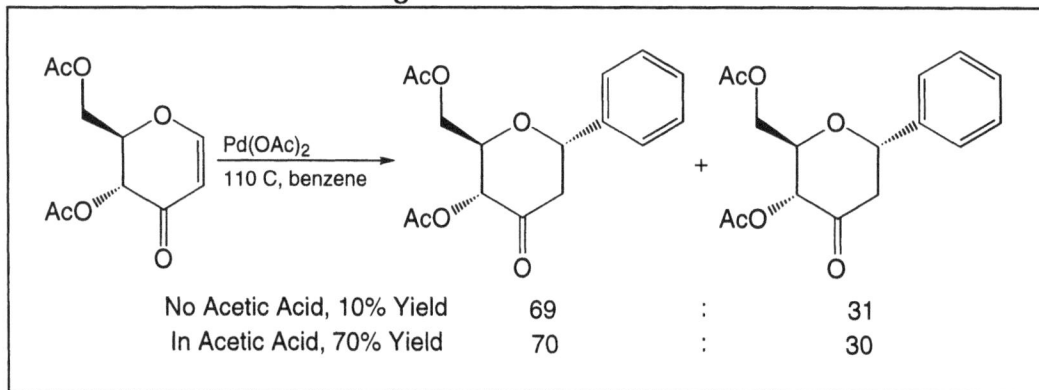

| | No Acetic Acid, 10% Yield | 69 | : | 31 |
| | In Acetic Acid, 70% Yield | 70 | : | 30 |

Complimentary to the use of organocuprates, silylenol ethers have been useful as agents capable of adding to sugar-derived enones. Already discussed in detail in section 2.5, silylenol ethers have been exploited in *C*-glycoside technology with respect to their combination with anomerically activated sugars. As sugar-derived enones are highly activated sugars, the incorporation of silylenol ethers as reagents for effecting *C*-glycosidations is logical and complimentary to methods previously addressed. As shown in Scheme 2.10.3, Kunz, *et al.*,[142] effected the use of silylenol ethers in the formation of a 90% yield of the *C*-glycoside shown. Compared to the use of organocuprates, the use of silylenol ethers allows the formation of β anomers further demonstrating advantages to having a variety of available *C*-glycosidation methods.

As a final example of the use of sugar-derived enones as substrates for the formation of *C*-glycosides, the use of transition metal reagents will be addressed. As shown in Scheme 2.10.4, Benhaddou, *et al.*,[143] effected the formation of α-1-arylglycosides in addition to the substituted enones formed as the major products. As illustrated, the specific reaction proceeded in 70% yield only when acetic acid was used in the reaction. As shown in Figure, 2.10.3, this reaction proceeds through the intermediate 2-metallo compound capable of isomerizing between axial and equatorial orientations. Elimination thus liberates the enone while reduction affords the α-1-arylglycoside.

Figure 2.10.3   Palladium Mediated *C*-Glycosidations of Sugar-Derived Enones

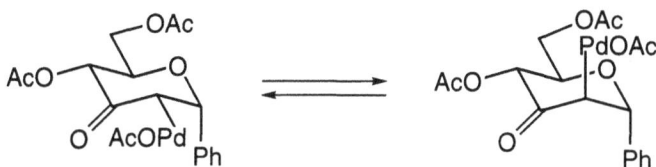

Technically different from the chemistry discussed in this chapter, this final example was chosen to illustrate the versatility of sugar-derived enones. The use of transition metals as agents mediating the formation of *C*-glycosides will be addressed in detail in Chapter 4.

2.11 Transition Metal Mediated Carbon Monoxide Insertions

As alluded to at the end of section 2.10 and to be discussed in Chapter 4, transition metals are capable of mediating reactions that produce *C*-glycosides. One such type of reaction is the transition metal mediated insertion of carbon monoxide generalized in Figure 2.11.1. As a means of preparing *C*-glycosides this technology has been, until recently, relatively unexplored. The examples that have been reported demonstrate a propensity for β-selectivity. These reactions can be executed under catalytic conditions utilizing both benzylated and acylated sugars. Additionally, anomeric activation is not a prerequisite for reactions to proceed.

**Figure 2.11.1   Example of Transition Metal Mediated
                   Carbon Monoxide Insertion Reactions**

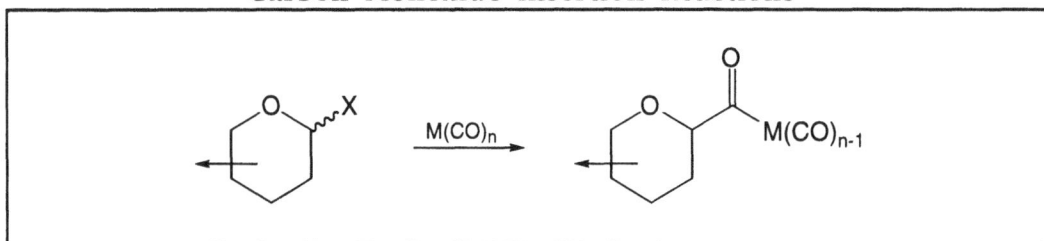

2.11.1   Manganese Glycosides

**Scheme 2.11.1   C-Glycosides from
                     Carbon Monoxide Insertions**

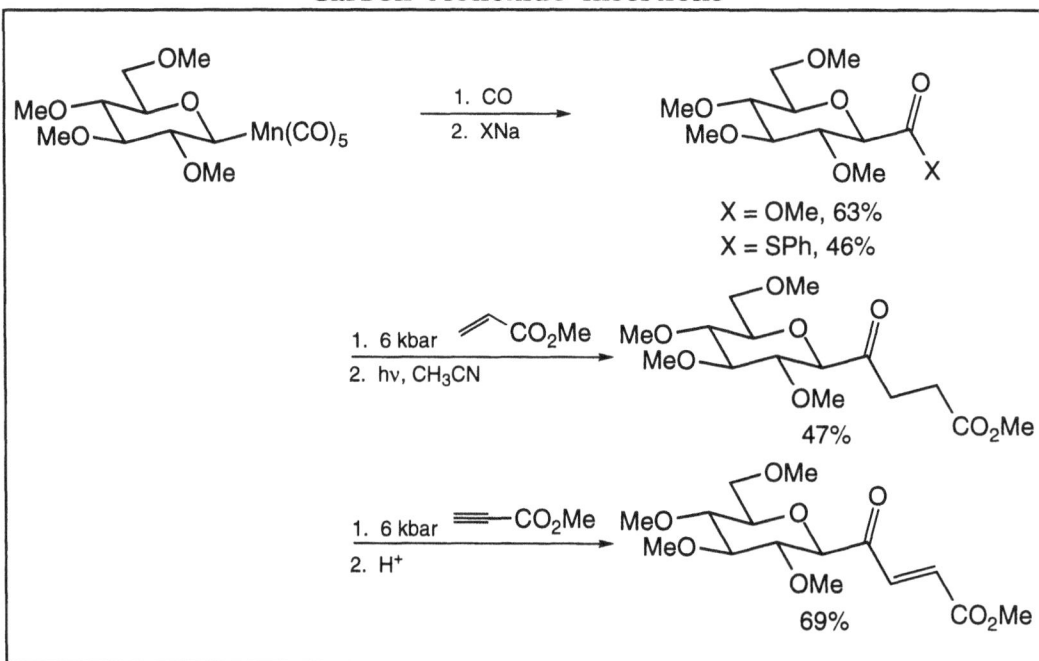

In one of the earliest examples of carbon monoxide insertion reactions for the formation of *C*-glycosides, Deshong, *et al.*,[144] utilized the manganese glycoside shown in Scheme 2.11.1. This compound was reacted with carbon monoxide followed by the sodium salts of either methanol or thiophenol. The resulting products were the corresponding ester and thioester. Additional examples demonstrated that vinylic and acetylenic esters were also substrates for these reactions yielding the corresponding Michael products.

The manganese glycosides utilized in the above study are easily prepared. As shown in Scheme 2.11.2, Deshong, *et al.*,[145] prepared these compounds from bromoglycosides. Interestingly, from pyranose sugars, the β anomer could be obtained. However, when furano sugars were used, the α anomer was the predominant product. In all cases, the isolated yields ranged from 50% to 90%.

Scheme 2.11.2   *C*-Glycosides from
                    Carbon Monoxide Insertions

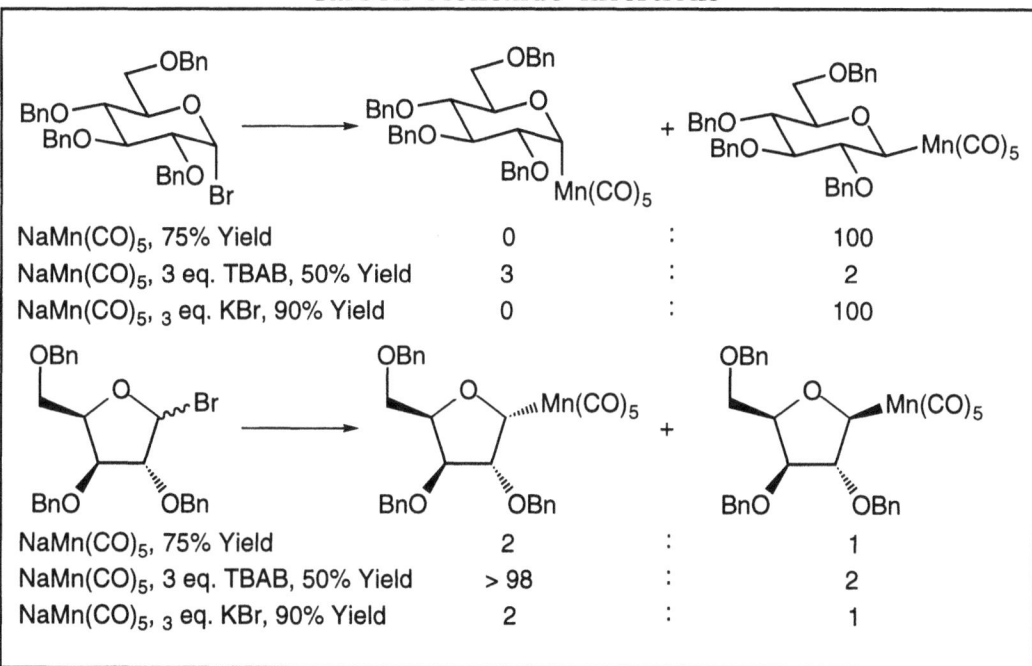

NaMn(CO)₅, 75% Yield	0	:	100
NaMn(CO)₅, 3 eq. TBAB, 50% Yield	3	:	2
NaMn(CO)₅, 3 eq. KBr, 90% Yield	0	:	100

NaMn(CO)₅, 75% Yield	2	:	1
NaMn(CO)₅, 3 eq. TBAB, 50% Yield	> 98	:	2
NaMn(CO)₅, 3 eq. KBr, 90% Yield	2	:	1

2.11.2   Cobalt Mediated Glycosidations

Manganese is not the only metal capable of effecting carbon monoxide insertion reactions. In 1988, Chatani, *et al.*,[146] demonstrated the utility of cobalt in the generation of *C*-glycosides. Specific examples, shown in Scheme 2.11.3, involve the conversion of glucose and galactose pentaacetates to the trimethylsilyloxymethylglycosides with β-selectivity. In the case of glucose pentaacetate, the use of trimethylsilylcobalt tetracarbonyl was illustrated.

As a final example, in 1992, Luengo, *et al.*,[147] applied cobalt methodology to fucose tetraaectate. The results, shown in Scheme 2.11.4, are interesting in that formation of the methyldiethylsilyloxymethyl fucoside proceeded with near exclusive formation of the β-isomer while employing allyltrimethylsilane in the reaction yielded the α-isomer in a 14 : 1 ratio. In applying this methodology to a practical synthesis, *C*-GDP-fucose, discussed in section 1.4 and shown in Figure 2.11.2, was prepared as a fucosyltransferase inhibitor.

Scheme 2.11.3   **C*-Glycosides from**
Carbon Monoxide Insertions

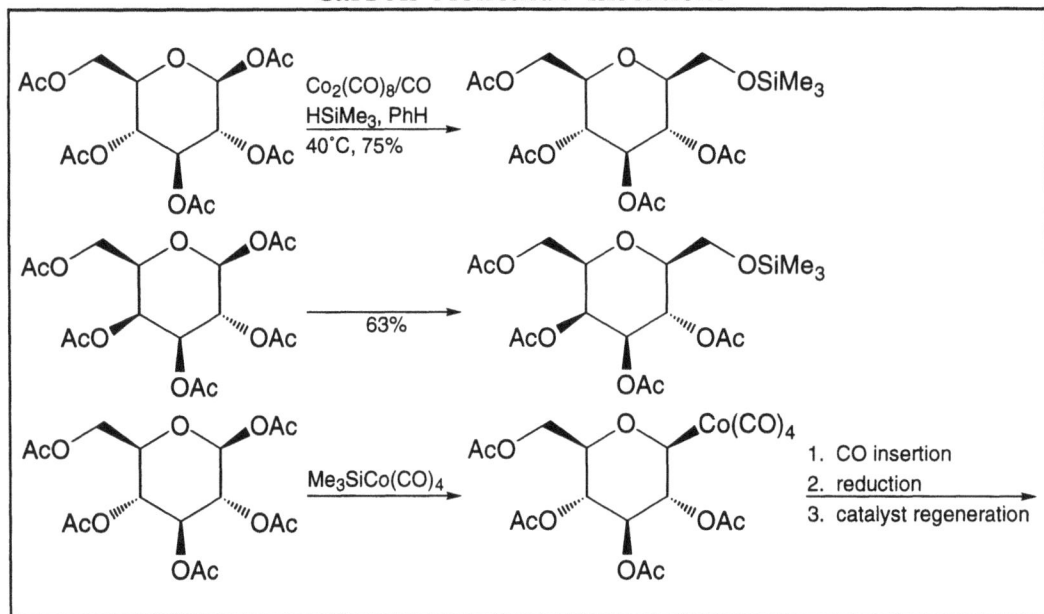

Scheme 2.11.4   **C*-Glycosides from**
Carbon Monoxide Insertions

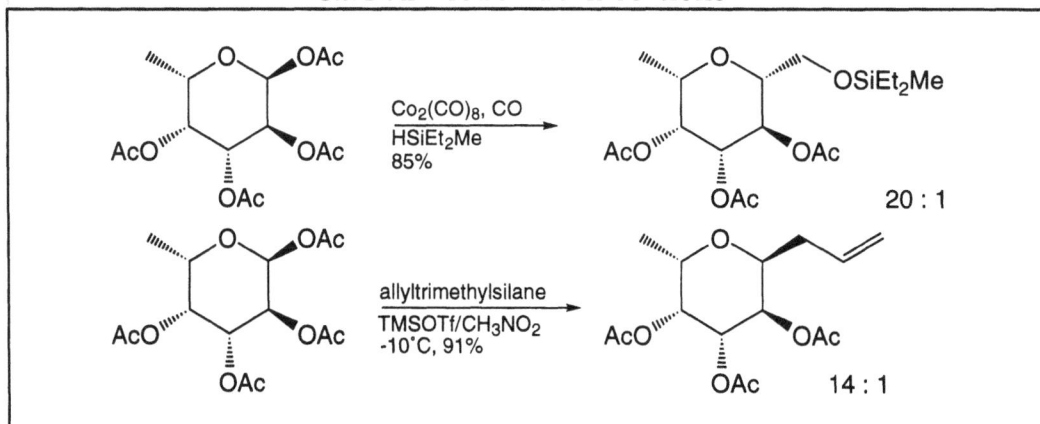

Figure 2.11.2   ***C*-GDP Fucose and Analogs**

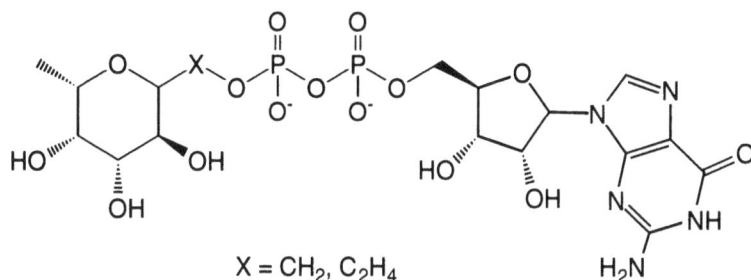

X = CH$_2$, C$_2$H$_4$

2.12 Reactions Involving Anomeric Carbenes

The incorporation of anomeric carbenes in *C*-glycoside technology has not received much attention. However, anomeric carbenes, as shown in Figure 2.12.1, provide, to date, the only known method for the direct cyclopropanation of anomeric centers. Advantages to this method include general synthetic accessibility of the carbenes when benzylated sugars are utilized.

Figure 2.12.1   **Example of Glycosidic**
Cyclopropanations

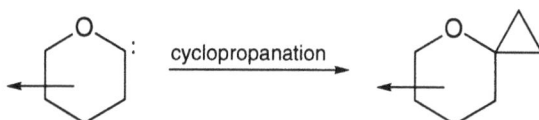

cyclopropanation

Two examples of the applicability of anomeric carbenes were reported by Vasella, *et al.*[148,149] The first demonstrated the utility of olefinic substrates such as *N*-phenylmaleimide, acrylonitrile, and dimethylmaleate for the formation of glycosidic cyclopropanes. The carbene precursor, in this example, was the glucose-derived diazirine. As shown in Scheme 2.12.1, the use of dimethylmaleate produced a mixture of diastereomers with a combined yield of 72%.

As shown in Scheme 2.12.2, the more dramatic example of cyclopropanations involving fullerenes is illustrated. The reaction proceeded with both benzyl and pivalyl protecting groups illustrating the versatility of this reaction under a variety of electronic conditions. The yields in both cases were approximately 55%. A particularly important aspect of this chemistry centers around the desired ability to functionalize the fullerenes and, especially,

C_{60} or buckminsterfullerene.   Through the use of glycosidic carbenes, this technology has been demonstrated to be both efficient and capable of producing a variety of products not readily available through other methods.

Scheme 2.12.1   Glycosidic Cyclopropanations

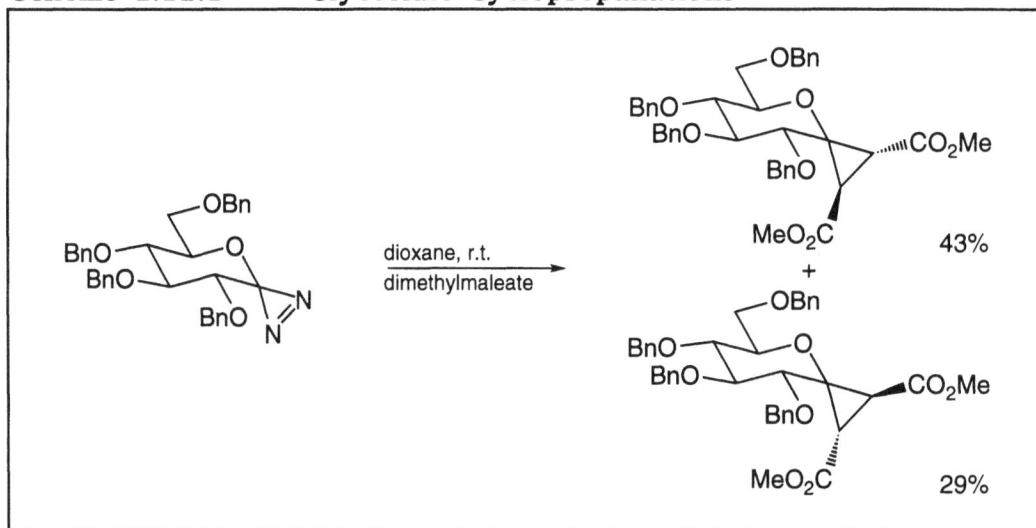

Scheme 2.12.2   Glycosidic Cyclopropanations

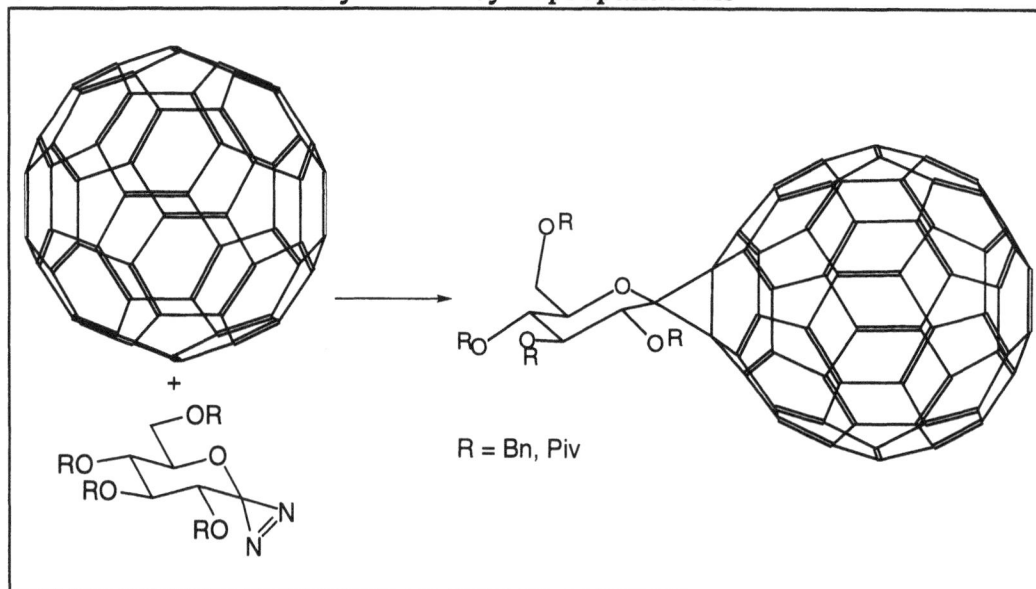

2.13 Reactions Involving Exoanomeric Methylenes

In the final section of this chapter, the elaboration of exoanomeric methylenes to more complex *C*-glycosides is addressed. These reactions, generalized in Figure 2.13.1, involve the addition of electrophiles to the anomeric olefin. Utilizing this technology makes double *C*-glycosidations possible. These reactions prefer benzyl protecting groups on the sugars. However, the olefins are easily generated *in situ* and reactions at the intermediate tertiary carbocation may be induced.

Figure 2.13.1   Example of Electrophilic Additions to Exoanomeric Methylene Groups

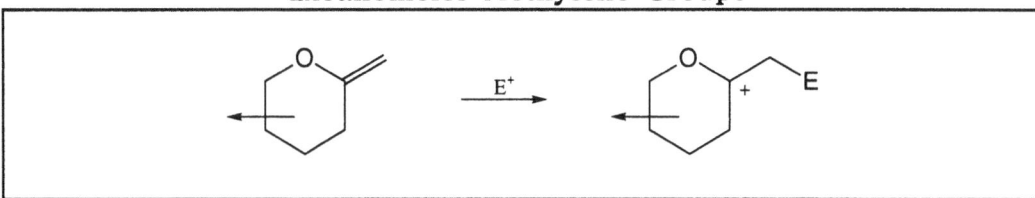

Scheme 2.13.1   Formation of Exoanomeric Methylene Groups

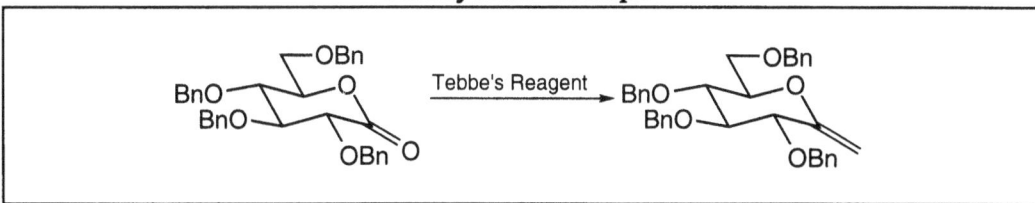

Technically *C*-glycosides, themselves, sugars bearing exoanomeric methylene groups are easily prepared by reacting sugar lactones with Tebbe's reagent. As shown in Scheme 2.13.1, RajanBabu, *et al.*,[150] utilized this technology as applied to pyranose sugars. The resulting products, as shown in Scheme 2.13.2, were reacted under a variety of conditions yielding a broad spectrum of products. As demonstrated, alcohols may be introduced in high yields and high stereoselectivity. Interestingly, when less bulky agents such as borane-THF complex were used, no stereoselectivity was observed. Furthermore, 1,3-dipolar cycloaddition reactions and arylations utilizing heavy metal reagents were demonstrated. In the later, non-specific delivery of the phenyl group resulted in the formation of a mixture of products. However, the ring-opened compound will cyclize under reductive oxymercuration conditions.

Utilizing similar methodology, Wilcox, *et al.*,[54] derivitized furanose sugars. One specific example is shown in Scheme 2.13.3 and demonstrates the ability to form hemiketal analogs from these sugar derivatives. The advantage of this transformation is apparent considering the ability to further derivatize these compounds as previously illustrated in Scheme 2.3.23.

Scheme 2.13.2   Reactions with Exoanomeric
Methylene Groups

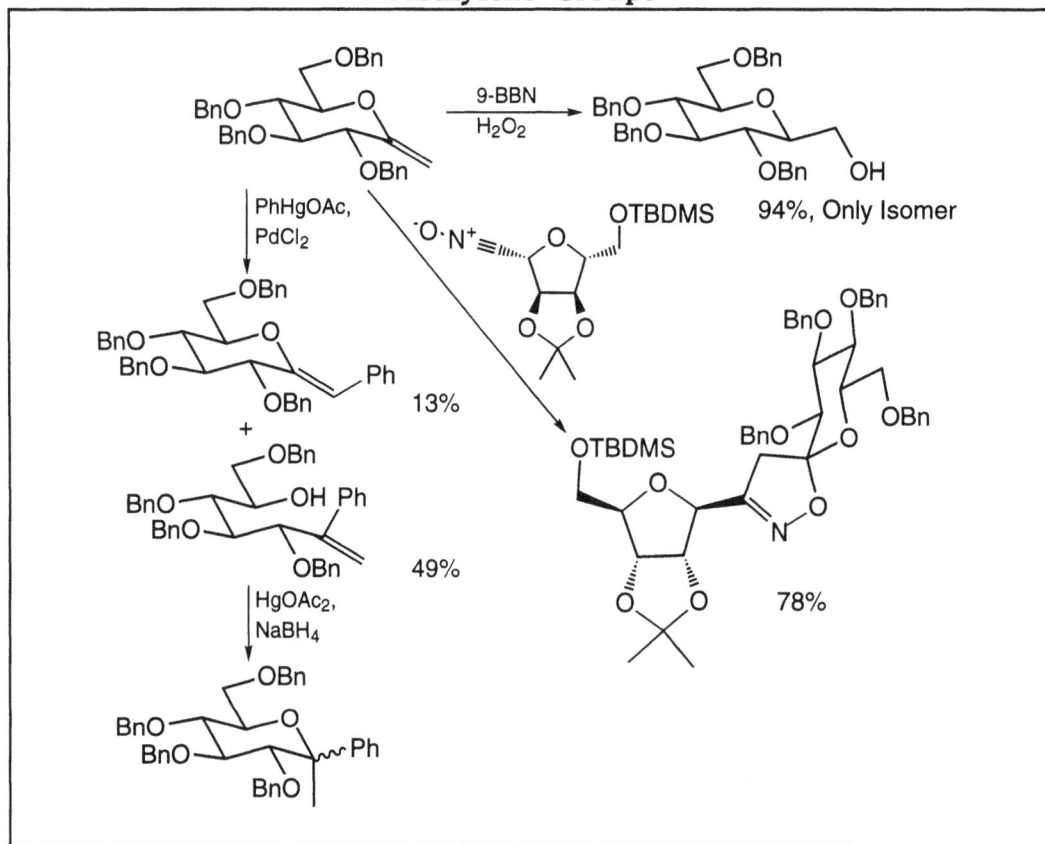

Scheme 2.13.3   Reactions with Exoanomeric
Methylene Groups

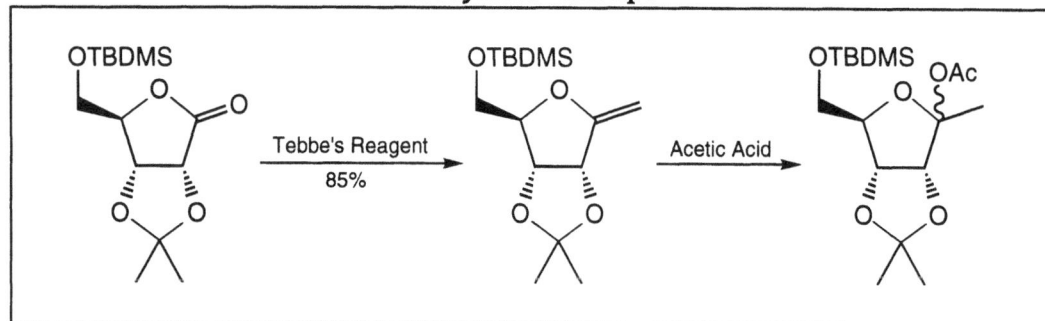

As a final example closing this chapter, the formation of difluoro olefins is discussed. As shown in Scheme 2.13.4, Motherwell, et al.,[151,152] utilized phosphorus chemistry to effect the formation of these species in yields ranging from 55% to 65%. Furthermore, reduction stereospecifically converted these difluoro olefins to difluoromethylglycosides. In the last example, reduction was effected with thiophenol and allyltin induced the chain elongation to form a difluorobutenyl glycoside.

Scheme 2.13.4   Reactions with Exoanomeric Methylene Groups

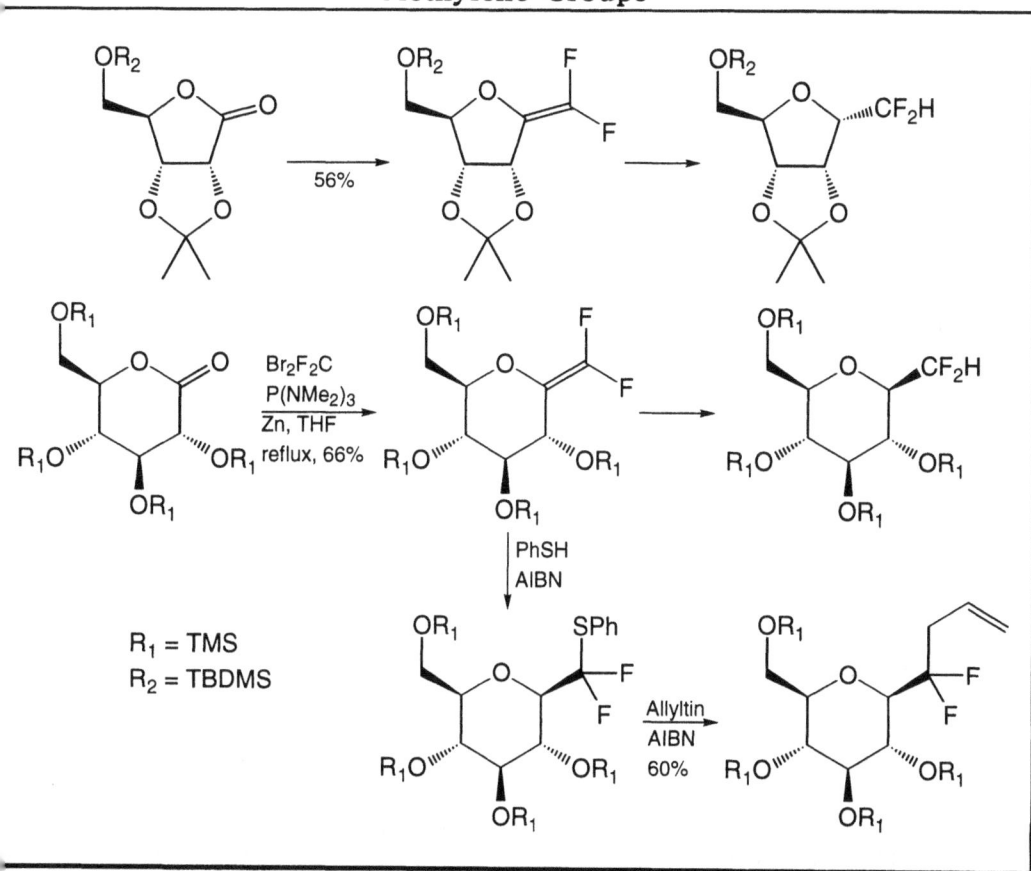

This section has attempted to introduce the use of exoanomeric methylene groups as a means for the formation of more complex C-glycosides. Such species are also apparent as intermediates in the formation of C-disaccharides and will be addressed in Chapter 8. For now, let it suffice that exoanomeric methylene groups, considered within other means of C-glycoside syntheses are quite complimentary and beneficial to the general chemical knowledge surrounding C-glycosides.

2.14 References

1.   Lopez, M. T. G.; Heras, F. G. D.; Felix, A. S. *J. Carbohydrate Chem.* **1987**, *6*, 273.
2.   Kini, G. D.; Petrie, C. R.; Hennen, W. J.; Dalley, N. K.; Wilson, B. E.; Robbins, R. K. *Carbohydrate Res.* **1987**, *159*, 81.
3.   Hoffman, M. G.; Schmidt, R. R. *Liebigs Ann. Chem.* **1985**, 2403.
4.   Nicolaou, K. C.; Dolle, R. E.; Chucholowski, A.; Randall, J. L. *J. Chem. Soc. Chem. Comm.* **1984**, 1153.
5.   Araki, Y.; Kobayashi, N.; Watanabe, K.; Ishido, Y. *J. Carbohydrate Chem.* **1985**, *4*, 565.
6.   Grynkiewicz, G.; Bemiller, J. N. *Carbohydrate Res.* **1982**, *108*, 229.
7.   Nicolaou, K. C.; Hwang, C. K.; Duggan, M. E. *J. Chem. Soc. Chem. Comm.* **1986**, 925.
8.   Smolyadove, I. P.; Smit, W. A.; Zal'chenko, E. A.; Chizhov, O. S.l Shashakov, A. S. *Tetrahedron Lett.* **1993**, *34*, 3047.
9.   Mukaiyama, T.; Kobayashi, S.; Shoda, S. I. *Chemistry Lett.* **1984**, 1529.
10. Grierson, D. S.; Bonin, M.; Husson, H. P. *Tetrahedron Lett.* **1984**, *25*, 4645.
11. Drew, K. N.; Gross, P. H. *J. Org. Chem.* **1991**, *56*, 509.
12. Grynkiewicz, G.; BeMiller, J. N. *Carbohydr. Res.* **1983**, *112*, 324.
13. Tatsuta, K.; Hayakawa, J.; Tatsuzawa, Y. *Bull. Chem. Soc. Jpn.* **1989**, *62*, 490.
14. BeMiller, J. N.; Yadav, M. P.; Kalabokis, V. N.; Myers, R. M. *Carbohydr. Res.* **1990**, *200*, 111.
15. Myers, R. W.; Lee, Y. C. *Carbohydr. Res.* **1986**, *152*, 143.
16. Pochet, S.; Allard, P.; Tam, H. D.; Igolen, J. *J. Carbohydrate Chem.* **1982-83**, *1*, 277.
17. Albrecht, H. P.; Repke, D. B.; Moffat, J. G. *J. Org. Chem.* **1973**, *38*, 1836.
18. Trummlitz, G.; Moffat, J. G. *J. Org. Chem.* **1973**, *38*, 1841.
19. Ugi, I. *Angew. Chem.* **1962**, *74*, 9.
20. Poonian, M. S.; Nowoswiat, E. F. *J. Org. Chem.* **1980**, *45*, 203.
21. El Khadem, H. S.; Kewai, J. *Carbohydr. Res.* **1986**, *153*, 271.
22. Hanessian, S.; Pernet, A. G. *Can. J. Chem.* **1974**, *52*, 1268.
23. Cristol, S. J.; Firth, W. C. *J. Org. Chem.* **1960**, *26*, 280.
24. Bihovsky, R.; Selick, C.; Giusti, I. *J. Org. Chem.* **1988**, *53*, 4026.
25. Bellosta, V.; Czernecki, S. *Carbohydrate Res.* **1993**, *244*, 275.
26. Bellosta, V.; Czernecki, S. *J. Chem. Soc. Chem. Comm.* **1989**, 199.
27. Shulman, M. L.; Shiyan, S. D.; Khorlin, A. Y. *Carbohydrate Res.* **1974**, *33*, 229.
28. Hanessian, S.; Liak, T. J.; Dixit, D. M. *Carbohydr. Res.* **1981**, *88*, C14.
29. Bolitt, V.; Mioskowski, C.; Falck, J. R. *Tetrahedron Lett.* **1989**, *30*, 6027.
30. Brown, D.; Ley, S. V. *Tetrahedron Lett.* **1988**, *29*, 4869.
31. Ley, S. V.; Lygo, B.; Wonnacott, A. *Tetrahedron Lett.* **1985**, *29*, 535.
32. Ley, S. V.; Lygo, B.; Sternfeld, F.; Wonnacott, A. *Tetrahedron Lett.* **1986**, *42*, 4333.

33. Beau, J.-M.; Sinay, P. *Tetrahedron Lett.* **1985**, *26*, 6185.
34. Cassidy, J. F.; Williams, J. M. *Tetrahedron Lett.* **1986**, *27*, 4355.
35. Orsini, F.; Pelizzoni, F. *Carbohydrate Res.* **1993**, *243*, 183.
36. Kozikowski, A. P.; Konoike, T.; Ritter, A. *Carbohydrate Res.* **1987**, *171*, 109.
37. Bols, M.; Szarek, W. A. *J. Chem. Soc. Chem. Commun.* **1992**, 445.
38. Koll, P.; Oelting, M. *Tetrahedron Lett.* **1986**, *27*, 2837.
39. Lewis, M. D.; Cha, J. K.; Kishi, Y. *J. Am. Chem. Soc.* **1982**, *104*, 4976.
40. Cupps, T. L.; Wise, D. S.; Townsend, L. B. *Carbohydrate Res.* **1983**, *115*, 59.
41. Herscovici, J.; Muleka, K.; Antonakis, K. *Tetrahedron Lett.* **1984**, *25*, 5653.
42. Herscovici, J.; Muleka, K.; Boumaiza, L.; Antonakis, K. *J. Chem. Soc. Perkin Trans. 1* **1990**, 1995.
43. Levy, D. E.; Dasgupta, F.; Tang, P. C. *Tetrahedron Asymmetry* **1994**, *in press*.
44. Adams, D. R.; Bhatnagar, S. P. *Synthesis* **1977**, *10*, 661.
45. Cooper, A. J.; Salomon, R. G. Tetrahedron Lett. 1990, 31, 3813.
46. Newcombe, N. J.; Mahon, M. F.; Molloy, K. C.; Alker, D.; Gallagher, T. *J. Am. Chem. Soc.* **1993**, *115*, 6430.
47. Terahara, A.; Haneishi, T.; Arai, M.; Hata, T. *J. Antibiot.* **1982**, *35*, 1711.
48. Yoshikawa, H.; Takiguchi, Y.; Terao, M. *J. Antibiot.* **1983**, *36*, 30.
49. Kozaki, S.; Sakanaka, O.; Yasuda, T.; Shimizu, T.; Ogawa, S.; Suami, T. *J. Org. Chem.* **1988**, *53*, 281.
50. Cabaret, D.; Wakselman, M. *Carbohydr. Res.* **1989**, *189*, 341.
51. De Mesmaeker, A.; Waldner, A.; Hoffman, P.; Hug, P.; Winkler, T. *Synlett* **1992**, 285.
52. Giannis, A.; Sandhoff, K. *Tetrahedron Lett.* **1985**, *26*, 1479.
53. Allevi, P.; Anastasia, M.; Ciuffreda, P.; Fiecchi, A.; Scala, A. *J. Chem. Soc. Chem. Comm.* **1987**, 1245.
54. Wilcox, C. S.; Long, G. W.; Suh, H. S. *Tetrahedron Lett.* **1984**, *25*, 395.
55. Acton, E. M.; Ryan, K. J.; Tracy, M. *Tetrahedron Lett.* **1984**, *25*, 5743.
56. Martin, M. G. G.; Horton, D. *Carbohydrate Res.* **1989**, *19*, 223.
57. Panek, J. S.; Sparks, M. A. *J. Org. Chem.* **1989**, *54*, 2034.
58. Babirad, S. A.; Wang, Y.; Kishi, Y. *J. Org. Chem.* **1987**, *52*, 1370.
59. Nicotra, F.; Panza, L.; Russo, G. *J. Org. Chem.* **1987**, *52*, 5627.
60. Martin, O. R.; Rao, S. P.; Kurz, K. G.; El-Shenawy, H. A. *J. Am. Chem. Soc.* **1988**, *110*, 8698.
61. Bennek, J.; Gray, G. R. *J. Org. Chem.* **1987**, *52*, 892.
62. Hosomi, A.; Sakata, Y.; Sakurai, H. *Carbohydrate Res.* **1987**, *171*, 223.
63. Araki, Y.; Kobayashi, N.; Ishido, Y. *Carbohydrate Res.* **1987**, *171*, 125.
64. Bertozzi, C. R.; Bednarski, M. D. *Tetrahedron Lett.* **1992**, *33*, 3109.
65. Ichikawa, Y.; Isobe, M.; Konobe, M.; Goto, T. *Carbohydrate Res.* **1987**, *171*, 193.

66. de Raddt, A.; Stutz, A. E. *Carbohydrate Res.* **1991**, *220*, 101.
67. Ferrier, R. J.; Petersen, P. M. *J. Chem. Soc. Perkin Trans. 1* **1992**, 2023.
68. Tsukiyama, T.; Isobe, M. *Tetrahedron Lett.* **1992**, *33*, 7911.
69. Keck, G. E.; Enholm, E. J.; Kachensky, D. F. *Tetrahedron Lett.* **1984**, *25*, 1867.
70. Nagy, J. O.; Bednarski, M. D. *Tetrahedron Lett.* **1991**, *32*, 3953.
71. Waglund, T.; Cleasson, A. *Acta Chemica Scandinavica* **1992**, *46*, 73.
72. Zhai, D.; Zhai, W.; Williams, R. M. *J. Am. Chem. Soc.* **1988**, *110*, 2501.
73. Tolstikov, G. A.; Prokhorova, N. A.; Spivak, A. Y.; Khalilov, L. M.; Sultanmuratova, V. R. *Zh. Org. Khim.* **1991**, *27*, 2101.
74. Posner, G. H.; Haines, S. R. *Tetrahedron Lett.* **1985**, *26*, 1823.
75. Bellosta, V.; Chassagnard, C.; Czernecki, S. *Carbohydrate Res.* **1991**, *219*, 1.
76. Macdonald, S. J. F.; Huizinga, W. B.; McKenzie, T. C. *J. Org. Chem.* **1988**, *53*, 3371.
77. Outten, R. A.; Daves, D. G. Jr. *J. Org. Chem.* **1987**, *52*, 5064.
78. Outten, R. A.; Daves, D. G. Jr. *J. Org. Chem.* **1989**, *54*, 29.
79. Hamamichi, N.; Miyasaka, T. *J. Org. Chem.* **1991**, *56*, 3731.
80. Kwok, D. I.; Outten, R. A.; Huhn, R.; Daves, G. D. Jr. *J. Org. Chem.* **1988**, *53*, 5359.
81. Farr, R. N.; Kwok, D. I.; Daves, G. D. Jr. *J. Org. Chem.* **1992**, *57*, 2093.
82. Cai, M. S.; Qiu, D. X. *Carbohydrate Res.* **1989**, *191*, 125.
83. Cai, M. S.; Qiu, D. X. *Syn. Comm.* **1989**, *19*, 851.
84. Frick, W.; Schmidt, R. R. *Carbohydrate Res.* **1991**, *209*, 101.
85. Schmidt, R. R.; Hoffman, M. *Tetrahedron Lett.* **1982**, *23*, 409.
86. Schmidt, R. R.; Effenberger, G. *Liebigs Ann. Chem.* **1987**, 825.
87. El-Desoky, E. I.; Abdel-Rahman, H. A. R.; Schmidt, R. R. *Liebigs Ann. Chem.* 1990, 877.
88. Matsumoto, T.; Katsuki, M.; Suzuki, K. *Tetrahedron Lett.* **1989**, *30*, 833.
89. Allevi, P.; Anastasia, M.; Ciuffreda, P.; Fiecchi, A.; Scala, A. *J. Chem. Soc. Chem. Comm.* **1987**, 101.
90. Casiraghi, G.; Cornia M.; Colombo, L.; Rassu, G.; Fava, G. G.; Belicchi, M. F.; Zetta, L. *Tetrahedron Lett.* **1988**, *29*, 5549.
91. Araki, Y.; Mokubo, E.; Kobayashi, N.; Nagasawa, J. *Tetrahedron Lett.* **1989**, *30*, 1115.
92. Martin, O. R.; Hendricks, C. A.; Desphpande, P. P.; Cutler, A. B.; Kane, S. A.; Rao, S. P. *Carbohydrate Res.* **1990**, *198*, 41.
93. Martin, O. R.; Mahnken, R. E. *J. Chem. Soc. Chem. Comm.* **1986**, 497.
94. Anastasia, M.; Allevi, P.; Ciuffreda, P.; Fiecchi, A.; Scala, A. *Carbohydrate Res.* **1990**, *208*, 264.
95. Ramesh, N. G.; Balasubramanian, K. K. *Tetrahedron Lett.* **1992**, *33*, 3061.
96. Casiraghi, G.; Cornia, M.; Rassu, G.; Zetta, L.; Fava, G. G.; Belicchi, M. F. *Tetrahedron Lett.* **1988**, *29*, 3323.

97. Casiraghi, G.; Cornia, M.; Rassu, G.; Zetta, L.; Fava, G.; Belicchi, F. *Carbohydrate Res.* **1989**, *191*, 243.
98. Matsumoto, T.; Hosoya, T.; Suzuki, K. *Tetrahedron Lett.* **1990**, *32*, 4629.
99. Toshima, K.; Matsuo, G.; Ishizuka, T.; Nakata, M.; Kninoshita, M. *J. Chem. Soc. Chem. Comm.* **1992**, 1641.
100. Toshima, K.; Matsuo, G.; Tatsuta, K. *Tetrahedron Lett.* **1992**, *33*, 2175.
101. Matsumoto, T.; Hosoya, T.; Suzuki, K. *J. Am. Chem. Soc.* **1992**, *114*, 3568.
102. Matsumoto, T.; Katsuki, M.; Jona, H.; Suzuki, K. *Tetrahedron Lett.* **1989**, *30*, 6185.
103. Gonzales, M. A.; Requejo, J.; Albarran, P.; Perez, J. A. G. *Carbohydrate Res.* **1986**, *158*, 53.
104. Yamaguchi, M.; Horiguchi, A.; Ikegura, C.; Minami, T. *J. Chem. Soc. Chem. Comm.* **1992**, 434.
105. Narasaka, K.; Ichikawa, Y. I.; Kubota, H. *Chemistry Lett.* **1987**, 2139.
106. Ichikawa, Y.; Kubota, H.; Fujita, K.; Okauchi, T.; Narasaka, K. *Bull Chem. Soc. Jpn.* **1989**, *62*, 845.
107. Araki, Y.; Watanabe, K.; Kuan, F. H.; Itoh, K.; Kobayashi, N.; Ishido, Y. *Carbohydrate Res.* **1984**, *C5*, 127.
108. Allevi, P.; Anastasia, M.; Ciuffreda, P.; Fiecchi, A.; Scala, A. *J. Chem. Soc. Perkin Trans. I* **1989**, 1275.
109. Craig, D.; Munasinghem, V. R. N. *J. Chem. Soc. Chem. Comm.* **1993**, 901.
110. Dawe, R. D.; Fraser-Reid, B. *J. Chem. Soc. Chem. Comm.* **1981**, 1180.
111. Kunz, H.; Muller, B.; Weissmuller, J. *Carbohydrate Res.* **1987**, *171*, 25.
112. Allevi, P.; Anastasia, M.; Ciuffreda, P.; Eicchim, A.; Scala, A. *J. Chem. Soc. Chem. Comm.* **1988**, 57.
113. Drew, K. N.; Gross, P. H. *Tetrahedron* **1991**, *47*, 6113.
114. Koll, P.; Kopf, J.; Wess, D.; Brandenburg, H. *Liebigs Ann. Chem.* **1988**, 685.
115. Aebischer, B.; Vasella, A. *Helv. Chim. Acta* **1983**, *66*, 789.
116. Baumberger, F.; Vasella, A. *Helv. Chim. Acta* **1983**, *66*, 2210.
117. Aebischer, B.; Meuwly, R.; Vasella, A. *Helv. Chim. Acta* **1984**, *67*, 2236.
118. Herscovici, J.; Delatre, S.; Antonakis, K. *J. Org. Chem.* **1987**, *52*, 5691.
119. Herscovici, J.; Delatre, S.; Antonakis, K. *Tetrahedron Lett.* **1991**, *32*, 1183.
120. Herscovici, J.; Boumaiza, L.; Antonakis, K. *J. Org. Chem.* **1992**, *57*, 2476.
121. Acton, E. M.; Ryan, K. J.; Smith, T. H. *Carbohydrate Res.* **1981**, *97*, 235.
122. Nicotra, F.; Russo, G.; Ronchetti, F.; Toma, L. *Carbohydrate Res.* **1983**, *124*, C5.
123. Giannis, A.; Sandhoff, K. *Carbohydrate Res.* **1987**, *171*, 201.
124. Dheilly, L.; Frechou, C.; Beaupere, D.; Uzan, R.; Demailly, G. *Carbohydrate Res.* **1992**, *224*, 301.
125. Shen, Y.; Xin, Y.; Zhao, J. *Tetrahedron Lett.* **1988**, *29*, 6119.
126. Sun, K. M.; Dawe, R. D.; Fraser-Reid, B. *Carbohydrate Res.* **1987**, *171*, 36.
127. Bandzouzi, A.; Chapleur, Y. *Carbohydrate Res.* **1987**, *171*, 13.
128. Allevi, P.; Ciuffreda, P.; Colombo, D.; Monti, D.; Speranza, G.; Manitto, P. *J. Chem. Soc. Perkin Trans. I* **1989**, 1281.

129. Barnes, J. J.; Davidson, A. H.; Hughes, L. R.; Procter, G. *J. Chem. Soc. Chem. Comm.* **1985**, 1292.
130. Davidson, A. H.; Hughes, L. R.; Qureshi, S. S.; Wright, B. *Tetrahedron Lett.* **1988**, *29*, 693.
131. Fréchou, C.; Dheilly, L.; Beaupere, D.; Demailly, R. U. *Tetrahedron Lett.* **1992**, *33*, 5067.
132. Horton, D.; Priebe, W. *Carbohydrate Res.* **1981**, *94*, 27.
133. Lancelin, J. M.; Zollo, P. H. A.; Sinay, P. *Tetrahedron Lett.* **1983**, *24*, 4833.
134. Kraus, G. A.; Molina, M. T. *J. Org. Chem.* **1988**, *53*, 752.
135. Czernecki, S.; Ville, G. *J. Org. Chem.* **1989**, *54*, 610.
136. Krohn, K.; Heins, H.; Wielckens, K. *J. Med. Chem.* **1992**, *35*, 511.
137. Wilcox, C. S.; Cowart, M. D. *Carbohydrate Res.* **1987**, *171*, 141.
138. Rouzaud, D.; Sinay, P. *J. Chem. Soc. Chem. Comm.* **1983**, 1353.
139. Preuss, R.; Schmidt, R. R. *J. Carbohydrate Chem.* **1991**, *10*, 887.
140. Goodwin, T. E.; Crowder, C. M.; White, R. B.; Swanson, J. S.; Evans, F. E.; Meyer, W. L. *J. Org. Chem.* **1983**, *48*, 376.
141. Bellosta, V.; Czernecki, S. *Carbohydrate Res.* **1987**, *171*, 279.
142. Kunz, H.; Weibuller, J.; Muller, B. *Tetrahedron Lett.* **1984**, *25*, 3571.
143. Benhaddou, R.; Czernecki, S.; Ville, G. *J. Org. Chem.* **1992**, *57*, 4612.
144. Deshong, P.; Slough, G. A.; Elango, V. *J. Am. Chem. Soc.* **1985**, *107*, 7788.
145. Deshong, P.; Slough, G. A.; Elango, V. *Carbohydrate Res.* **1987**, *171*, 342.
146. Chatani, N.; Ikeda, T.; Sano, T.; Sonoda, N.; Kurosawa, H.; Kawasaki, Y.; Murai, S. *J. Org. Chem.* **1988**, *53*, 3387.
147. Luengo, J. I.; Gleason, J. G. *Tetrahedron Lett.* **1992**, *33*, 6911.
148. Vasella, A.; Waldraff, C. A. A. *Helv. Chim. Acta.* **1991**, *74*, 585.
149. Vasella, A.; Uhlmann, P.; Waldraff, C. A. A.; Diederich, F. *Angew. Chem. Int. Ed. Eng.* **1992**, *31*, 1388.
150. RajanBabu, T. V.; Reddy, G. S. *J. Org. Chem.* **1986**, *51*, 5458.
151. Motherwell, W. B.; Tozer, M. J.; Ross, B. C. *J. Chem. Soc. Chem. Comm.* **1989**, 1437.
152. Motherwell, W. B.; Ross, B. C.; Tozer, M. J. *Synlett* **1989**, 68.

3   Introduction

As thoroughly covered in Chapter 2, the use of electrophilic substitutions at the anomeric center have provided a substantially diverse variety of methods for the preparation of *C*-glycosides. However, a complimentary approach to the preparation of these compounds, for which there is no counterpart in *O*-glycoside chemistry, is found through the use of sugar-derived nucleophiles. The formation of C_1 lithiated sugar derivatives can be accomplished through a variety of methods including direct hydrogen-metal exchange, metal-metal exchange, and sulfone reductions.

Scheme 3.0.1   ***C*-Glycosides from**
Anomeric Nucleophiles

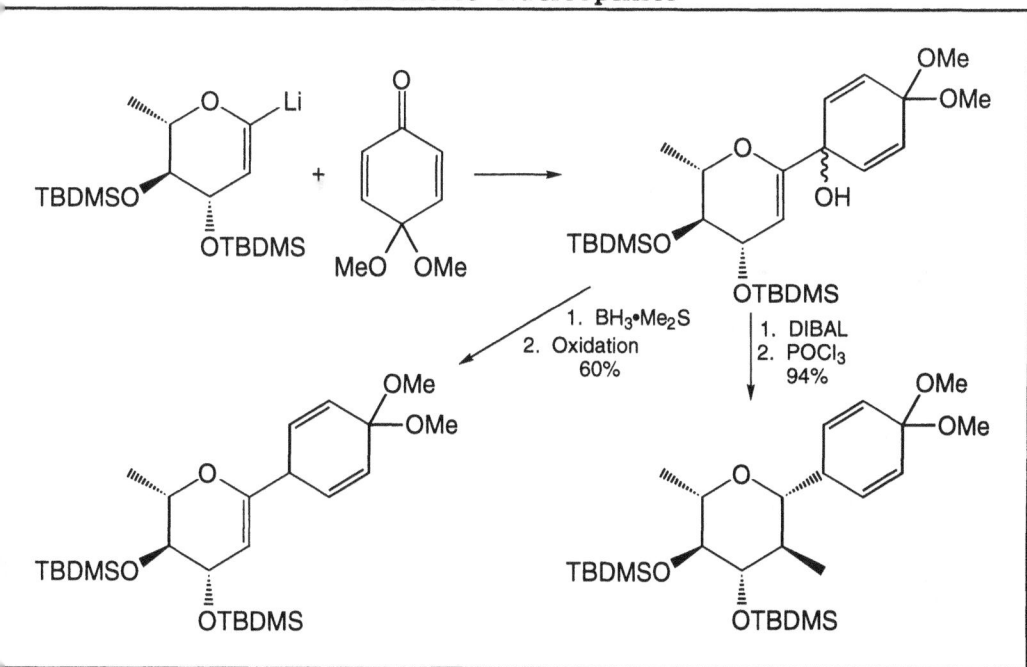

As shown in Scheme 3.0.1, C_1 lithiated carbohydrate analogs can act as nucleophiles in the formation of *C*-glycosides. This particular example, reported by Parker, *et al.*,[1] produced an intermediate compound capable of being converted to either a substituted glycal or a *C*-aryl glycoside. In this chapter the formation of C_1 lithiated sugar derivatives and their incorporation into the preparation of *C*-glycosides are addressed. Additionally, the use of electron withdrawing stabilizing groups are discussed.

3.1 C_1 Lithiated Anomeric Carbanions by Direct Metal Exchange

The direct metallation of sugar derivatives, shown in Figure 3.1.1, is a useful method for the preparation of anomeric nucleophiles. Anomeric halides and glycals are excellent substrates for this reaction. Additionally, selective generation of α and β anomers is possible.

Figure 3.1.1   **General Metallation Examples**

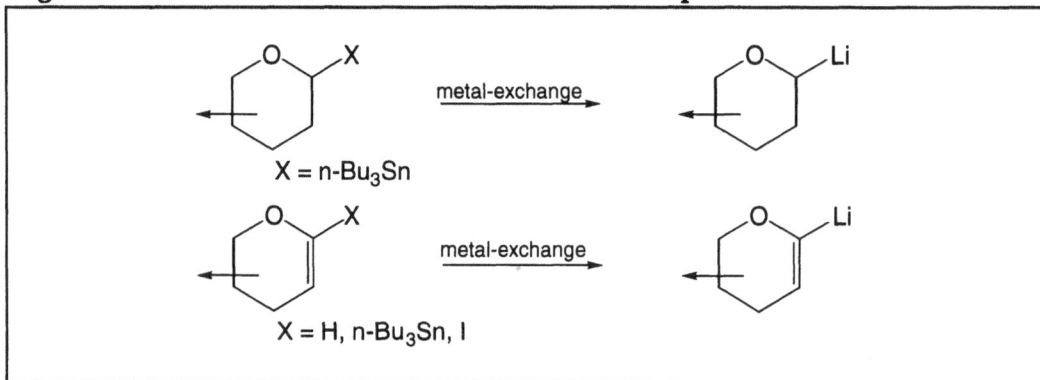

3.1.1   *Hydrogen-Metal Exchanges*

Scheme 3.1.1   **Glycal Lithiations**

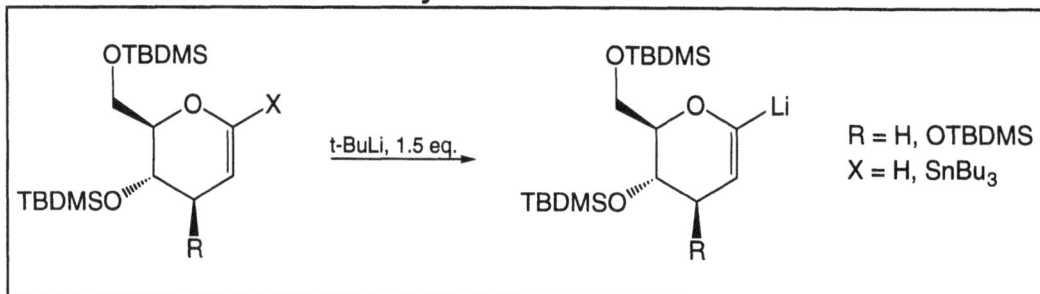

Direct applications of this methodology to the preparation of *C*-glycosides was demonstrated as early as 1986 when Sinay, *et al.*,[2] converted glycals and 1-stannyl glycals to 1-lithioglycals on treatment with *tert*-butyllithium. These results, shown in Scheme 3.1.1, were complemented by Nicolau's subsequent application to the generation of *C*-glycosides on treatment with allyl bromide, shown in Scheme 3.1.2.[3] As illustrated, the product of this addition was the 1-allylglycal shown. Utilizing a variety of methods described in Chapter 2, these compounds are easily transformed to a variety of true *C*-glycosides.

Scheme 3.1.2   *C*-**Glycosides from Glycals**

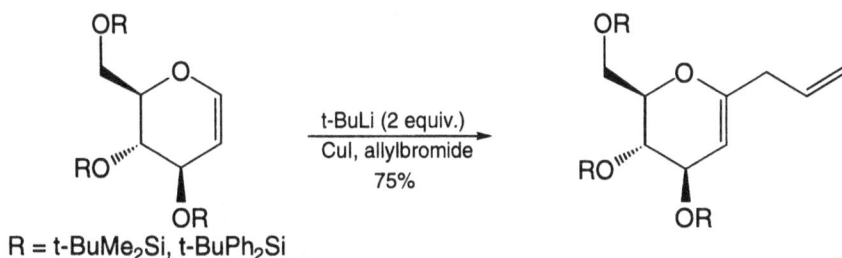

Scheme 3.1.3   **Glycal Stannylations**

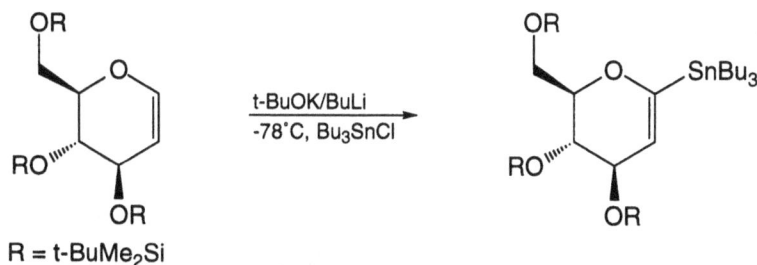

Scheme 3.1.4   **Glycal Lithiations**

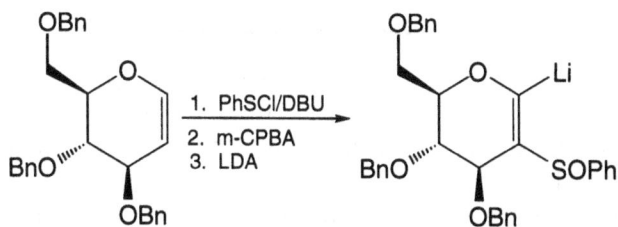

Later work elaborating on the chemistry of glycals demonstrated the ease of formation of 1-stannyl glycals. These compounds, introduced in Scheme 3.1.1, are useful substrates for the direct formation of *C*-glycosides as well as for metal-metal exchanges with lithium to be discussed later in this chapter. As shown in Scheme 3.1.3, Hanessian, *et al.*,[4] utilized potassium *tert*-butoxide and butyllithium to effect the formation of 1-stannyl glycals.

Continuing with the direct metallation of glycals, 2-phenylsulfinyl derivatives have found utility. Their formation and subsequent lithiation, shown in Scheme 3.1.4, is accomplished on reaction of glycals with phenylsulfenyl chloride under basic conditions. Subsequent oxidation with mCPBA yields the sulfinyl compound ready for lithiation on treatment with lithium diisopropylamide. Advantageous to the formation of this species is the stabilization of the anion by chelation of the sulfoxide to the metal. This procedure reported by Schmidt, *et al.*,[5] was utilized in the preparation of *C*-disaccharides, discussed in Chapter 8.

Similar reactions involving the lithiation of 2-phenylsulfinyl glycals, again reported by Schmidt, *et al.*,[6] are shown in Scheme 3.1.5. The actual couplings involved the addition of benzaldehyde to the lithiated species. The coupling proceeded in approximately 86%. An important observation is the induction of stereochemistry at the newly formed center with a diastereomeric ratio of approximately 4 : 1. Utilizing this methodology, the compounds shown in Figure 1.4.1 were prepared as potential β-glucosidase inhibitors.

Scheme 3.1.5   **Glycal Lithiations**

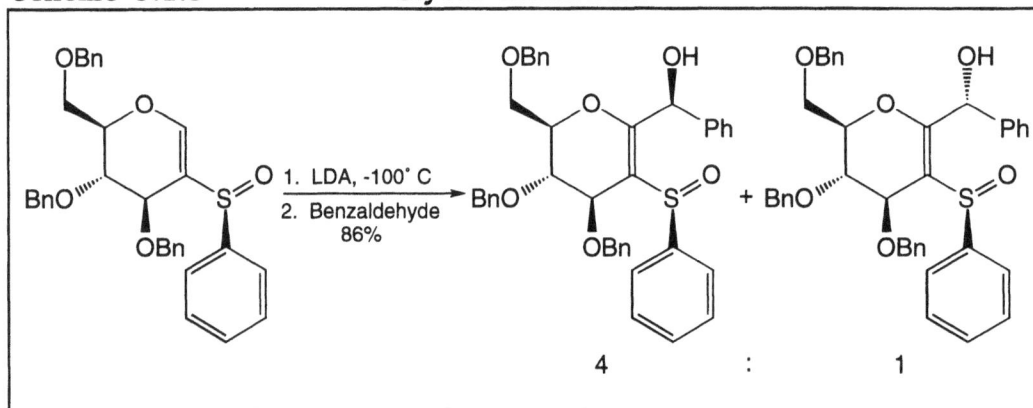

4   :   1

Before moving to other means used in the preparation of C_1 lithioglycals mention of a final example utilizing a hydrogen-metal exchange is warranted. The example of interest, shown in Scheme 3.1.6, was reported by Friesen, *et al.*,[7] and involves the formation of C_1 iodinated glycals. As shown, direct lithiation was accompanied by treatment with tri-*n*-butyl stannylchloride. Subsequent treatment of the stannylated glycal with iodine gave the iodinated compound in 85% overall yield. These compounds have subsequently found use in coupling reactions with arylzinc, vinyltin, and arylboron compounds. Thus, a compliment of unique *C*-glycosides are available through this methodology.

Scheme 3.1.6   *C*-Glycosides from C_1 Iodo Glycals

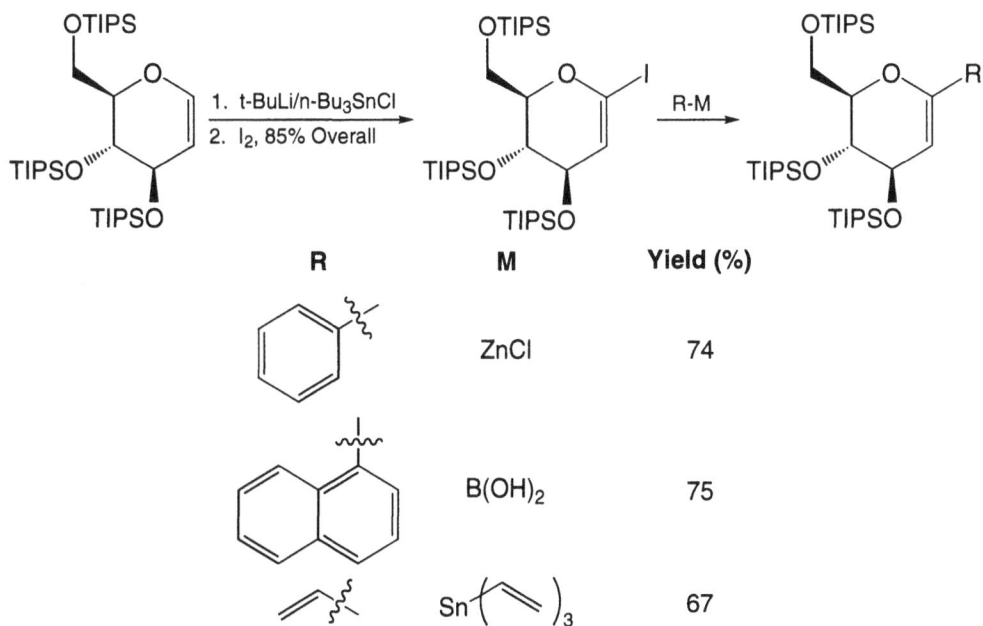

	R	M	Yield (%)
	phenyl	ZnCl	74
	naphthyl	B(OH)$_2$	75
	allyl	Sn(CH=CH$_2$)$_3$	67

3.1.2   Metal-Metal Exchanges

Although the hydrogen-metal exchange has found utility in the formation of C_1 lithiated glycals, sometimes, the best method for the formation of these compounds rests in the ability to trans-metallate from a different metallo glycal. These metal-metal exchanges have been extensively exploited in the derivitization of glycals as well as providing an efficient means of forming true anomeric anions from stannyl glycosides.

As already discussed in Scheme 3.1.3, Hanessian, *et al.*,[8] prepared the C_1 stannyl glycal shown. Further work by this group demonstrated the ability to transform this glycal analog to a lithiated species in preparation for coupling with an aldehyde. The reaction, shown in Scheme 3.1.7, produced a 68% yield of the *C*-glycoside as a mixture of isomers at the newly formed stereogenic center.

Electrophiles other than aldehydes have been demonstrated to effect the formation of *C*-glycosides under conditions involving metal-metal exchanges. As shown in Scheme 3.1.8, such reactions have been accomplished utilizing allyl halides. The reaction shown, reported by Grondin, *et al.*,[9] involves the formation of the lithioglycal on reaction with butyllithium. The resulting anionic glycal was coupled with allyl halides utilizing copper bromide-dimethyl sulfide complex to assist the progression of the reaction *via* a tandem cuprate formation.

Scheme 3.1.7   Glycal Trans-Metallations

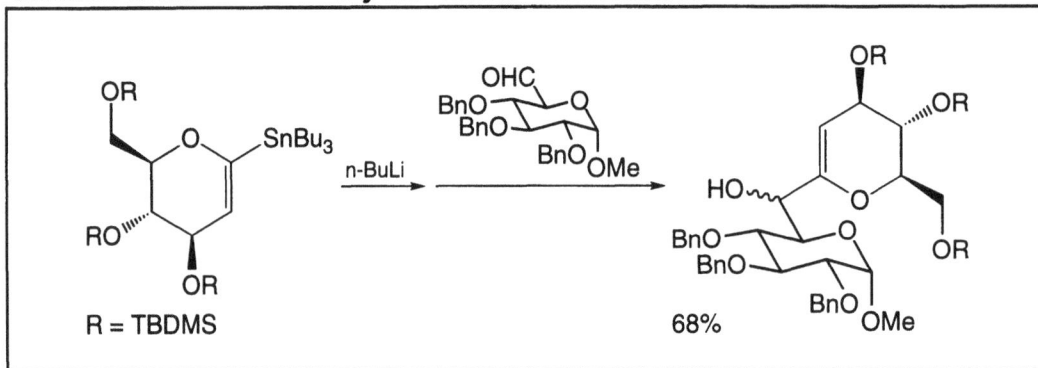

R = TBDMS   68%

Scheme 3.1.8   Glycal Trans-Metallations

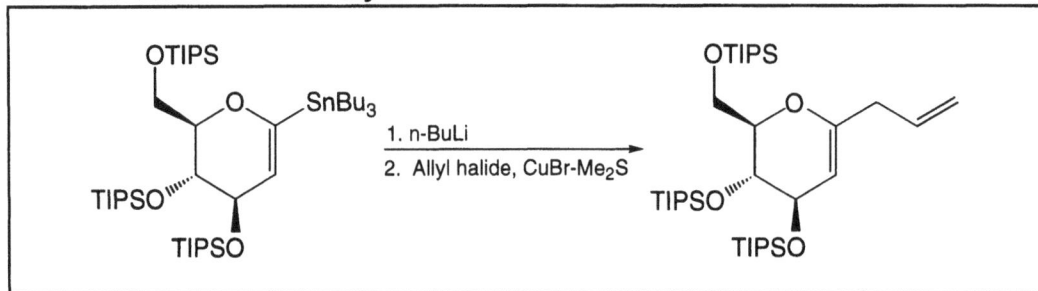

Applying this methodology to stannyl glycosides, the preparation of which was discussed in Chapter 2, Michael acceptors have found utility in the direct formation of *C*-glycosides. As shown in Scheme 3.1.9, Hutchinson, *et al.*,[10] treated the protected 2-deoxyglucose-derived β-stannyl glycoside with butyllithium followed by copper bromide-dimethyl sulfide complex. Once formed, the cuprate was exposed to methyl vinyl ketone producing a 55% yield of the β-*C*-glycoside. Furthermore, utilizing the α-stannyl glycoside and comparable reaction conditions, a 75% yield of the desired α-*C*-glycoside was formed. One may infer, from these results, that the stereochemistry of anomeric anions is stable and remains intact through the duration of these reactions.

Corroboration of the above described results was accomplished numerous times. However, one particular example deserved mention simply to illustrate the versatility available not only in the electrophile used but also in the nature

of the cuprate formed. As shown in Scheme 3.1.10, Prandi, et al.,[11] utilized the
same stannyl glycoside and formed the mixed cuprate from the lithioglycoside.
The cuprate was formed utilizing lithium thienyl copper cyanide. Subsequent
condensation with the epoxide, shown, produced the desired C-glycosides in
50% to 67% yields with retention of the anomeric stereochemistry.

Scheme 3.1.9   Glycosidic Trans-Metallations

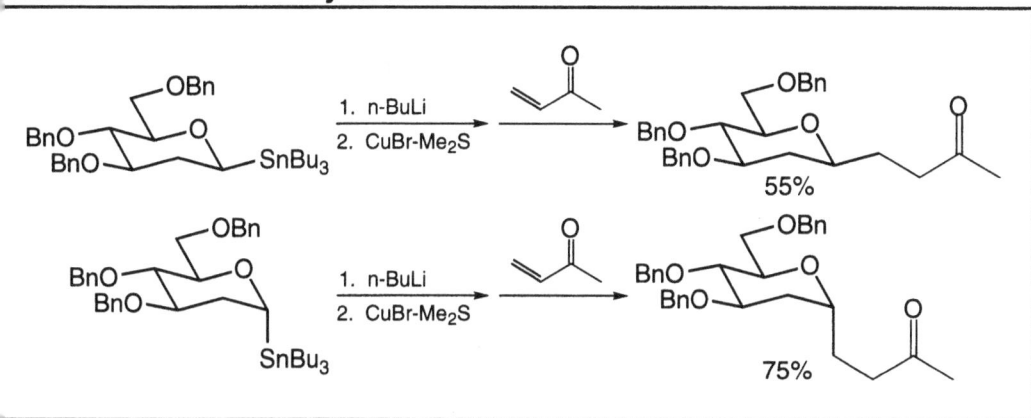

Scheme 3.1.10   Glycosidic Trans-Metallations

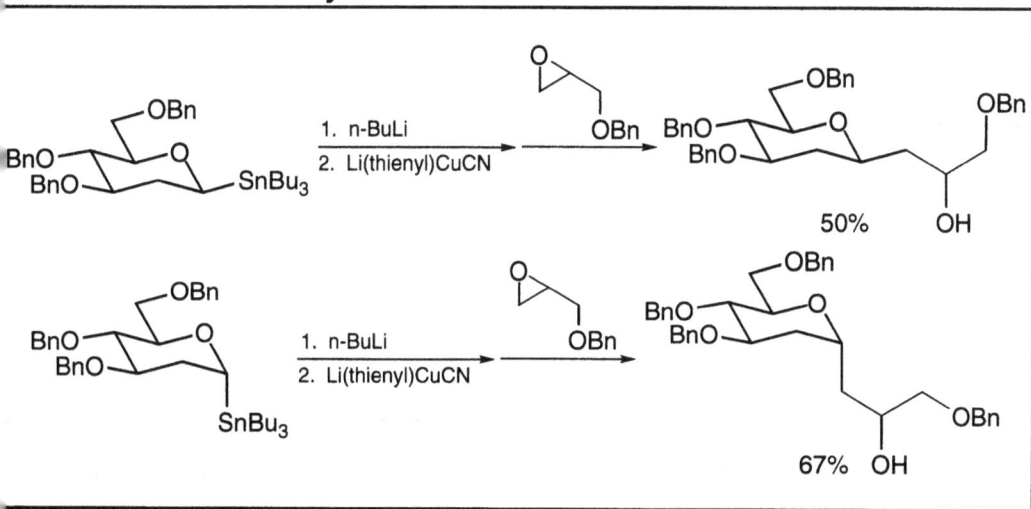

3.1.3   Halogen-Metal Exchanges

Complimentary to hydrogen-metal exchanges and metal-metal exchanges
are halogen-metal exchanges. This method has particularly been useful in
conversions involving glycosyl chlorides. As shown in Scheme 3.1.11, Lesimple,

et al.,[12] treated 2-deoxy-3,4,6-tri-*O*-benzyl-D-glucopyranosyl chloride with tri-*n*-butyl stannyllithium. The result was the formation of the desired stannyl glycoside in 85% yield with inversion of the stereochemistry at the anomeric center. Thus, the α-pyranosyl chloride yielded the β-stannyl glycoside.

Scheme 3.1.11   Glycosidic Halogen-Metal Exchanges

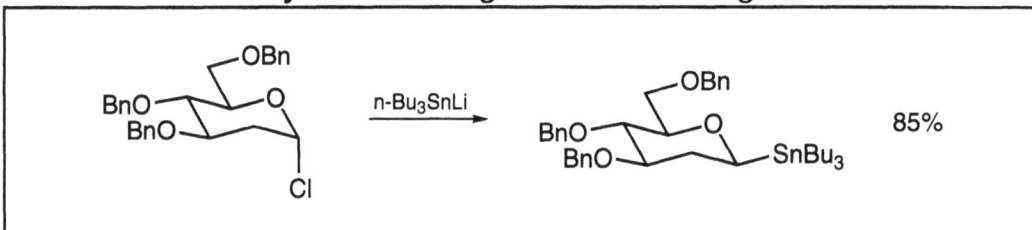

3.2   C₁ Lithiated Anomeric Carbanions by Reduction

Scheme 3.2.1   Glycosidic Halogen-Metal Exchanges

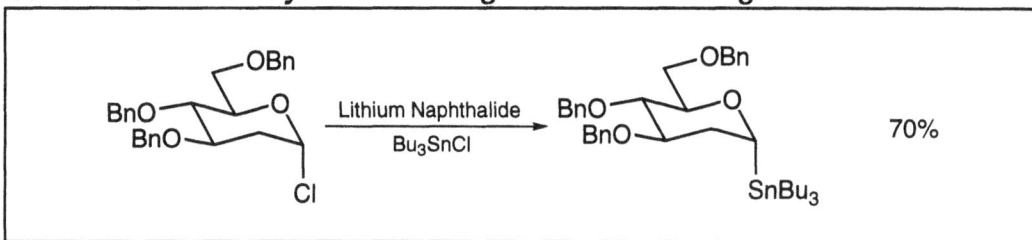

Figure 3.2.1   Stable Lithioglycosides

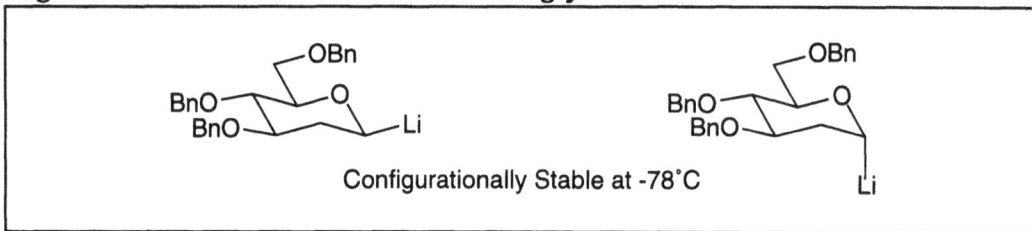

Complementary to the reaction shown in Scheme 3.1.11 is a method reported by the same group demonstrating the formation of the corresponding α stannyl glycoside.[12] As shown in Scheme 3.2.1, this reaction provided a 70% yield of the desired product utilizing lithium naphthalide to form the intermediate C₁ lithiated sugar. Furthermore, the intermediate lithioglycosides relevant to Schemes 3.1.11 and 3.2.1, shown in Figure 3.2.1, were demonstrated to be configurationally stable at -78° C. These observations are in complete agreement with the observed stereochemical stability of anomeric cuprates reported by Hutchinson, *et al.*,[10] and illustrated in Scheme 3.1.9 as reagents effecting the formation of *C*-glycosides on reaction with Michael acceptors.

Figure 3.2.2   Reductive Halogen-Metal Exchanges

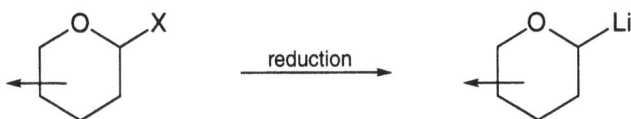

Scheme 3.2.2   *C*-Glycosidations Following Reductive Halogen-Metal Exchanges

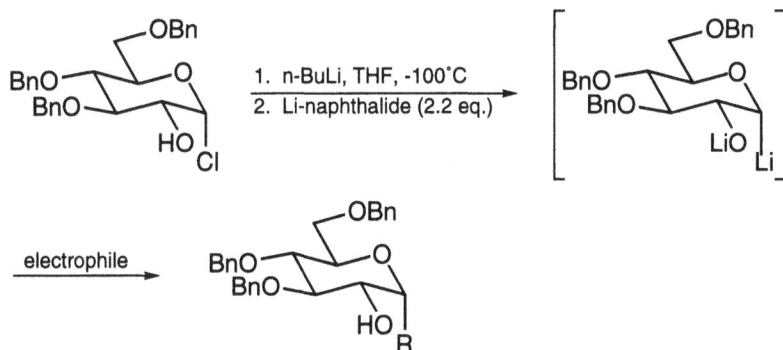

Table 3.2.1   Electrophiles, Products, and Yields from Scheme 3.2.2

Electrophile	R	% Yield
MeOD	D	75
MeOH	H	82
MeCHO	CH(OH)Me	62
PhCHO	CH(OH)Ph	70
iPrCHO	CH(OH)iPr	59
HCHO	CH$_2$OH	17
MeI/CuI	Me	72

The transient anomeric lithiation shown in Scheme 3.2.1 differs from standard halogen-metal exchanges in that a reductive mechanism is involved. The use of reductive methods, illustrated in Figure 3.2.2, is complimentary to the exploitation of metal exchange reactions for the formation of glycosidic nucleophiles in that strategies towards selectivity in the anomeric stereochemistry become available.

As shown in Scheme 3.2.2, Wittman, *et al.*,[13] utilized reductive anomeric lithiations in the preparation of a variety of glucose analogs. The specific procedure involved the use of 3,4,6-tri-*O*-benzyl-α-D-pyranosyl chloride. Initial treatment with butyllithium deprotonated the 2-hydroxyl group. Subsequent treatment with lithium naphthalide effected formation of the dianionic intermediate shown.   Notably, this species possessed the α configuration at the anomeric center.   Treatment of this species with a variety of electrophiles provided a variety of compounds predominantly formed in greater than 60% yield.   Additionally, reflecting the stereochemistry of the glycosidic anion, the products reported possessed the α configuration.   The actual yields of the isolated α anomers obtained from various electrophiles are shown in Table 3.2.1.   As a final point, formation of side products resulting from β-elimination is unfavorable due to O^{2-} being the leaving group.

Similar reductive lithiations are available from 2-deoxy sugars. As shown in Scheme 3.2.3, tri-*O*-benzyl glucal was converted to the desired 2-deoxy glucopyranosyl chloride on treatment with hydrochloric acid.   Following formation of the chloride, Lancelin, *et al.*,[14] effected formation of the α lithiated sugar, *via* initial conversion to the axially oriented anionic radical, on treatment with lithium naphthalide.

Scheme 3.2.3   Reductive Halogen-Metal Exchanges

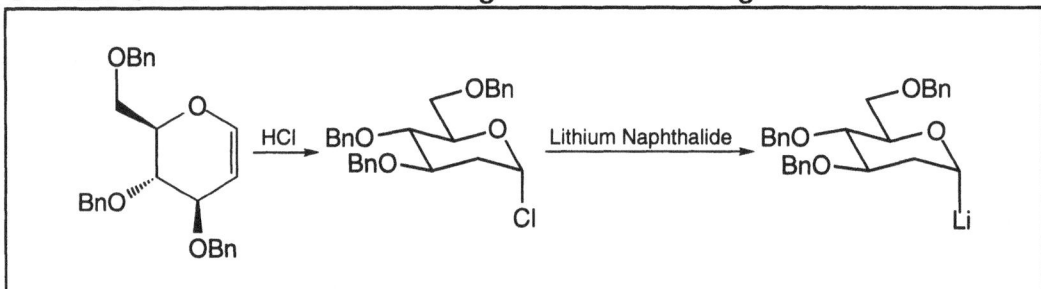

Scheme 3.2.4   Reductive Metallations from Sulfones

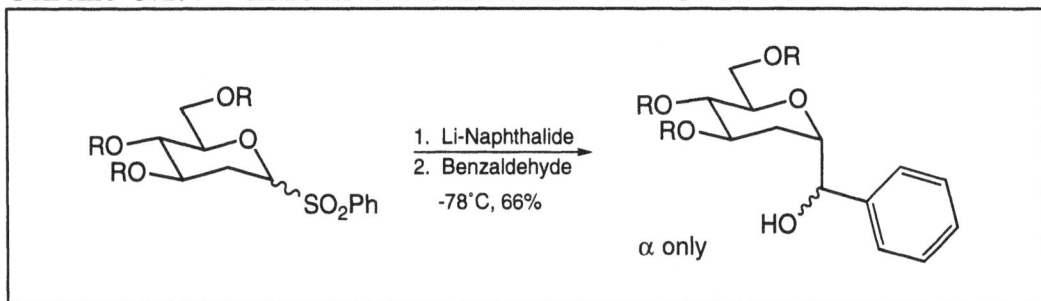

Complimentary to the use of pyranosyl chlorides as substrates for reductive lithiations is the use of pyranosyl phenylsulfones. These species are easily reduced to the corresponding lithioglycosides with equal propensities for the selective formation of α anomers. As shown in Scheme 3.2.4, Beau, et al.,[15] utilized this methodology in the formation of C-glycosides resulting from the treatment of the formed lithioglycosides with various aldehydes. As shown, the use of benzaldehyde effected the formation of the desired C-glycoside in 66% yield with exclusive formation of the α anomer.

One additional example of reductive glycosidic lithiations is warranted in that the specific example, shown in Scheme 3.2.5, results in a carboxyl stabilized glycosidic anion. As illustrated, Crich, et al.,[16] treated the 2-deoxy-D-glucose derivative, shown, with lithium naphthalide to effect reductive lithiation. The resulting stabilized anion was treated with allyl bromide thus effecting allylation in 44% yield. Hydrolysis and reductive decarboxylation under Barton conditions[17,18] gave the β C-glycoside in 50% yield. Notably, the resulting C-glycoside exhibits the opposite stereochemistry usually observed under reductive conditions. This is due to the preference of the intermediate radical anion for the α configuration. Furthermore, the stabilized nature of the intermediate anionic species facilitates retention of the observed stereochemical outcome. These observations will be addressed in detail in Section 3.3.

Scheme 3.2.5   Reductive Halogen-Metal Exchanges

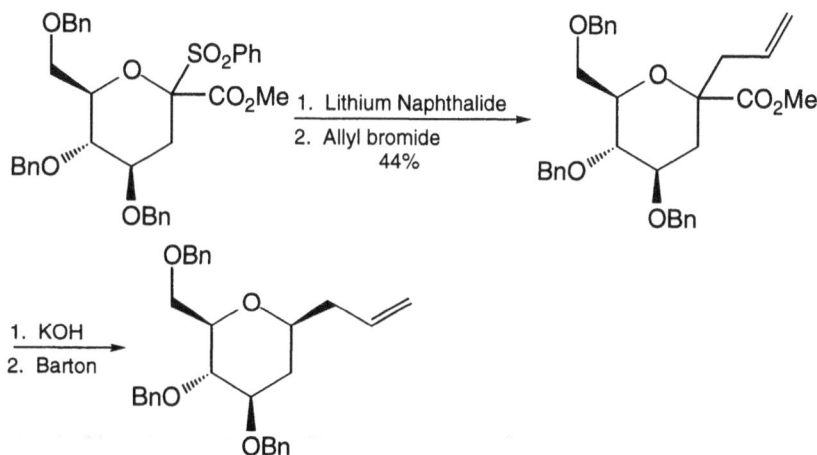

3.3 C$_1$ Carbanions Stabilized by Sulfones, Sulfides, Sulfoxides, Carboxyl and Nitro Groups

The use of carbanions in the preparation of C-glycosides is often dependent on the ability to generate stable lithiated species. Consequently, the use of electron withdrawing groups as a means for anion stabilization is often

employed.   Shown in Figure 3.3.1 are several representations of sugar
derivatives containing electron withdrawing groups in order to facilitate their
C_1 deprotonation and subsequent utility in the formation of *C*-glycosides.

Figure 3.3.1   Electron Deficient Sugar Derivatives

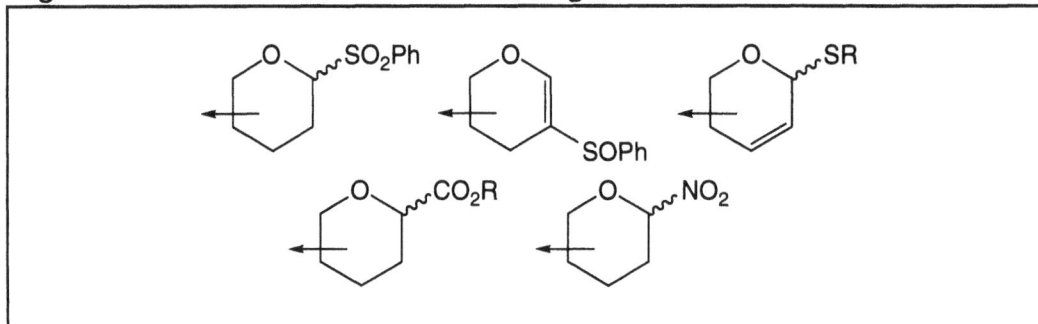

3.3.1   Sulfone Stabilized Anions

Scheme 3.3.1   Lithiation of Glycosyl Phenylsulfones

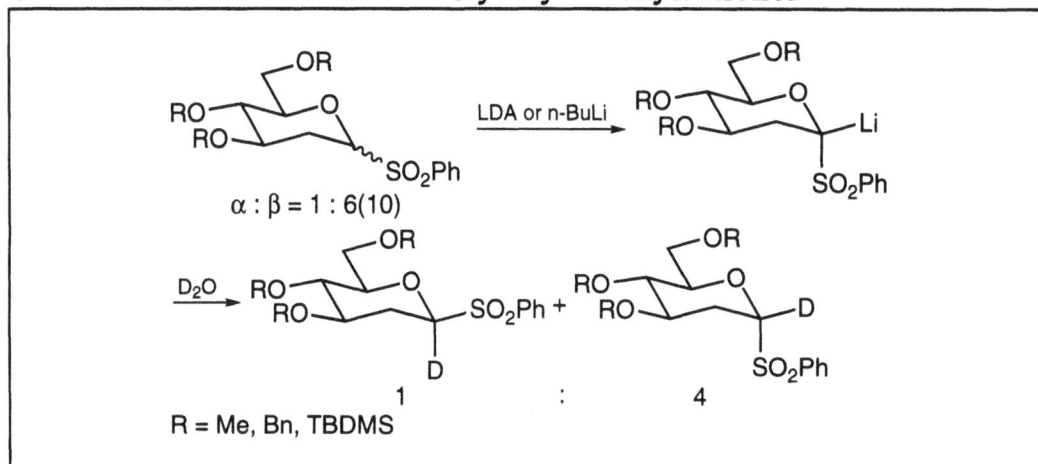

Of the examples shown in Figure 3.3.1, glycosyl phenylsulfones are
extremely useful, particularly in conjunction with reductive methods discussed
in Section 3.2. For example, Scheme 3.3.1 illustrates work reported by Beau, et
al.,[15] in which the illustrated phenylsulfone was utilized as follows.
Deprotonation with either LDA or *n*-butyllithium effected the direct formation
of the desired β-lithiated species.   Subsequent deuterium oxide quench
provided a 4 : 1 mixture of anomeric favoring that bearing the deuterium in the
β position. As reported, This methodology is useful with a variety of protecting
groups including methyl, benzyl, and *tert*-butyl dimethylsilyl ethers.

Scheme 3.3.2   Lithiation of Glycosyl Phenylsulfones

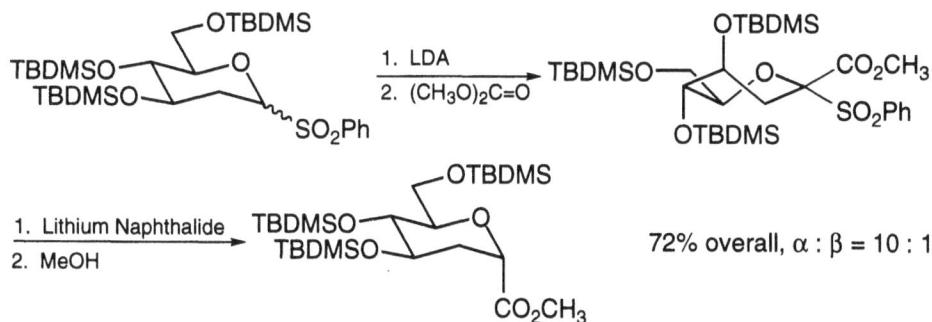

1. LDA
2. (CH₃O)₂C=O

1. Lithium Naphthalide
2. MeOH

72% overall, α : β = 10 : 1

Scheme 3.3.3   Mechanism for
C-Glycosidation of Sulfones

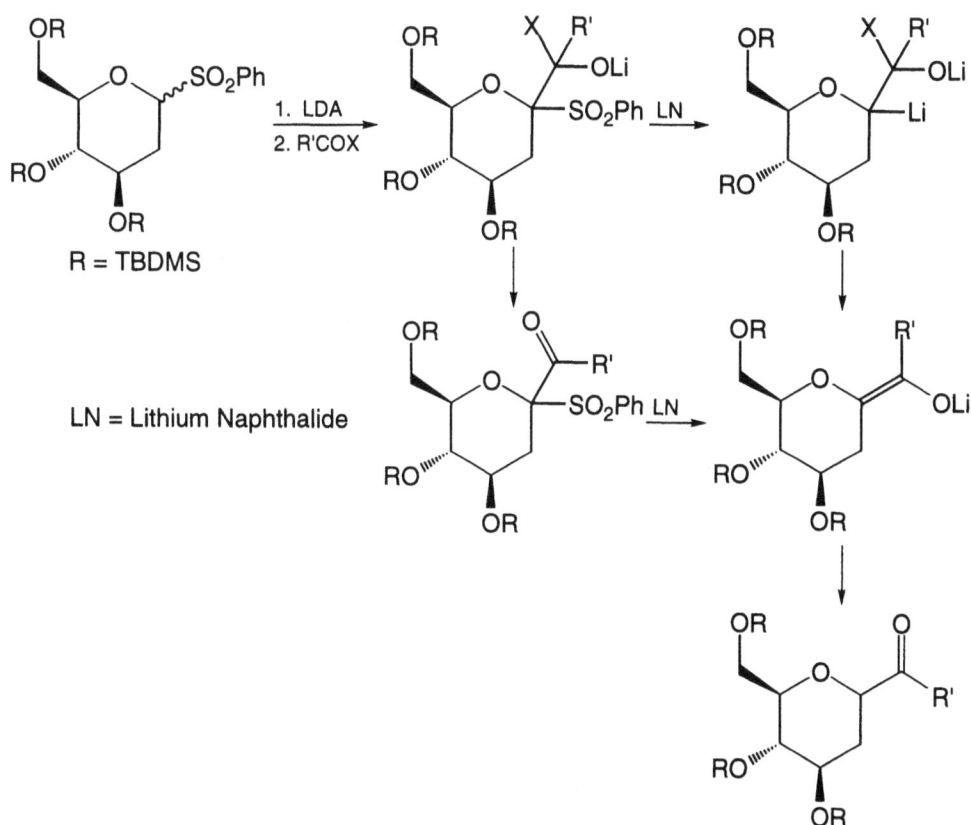

R = TBDMS

1. LDA
2. R'COX

LN = Lithium Naphthalide

In the application of this methodology to the preparation of *C*-glycosides, Beau, *et al.*,[19] utilized dimethyl carbonate as an electrophile. As shown in Scheme 3.3.2, the resulting product mixture favored the β oriented ester group in a ratio of 20 : 1. Subsequent reductive cleavage of the phenylsulfone was accomplished on treatment with lithium naphthalide giving the α-*C*-glycoside, shown, in 72% overall yield. Similar results were observed when phenyl benzoate was used as an electrophile.

At first glance, and in agreement with the observations presented in Scheme 3.2.5, the results shown in Scheme 3.3.2 seem contrary to the stereochemical consequences of the reactions discussed thus far. For example, although the deprotonation step proceeds as inferred in Scheme 3.3.1 with β-lithiation and formation of products possessing the β configuration at the anomeric center, the stereochemistry converts to the α configuration on reductive cleavage of the phenylsulfone. This may not seem consistent with the preferential formation of α lithioglycosides under reductive conditions. As shown in Scheme 3.3.3, this apparent inconsistency is explained. Once the *C*-glycosidation has been accomplished, lithium naphthalide effects the formation of the α lithioglycoside which conjugates into the carbonyl of the added group. Upon quenching, the preferred α configuration is adopted.

Additional work reported by Beau, *et al.*,[20] extended the use of stabilized anions and reductive cleavages to the preparation of *C*-disaccharides. As shown in Scheme 3.3.4, the phenylsulfone was deprotonated and treated with a sugar-derived aldehyde. The phenylsulfone was then cleaved from the resulting product utilizing lithium naphthalide. Unlike the example shown in Scheme 3.3.2, conjugation of the intermediate lithioglycoside resulting from cleavage of the phenylsulfone is not possible. Therefore, isomerization does not occur and the predominant product bears the β anomeric configuration.

Scheme 3.3.4   Lithiation of Glycosyl Phenylsulfones

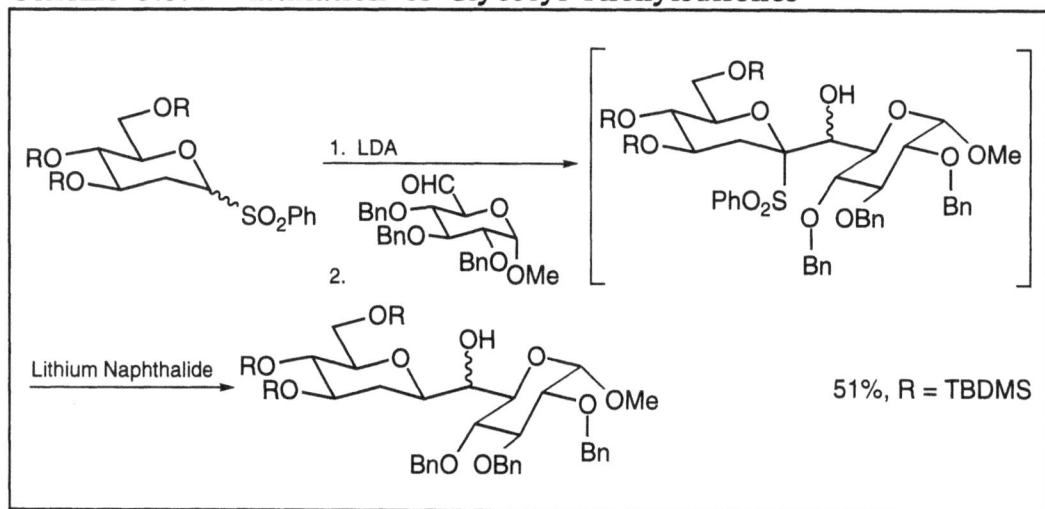

3.3.2   Sulfide and Sulfoxide Stabilized Anions

Scheme 3.3.5   **Glycal Metallations**

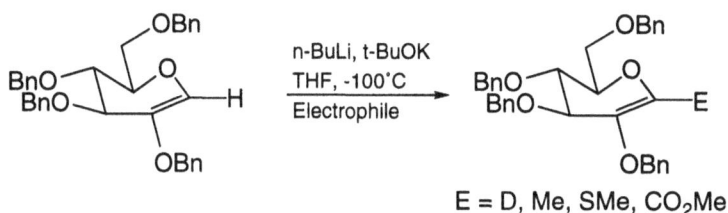

E = D, Me, SMe, CO$_2$Me

Scheme 3.3.6   **C$_2$ Activated Glycals
in C-Glycosidations**

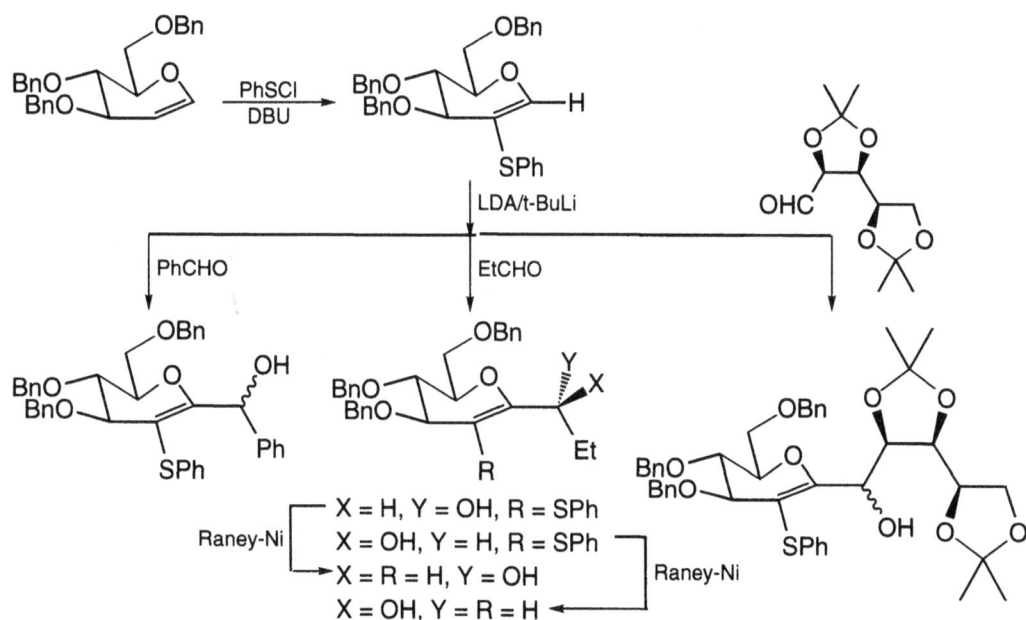

X = H, Y = OH, R = SPh
X = OH, Y = H, R = SPh
X = R = H, Y = OH
X = OH, Y = R = H

Until now, only glycosidic phenylsulfones have been discussed as substrates for the formation of glycosidic anions. However, as illustrated in Figure 3.3.1, other sugar derivatives are useful. Among these are glycal analogs. Early examples of the utilization of glycal analogs were reported by Schmidt, et al.,[21] and are shown in Scheme 3.3.5. In this report, the ability to directly lithiate protected glycals and introduce new glycosidic moieties was documented. As shown, metallations were accomplished utilizing butyllithium and potassium *tert*-butoxide at low temperatures.

The observations illustrated in Scheme 3.3.5 were complimented, in the same report, by the demonstration of the utility of 2-phenylsulfide activated glycals as *C*-glycosidation substrates.   As shown in Scheme 3.3.6, 2-phenylsulfide substituted glycals were easily prepared on treatment with DBU and chlorophenylsulfide. Subsequent metallations were then accomplished utilizing *tert*-butyllithium and LDA. Treatment of the anionic species with a variety of aldehydes effected the formation of various *C*-glycosides.

Scheme 3.3.7   Phenyl Sulfoxides in *C*-Glycosidations

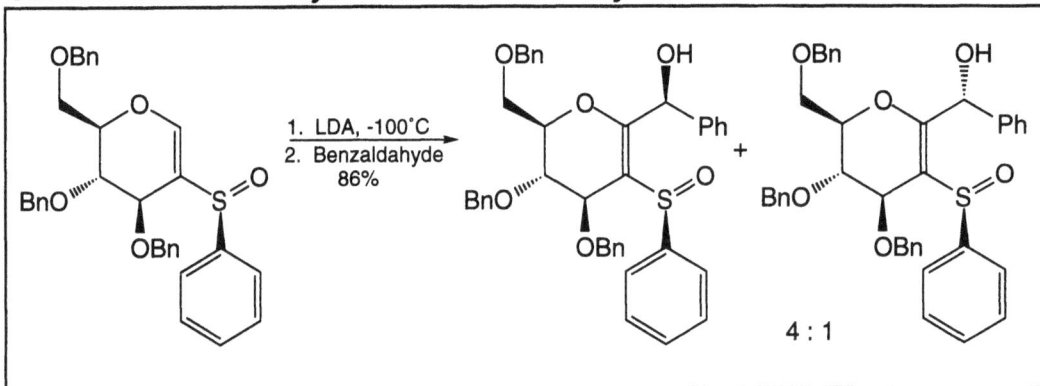

Scheme 3.3.8   Phenyl Sulfides in *C*-Glycosidations

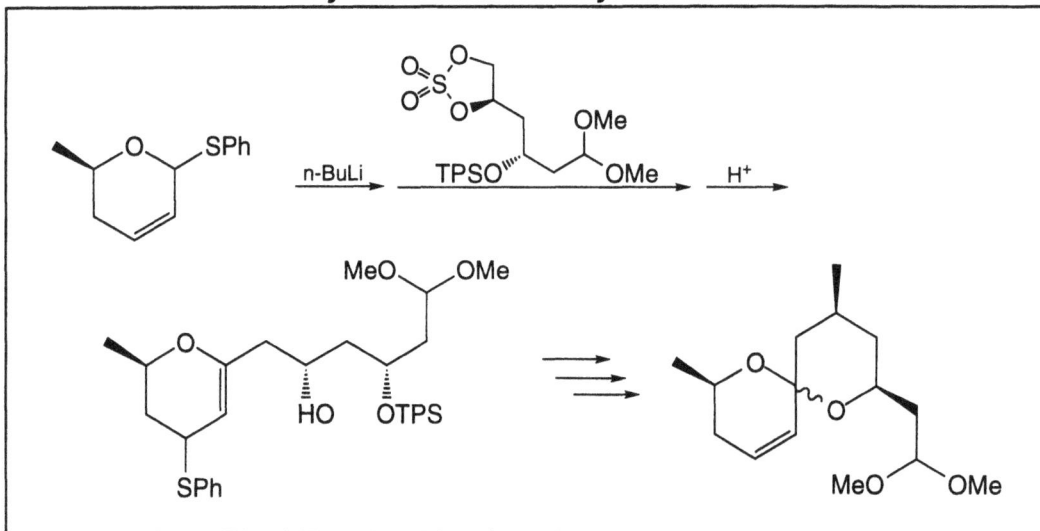

As a logical extension to the use of 2-phenylsulfide substituted glycals as substrates for the formation of *C*-glycosides is the use of 2-phenylsulfoxide substituted glycals. A particular advantage over the use of phenylsulfides is that incorporation of chirality to the phenylsulfoxide allows for stereospecificity in the addition of aldehydes to subsequently metallated species. As shown in Scheme 3.3.7, Schmidt, *et al.*,[6] incorporated the above described technology in the ultimate preparation of the potential β-glycosidase inhibitors illustrated in Figure 1.4.1. The key reaction, illustrated, involves LDA mediated deprotonation of the substituted glycal followed by addition of benzaldehyde to the resulting lithiated species. This particular example proceeded in 86% yield with the formation of a 4 : 1 ratio of diastereomers.

Concurrent with the development of phenylsulfoxide chemistry and complimentary to the use of 2-phenylsulfide substituted glycals came reports demonstrating the utility of glycosidic phenylsulfides of 2,3-di-deoxy sugars in the preparation of *C*-glycosides. As shown in Scheme 3.3.8, Gomez, *et al.*,[22] applied this chemistry to coupling reactions with cyclic sulfates. The immediate product of the reaction was a 3-phenylsulfide-1-substituted glycal. Further elaborations converted this product to an unsaturated spiroketal *via* hydrolysis of the sulfate and eventual cyclization.

3.3.3   Carboxy Stabilized Anions

Scheme 3.3.9   Carboxyglycosides in *C*-Glycosidations

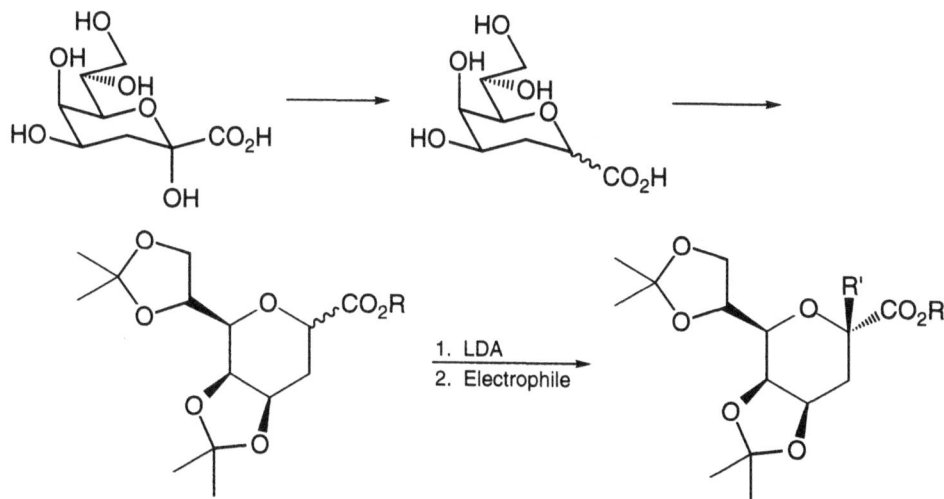

As already alluded to in Schemes 3.2.5 and 3.3.2, the carboxyl group is an excellent stabilizing group for anomeric anions. C_1 carboxyl glycosides are easily prepared from a number of methods discussed in Chapter 2. Additionally, many naturally occurring C_1 carboxyl glycosides are known.

Among these is 3-deoxy-D-manno-2-octulosonic acid. This sugar has been identified as a component of the lipopolysaccharides released by Gram-negative bacteria.[23,24] As shown in Scheme 3.3.9, Luthman, *et al.*,[25] demonstrated ability to derivatize this sugar through C_1 metallation and subsequent alkylation. The results, obtained from both the methyl and ethyl ester derivatives of this sugar demonstrate moderate yields and high stereoselectivities. For example, utilizing *tert*-butyl bromoacetate, the yield of the desired alkylated sugar was only 30% with a ratio of 95 : 5 favoring the β anomer. However, utilizing formaldehyde afforded a 62% yield of the desired hydroxymethyl analog exhibiting a ratio of 90 : 10 favoring the β anomer. These results, accompanied by additional observations are shown in Table 3.3.1. In all cases, metallation was accomplished utilizing LDA.

Table 3.3.1   Carboxyglycosides in *C*-Glycosidations

R	R'	% α	% β	Total Yield (%)
CN	CH_3	<5	>95	47
CN	CH_2CH_3	10	90	55
CH_2OH	CH_2CH_3	10	90	47
$COCH_3$	CH_2CH_3	5	95	62
CH_3	CH_2CH_3	<5	>95	50
$CH2C{\equiv}CH$	CH_2CH_3	10	90	50
$CH_2CO_2{}^tBu$	CH_2CH_3	<5	>95	30
CH_2Ph	CH_3	5	95	67
$CH_2CH_2CO_2CH_3$	CH_2CH_3	25	75	27

Scheme 3.3.10   Sialic Acid Derivatization

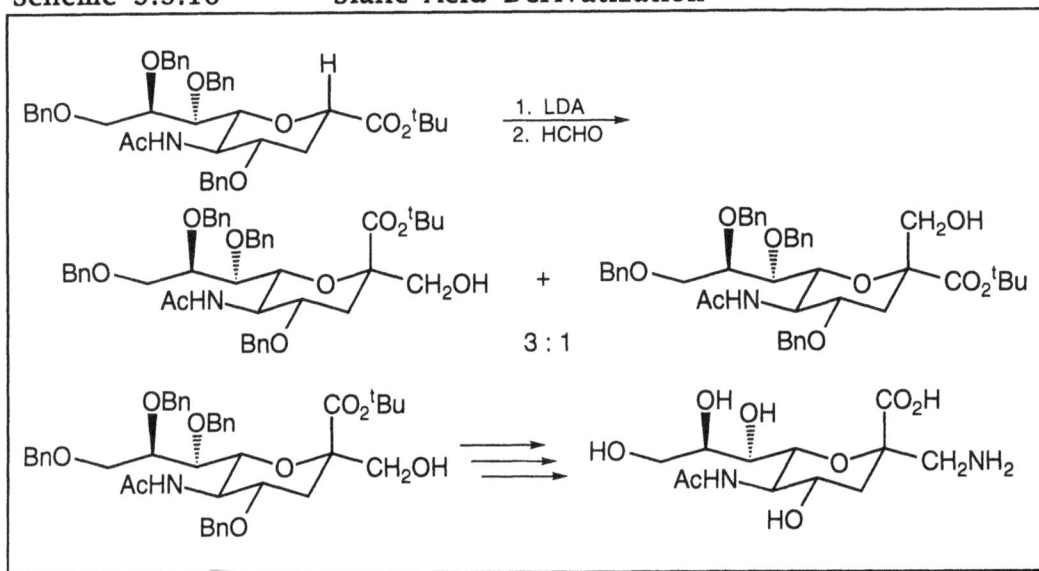

Among the most studied of the naturally occurring C_1 carboxyglycosides is *N*-acetyl neuraminic acid or sialic acid. The importance of this molecule centers around it being a component of sialyl Lewis^x, the natural ligand for the selectins and a key player in cell adhesion and leukocyte trafficking.[26] As shown in Scheme 3.3.10, Vasella, *et al.*,[27] prepared *C*-glycosidic sialic acid derivatives beginning with the 1-deoxy sialic acid analog shown. Deprotonation with LDA followed by treatment with formaldehyde produced a 3 : 1 ratio of the hydroxymethylated products favoring the β anomeric configuration for the added group. Following deprotection, these compounds were found to be weak inhibitors of Vibro cholerae sialidase. In the same report, the β-hydroxymethyl analog was converted to the corresponding aminomethyl derivative *via* the azide. As compared to the hydroxymethyl analogs, the aminomethyl analog was shown to stimulate Vibro cholerae sialidase.

3.3.4   Nitro Stabilized Anions

Scheme 3.3.11   C-Glycosides from Nitroglycosides

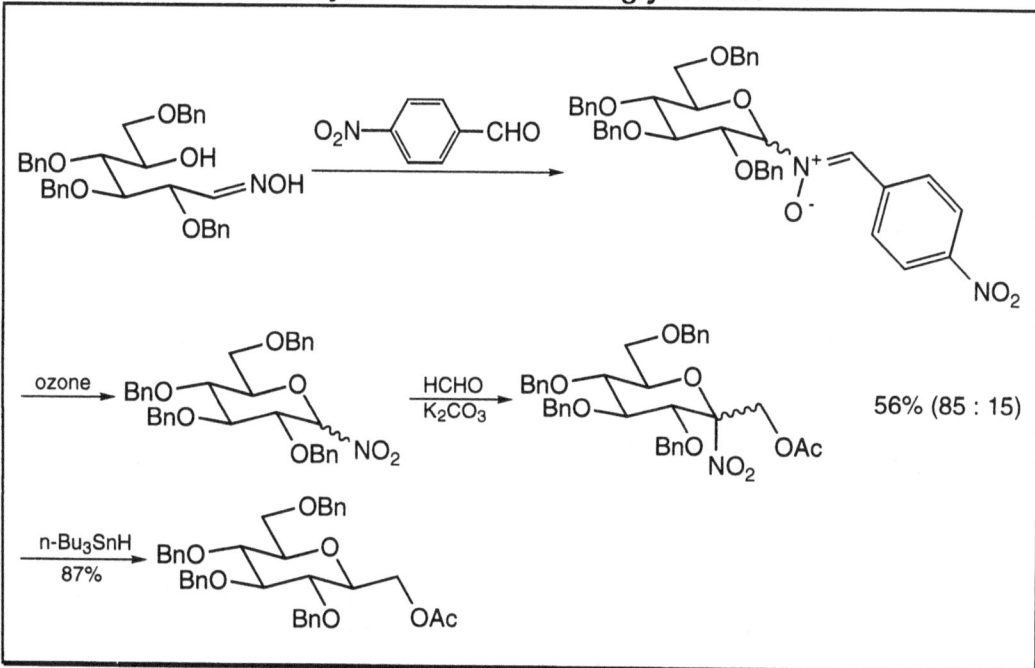

Of the anionic activating groups shown in Figure 3.3.1, the nitro group has yet to be discussed. This group was among the first to be utilized in the preparation of *C*-glycosides with its usage appearing as early as 1983. At that time, Vasella, *et al.*,[28,29] demonstrated the preparation of these compounds as shown in Scheme 3.3.11. Beginning with the oxime derived from 2,3,4,6-tetra-

O-benzyl-D-glucopyranose, condensation with *p*-nitrobenzaldehyde afforded the imine shown.   Subsequent ozonolysis afforded a 73% overall yield of the nitroglycoside as a mixture of anomers.   Further exploration of the utility of these compounds as applied to the preparation of *C*-glycosides showed their ease in deprotonation utilizing bases as mild as potassium carbonate. Furthermore, deprotonation in the presence of formaldehyde followed by acetylation gave the acetoxymethyl derivative shown.   Final cleavage of the nitro substituent was easily accomplished in 87% utilizing radical conditions and providing the desired *C*-glycoside exclusively as the β anomer.   As a final note, it is important to mention that utilizing nitro stabilized anomeric anions, β-elimination of the substituent at C_2 is rarely observed.

3.4   References

1.   Parker, K. A.; Coburn, C. A. *J. Am. Chem. Soc.* **1991**, *113*, 8516.
2.   Lesimple, P.; Beau, J. M.; Jaurand, G.; Sinay, P. *Tetrahedron Lett.* **1986**, *27*, 6201.
3.   Nicolau, K. C.; Hwang, C. K.; Duggan, M. E. *J. Chem. Soc. Chem. Comm.* **1986**, 925.
4.   Hanessian, S.; Martin, M.; Desai, R. C. *J. Chem. Soc. Chem. Comm.* **1986**, 926.
5.   Schmidt, R.; Preuss, R. *Tetrahedron Lett.* **1989**, *30*, 3409.
6.   Schmidt, R. R.; Dietrich, H. *Angew. Chem. Int. Ed. Engl.* **1991**, *30*, 1328.
7.   Friesen, R. W.; Loo, R. W. *J. Org. Chem.* **1991**, *56*, 4821.
8.   Hanessian, S.; Martin, M.; Desai, R. C. *J. Chem. Soc. Chem. Comm.* **1986**, 926.
9.   Grondin, R.; Leblanc, Y.; Hoogsteen, K. *Tetrahedron Lett.* **1991**, *32*, 5021.
10.   Hutchinson, D. K.; Fuchs, P. L. *J. Am. Chem. Soc.* **1987**, *109*, 4930.
11.   Prandi, J.; Audin, C.; Beau, J.-M. *Tetrahedron Lett.* **1991**, *32*, 769.
12.   Lesimple, P.; Brau, J. M.; Sinay, P. *Carbohydrate Res.* **1987**, *171*, 289.
13.   Wittman, V.; Kessler, H. *Angew. Chem. Int. Ed. Eng.* **1993**, *32*, 1091.
14.   Lancelin, J. M.; Morin-Allory, L.; Sinay, P. *J. Chem. Soc. Chem. Comm.* **1984**, 355.
15.   Beau, J. M.; Sinay, P. *Tetrahedron Lett.* **1985**, *26*, 6185.
16.   Crich, D.; Lim, L. B. L. *Tetrahedron Lett.* **1990**, *31*, 1897.
17.   Barton, D. H. R.; Crich, D.; Motherwell, W. B. *Tetrahedron* **1985**, *41*, 3901.
18.   Crich, D.; Quintero, L. *Chem. Rev.* **1989**, *89*, 1413.
19.   Beau, J. M.; Sinay, P. *Tetrahedron Lett.* **1985**, *26*, 6193.
20.   Beau, J. M.; Sinay, P. *Tetrahedron Lett.* **1985**, *26*, 6189.
21.   Schmidt, R. R.; Preuss, R.; Betz, R. *Tetrahedron Lett.* **1987**, *28*, 6591.
22.   Gomez, A. M.; Valverde, S.; Fraser-Reid, B. *J. Chem. Soc. Chem. Comm.* **1991**, 1207.
23.   Allen, N. E. *Annu. Rep. Med. Chem.* **1985**, *20*, 155.

24. Ray, P. H.; Kelsey, J. E.; Bigham, E. C.; Benedict, C. D.; Miller, T. A. "Bacterial Lipopolysaccharides", ACS Symposium Series 231, **1983**, 141.

25. Luthman, K.; Orbe, M.; Waglund, T.; Clesson, A. *J. Org. Chem.* **1987**, *52*, 3777.

26. Levy, D. E.; Tang, P. C.; Musser, J. H. *Annu. Rep. Med. Chem.* **1994**, *29*, 215.

27. Wallimann, K.; Vasella, A. *Helv. Chim. Acta* **1991**, *74*, 1520.

28. Acbischer, B.; Vasella, A. *Helv. Chim. Acta* **1983**, *66*, 789.

29. Baumberger, F.; Vasella, A. *Helv. Chim. Acta* **1983**, *66*, 2210.

4   Introduction

Whereas the combination of electrophiles with sugar nucleophiles provides routes to *C*-glycosides, these conditions are not applicable when the electrophiles bear centers prone to epimerization. However, where acid/base chemistry is not applicable due to reactive stereogenic centers, the use of transition metal mediated cross coupling reactions may often be utilized. Substantial research in this area has been applied to the couplings between glycals with aryl and other π-conjugated aglycones. Some sugar-derived substrates for this type of reaction are shown in Figure 4.0.1 and include glycals, metallated glycals, halogenated glycals, and sugar derivatives bearing allylic anomeric leaving groups. In this chapter the *C*-glycosidations introduced in Schemes 2.4.2, 2.13.2 and Figure 2.10.3 are elaborated upon. Additionally, mechanistic and stereochemical aspects of these reactions are addressed.

Figure 4.0.1   General Substrates for Transition Metal Mediated *C*-Glycosidations

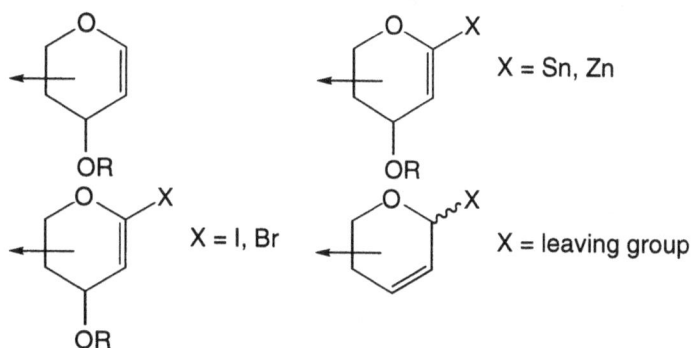

4.1   Direct Coupling of Glycals with Aryl Groups

The utility of glycals as substrates for *C*-glycosidations is emphasized by their availability from commercial sources as well as the examples presented thus far. Consequently, the actual coupling of these compounds with both activated and unactivated aryl groups is complimentary to the general chemistry surrounding *C*-glycosides. In this section, coupling reactions with unactivated, metallated, and halogenated aromatic rings are discussed.

4.1.1   Arylation Reactions with Unactivated Aromatic Rings

Scheme 4.1.1   Palladium Catalyzed Coupling of Glycals with Benzene

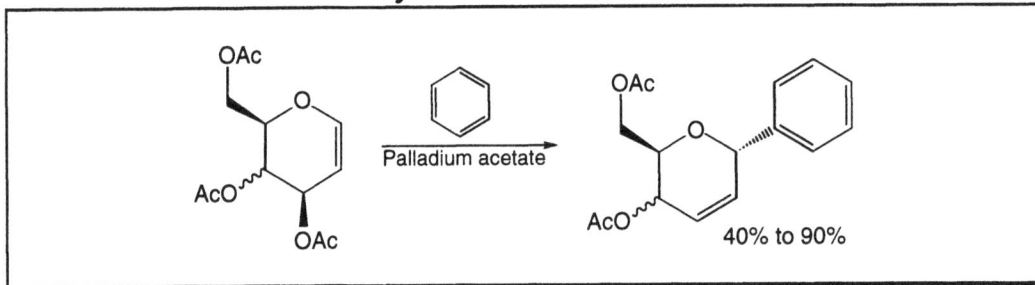

Scheme 4.1.2   Palladium Catalyzed Coupling of Glycals with Benzene Analogues

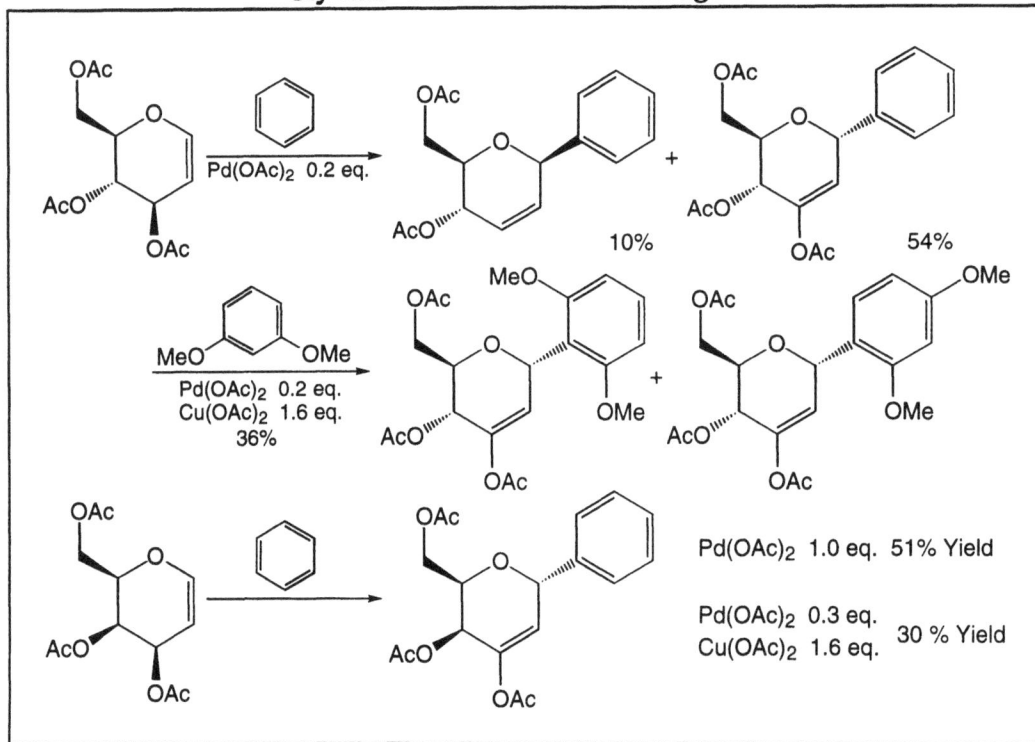

Beginning with the simplest of cases, the direct coupling of glycals with benzene is discussed.   As shown in Scheme 4.1.1, this reaction was accomplished by Czernecki, *et al.*,[1] utilizing palladium acetate as a catalyst. The yields for this reaction were generally in the range of 40% to 90% and resulted in the selective formation of the α configuration at the anomeric center.

Subsequent examples, reported by Czernecki, *et al.*,[2] elaborate upon this chemistry. As shown in Scheme 4.1.2, various acetate protected glycals were reacted with benzene and 1,3-dimethoxybenzene. In all cases, palladium acetate was used as the catalyst. However, varying equivalents were used. Consequently, the variances observed in the composition of the product mixture reflected the specific conditions used to induce reaction and the highest observed yields were approximately 50%.

4.1.2   *Arylation Reactions with Metallated Aromatic Rings*

In light of the relatively low yields for transition metal mediated *C*-glycosidations with unactivated aromatic aglycones, the use of substituted aryl groups was explored. Two particular examples, already presented in Schemes 2.4.2 and 2.4.3 involved the palladium mediated the couplings of glycals and aryltributyltin compounds.[3,4] While β anomeric selectivity was generally observed, the yields, some as high as 70%, were sensitive to the protecting groups used on the glycal substrates.

The example shown in Scheme 2.4.2 is particularly important in that, as illustrated in Scheme 4.1.3, aryl mercury reagents were shown to produce comparable results. Consequently, research into the nature of the aryl groups useful in these reactions have provided interesting results. Specifically, in addition to unactivated aryl groups, aryltin, arylmercury, and halogen substituted aryl compounds may be used. In the remainder of this section, examples utilizing these groups will be presented.

Scheme 4.1.3   *C*-**Glycosidation with**
Aryl Mercury Reagents

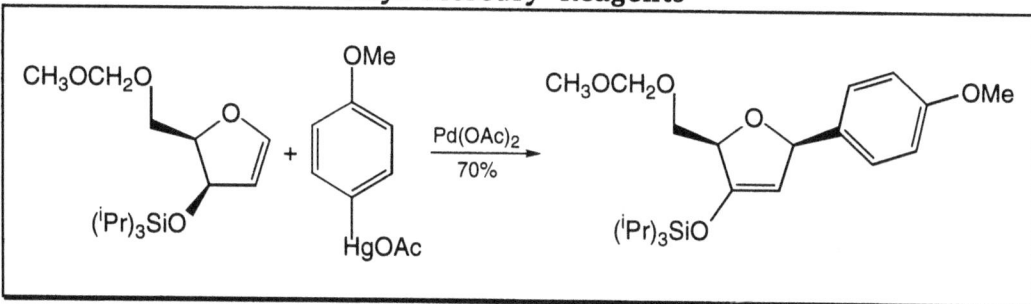

As illustrated in Scheme 4.1.2, salt additives have been used as a potential means of enhancing arylation reactions while decreasing the amount of palladium catalyst required. In Scheme 4.1.2, only copper acetate was considered. However, as shown in Schemes 4.1.4, 4.1.5, 4.1.6, and 4.1.7, the uses of lithium chloride, lithium acetate, and sodium bicarbonate were explored. Specifically, these salts were examined in conjunction with coupling reactions between glycals and pyrimidinyl mercuric acetates. The examples shown in the cited schemes were reported by Cheng, *et al.*,[5] and Arai, *et al.*,[6] and are segregated by glycal type. In Scheme 4.1.4, the reaction of tri-*O*-acetyl-D-glucal with a pyrimidinyl mercuric acetate is shown utilizing lithium chloride and palladium acetate as the catalytic mixture. Additionally, the complex product mixture and respective yields are shown. Scheme 4.1.5 elaborates upon these results utilizing a variety of reaction conditions in transformations involving tri-*O*-acetyl-D-glucal and other glycals. As illustrated, the products of these reactions are variations or subsets of the mixture presented in Scheme 4.1.4.

Scheme 4.1.4   *C*-Glycosidation with
                          Aryl Mercury Reagents

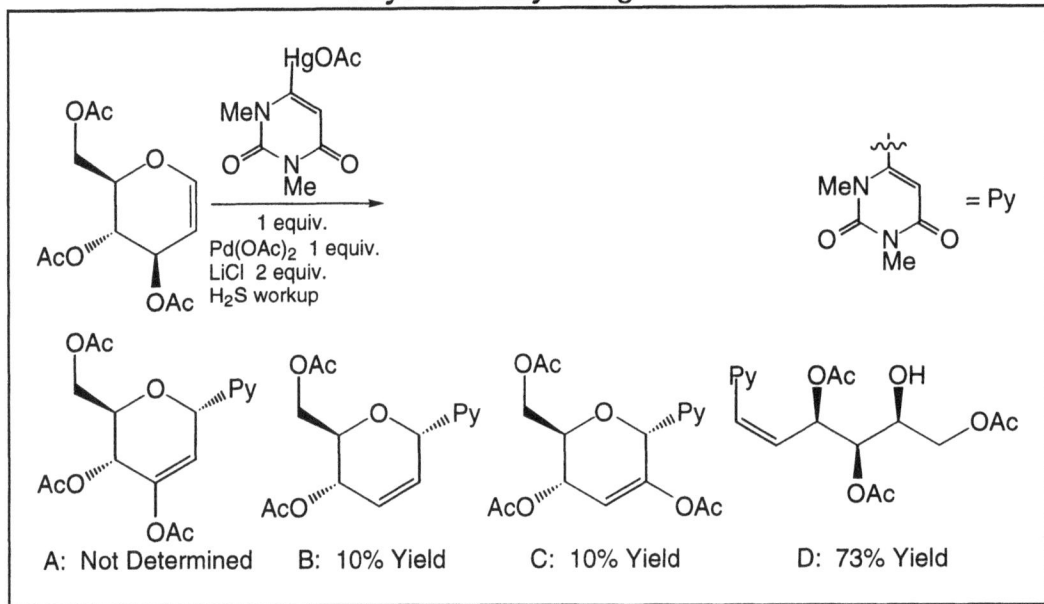

A: Not Determined   B: 10% Yield   C: 10% Yield   D: 73% Yield

In Scheme 4.1.6, variations from the acetate protecting groups used in Schemes 4.1.4 and 4.1.5 are illustrated. As shown, utilizing phenyl acetals, the isolated yields drop substantially. However, one consistency lies in the observation that the stereochemistry at C_3 adversely affects the stereochemistry of the product. This is interesting in light of the fact that under most circumstances, the stereochemistry at this center is either entirely lost due to conversion to a planar center or, alternately, the acetate group eliminates under the specified reaction conditions. These observations are further corroborated in Scheme 4.1.7 where the use of pentose-derived glycals are illustrated. As a final note, C3 keto derivatives become available when silyl groups are used to protect the alcohol at this center.

Scheme 4.1.5   **C-Glycosidation with**
Aryl Mercury Reagents

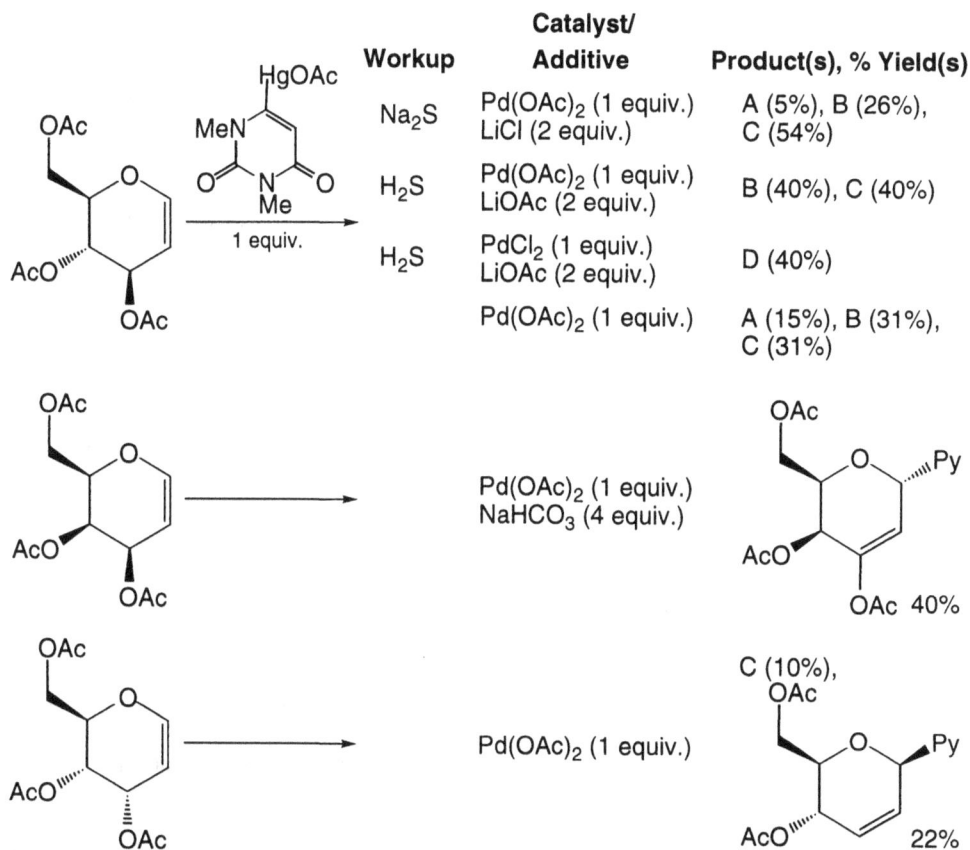

Pyranose-derived glycals are also useful as substrates in *C*-glycosidations with pyrimidinyl mercuric acetates. As shown in Scheme 4.1.8, Farr, *et al.*,[7] carried out such reactions utilizing 3-*O*-silyl protected glycals. Palladium acetate was used as the catalyst and no salt additives were used. Compared to the results presented in the previous four schemes, the illustrated coupling proceeded cleanly with the formation of an 84% yield of a β-*C*-glycoside.

Scheme 4.1.6   C-Glycosidation with
Aryl Mercury Reagents

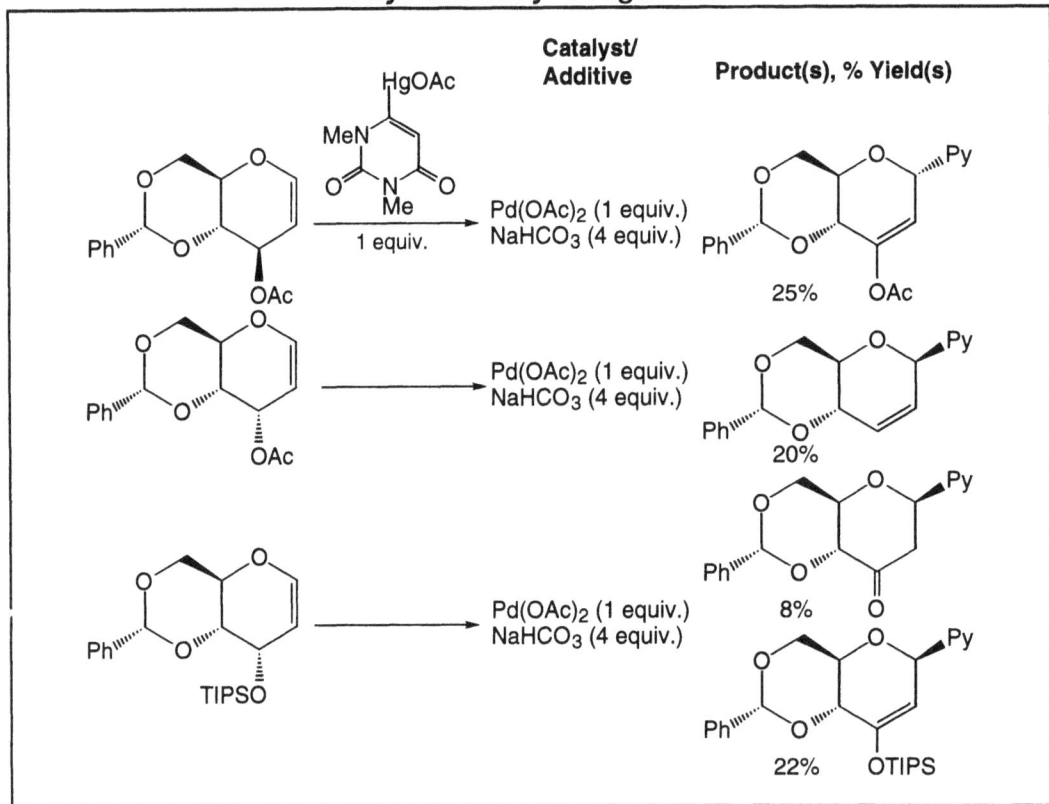

Scheme 4.1.7   C-Glycosidation with
Aryl Mercury Reagents

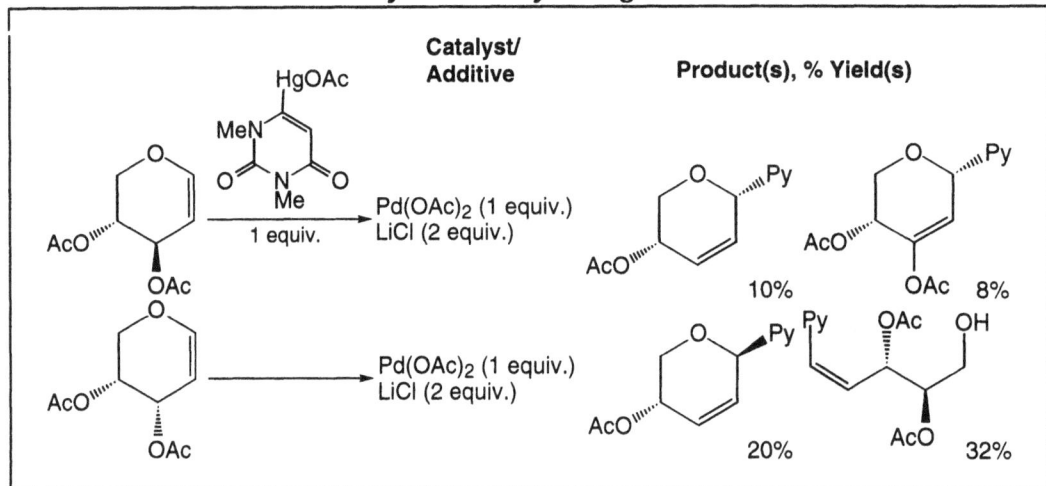

Scheme 4.1.8   *C*-Glycosidation with
                         Aryl Mercury Reagents

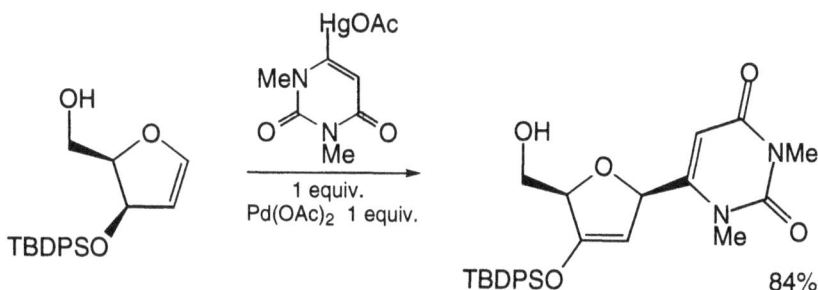

4.1.3   *Arylation Reactions with Halogenated Aromatic Rings*

Scheme 4.1.9   *C*-Glycosidation with Aryl
                       Mercury/Iodide Reagents

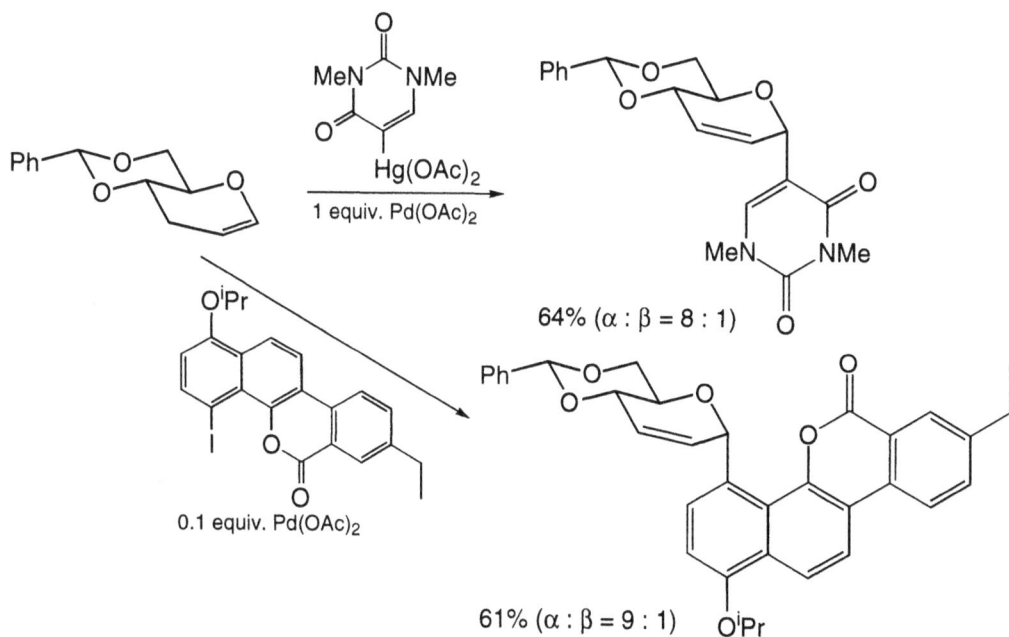

64% (α : β = 8 : 1)

0.1 equiv. Pd(OAc)₂

61% (α : β = 9 : 1)

Where mercury substituted aryl groups have proven useful in palladium
mediated *C*-glycosidations with glycals, aryl iodides have been shown to be
complimentary alternatives in these reactions. As shown in Scheme 4.1.9,

Kwok, *et al.*,[8] effected couplings with both mercury and iodo compounds. In both cases, the yields were comparable and the reactions showed a consistent preference for α anomers.

Scheme 4.1.10   **C-Glycosidation with Aryl Iodide Reagents**

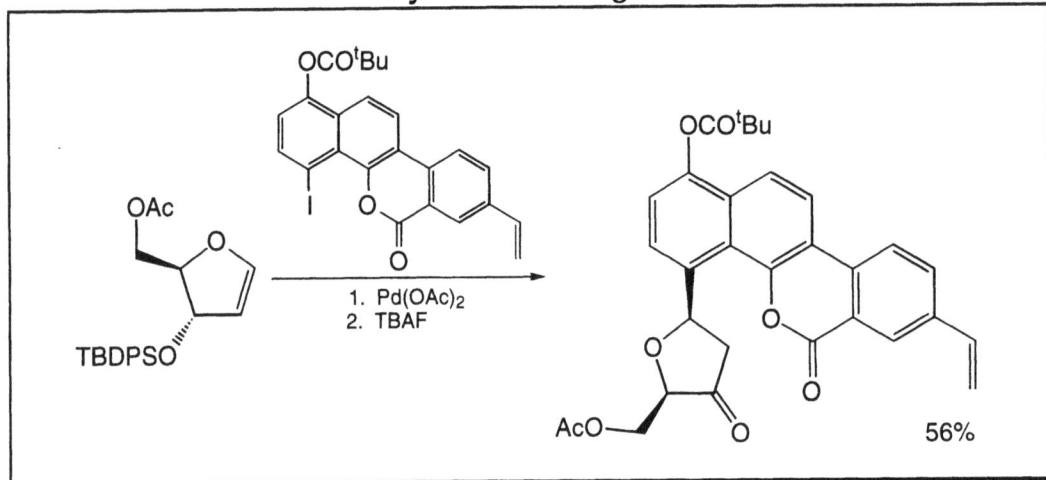

The above described comparison between aryl mercuric acetates and aryl iodides is applied specifically to pyranose-derived glycals. When applied to furanose-derived glycals, the aryl iodides react similarly with respect to reaction yields. However, unlike the pyranose glycals, reactions with furanose glycals appear to yield products bearing β anomeric configurations. These observations were reported by Farr, *et al.*,[9] and are illustrated in Scheme 4.1.10. As mentioned in the cited report, the products were utilized in the total synthesis of the antibiotics gilvocarcin, ravidomycin, and chrysomycin illustrated in Figure 2.4.3.

As illustrated in Schemes 4.1.4 through 4.1.9, the use of pyrimidinyl mercuric acetates have been instrumental in the development of *C*-nucleosides. Although the comparison in Scheme 4.1.9 illustrates the utility of aryl iodides in addition to aryl mercury compounds, the nature of the specific reagents were quite different. As shown in Scheme 4.1.11, Zhang, *et al.*,[10] utilized pyrimidinyl iodides in the preparation of *C*-nucleosides from furanose glycals. As illustrated, the product mixture composition was highly susceptible to the specific reaction conditions used. Thus, of the two products shown, either is available in comparable yields.

A final example illustrating the utility of pyrimidinyl iodides in the preparation of *C*-nucleosides is shown in Scheme 4.1.12. As shown, Zhang, *et al.*,[11] utilized glycals similar to those incorporated in Scheme 4.1.8. The actual coupling was accomplished utilizing palladium acetate with triphenyl arsine as

an additive. Following removal of the silyl group and reduction of the resulting
ketone, the product, 2'-deoxypseudouridine, was isolated in 63% overall yield.

Scheme 4.1.11   *C*-Glycosidation with
                                  Aryl Iodide Reagents

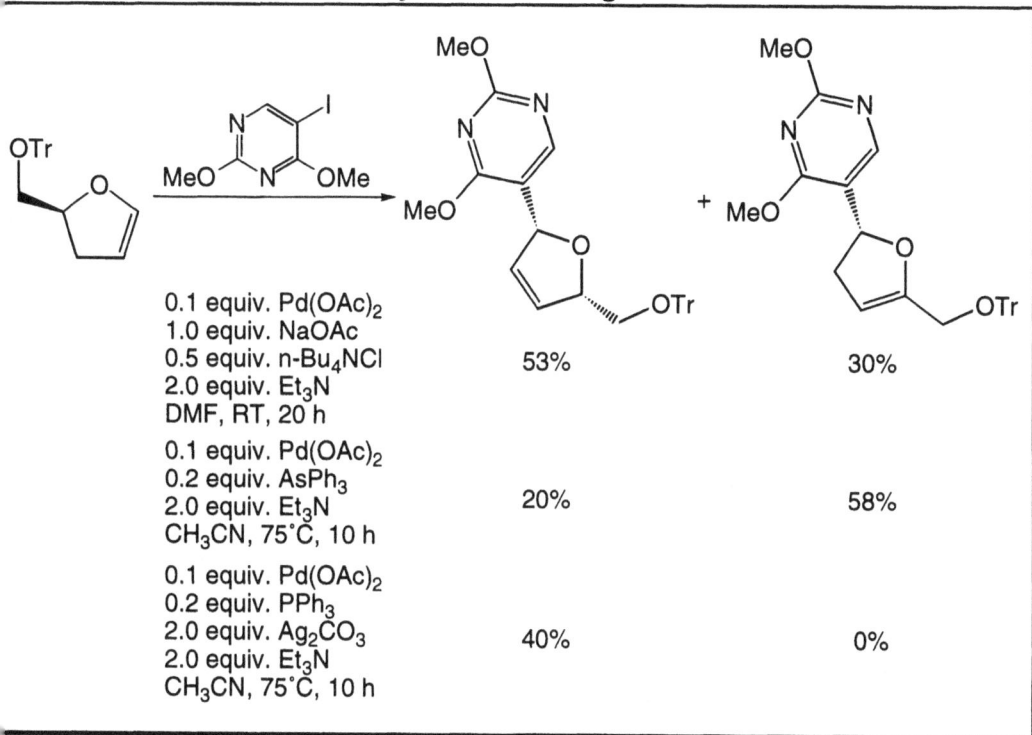

0.1 equiv. Pd(OAc)$_2$ 1.0 equiv. NaOAc 0.5 equiv. n-Bu$_4$NCl 2.0 equiv. Et$_3$N DMF, RT, 20 h	53%	30%
0.1 equiv. Pd(OAc)$_2$ 0.2 equiv. AsPh$_3$ 2.0 equiv. Et$_3$N CH$_3$CN, 75°C, 10 h	20%	58%
0.1 equiv. Pd(OAc)$_2$ 0.2 equiv. PPh$_3$ 2.0 equiv. Ag$_2$CO$_3$ 2.0 equiv. Et$_3$N CH$_3$CN, 75°C, 10 h	40%	0%

Scheme 4.1.12   *C*-Glycosidation with
                                  Aryl Iodide Reagents

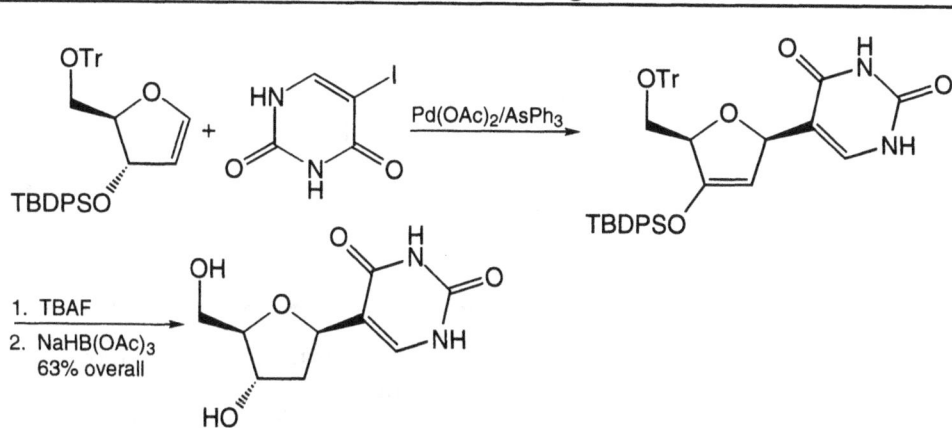

4.2   Coupling of Substituted Glycals with Aryl Groups

Although the direct use of glycals as *C*-glycosidation substrates has been fruitful, the results can often be improved upon utilizing C_1 activated glycals. Such activating groups include metals as well as halogens. Successful catalysts used to effect the coupling of these compounds with metal or halo substituted aryl compounds include nickel and palladium complexes. In this section, the formation of *C*-glycosides utilizing transition metal catalysts will be explored in the context of C_1 activated glycals as reaction substrates.

Beginning with a discussion of the utility of tin substituted glycals, Dubois, *et al.*,[12] utilized 2,3,6-tri-*O*-benzyl-1-tri-*n*-butylstannyl glucal in coupling reactions with various aromatic substrates. As shown in Scheme 4.2.1, tetrakis triphenylphosphinepalladium catalyzed the reaction with bromobenzene providing an 88% yield of the desired product. Additionally, when *p*-nitrobenzoyl chloride was used, dichloro dicyanopalladium effected the formation of the illustrated ketone in 71% yield. Unlike the reactions discussed in section 4.1, the products observed were C_1 substituted glycals. Further elaborations of this chemistry demonstrated its compatibility with both unprotected hydroxyl groups as well as very bulky aromatic bromides.[13] Additional reactions with dibromobenzene are addressed in Chapter 8.

Scheme 4.2.1   *C*-Glycosidation with Stannyl Glycals

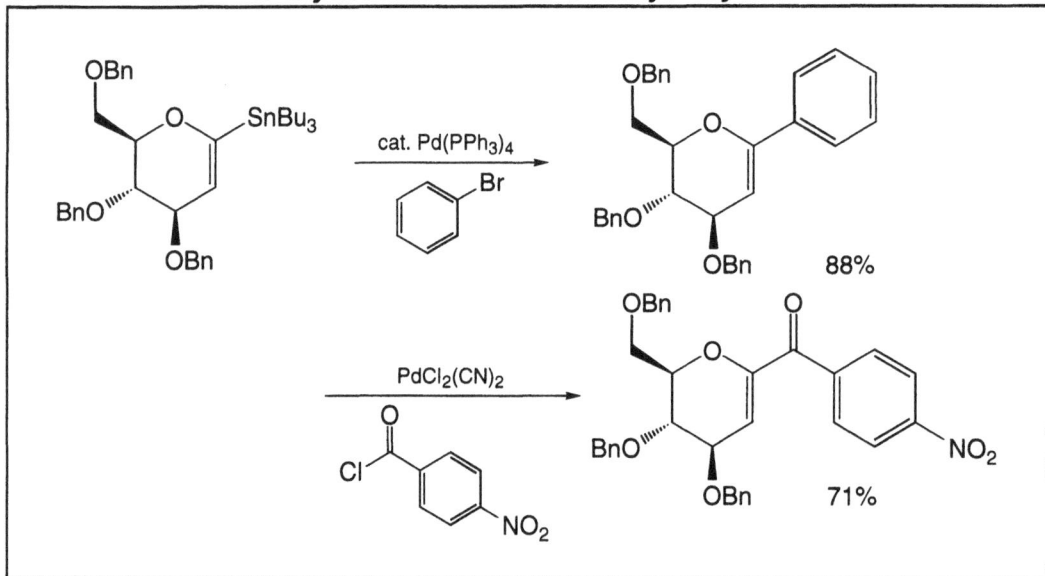

At approximately the same time, Friesen, *et al.*,[14] reported results in agreement with the work presented in Scheme 4.2.1. As shown in Scheme 4.2.2, the analogous silyl protected glycal was coupled to bromobenzene utilizing tetrakis triphenylphosphinepalladium as a catalyst. With tetrahydrofuran as the solvent, a 70% yield of the expected product was isolated. Complimentary to the examples cited in Scheme 4.2.1, the results presented in Scheme 4.2.2 illustrate the compatibility of silyl protecting groups applied to this type of reaction.

Scheme 4.2.2   *C*-Glycosidation with Stannyl Glycals

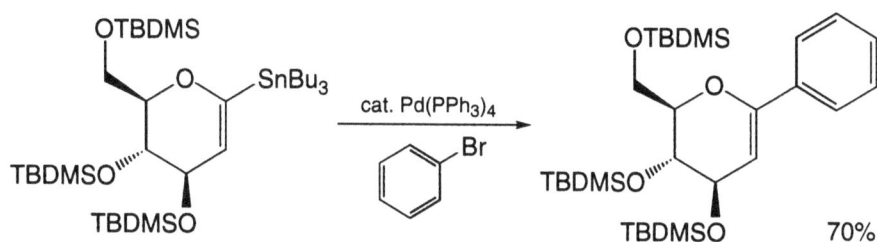

Scheme 4.2.3   *C*-Glycosidation with
C_1 Zinc and Iodo Glycals

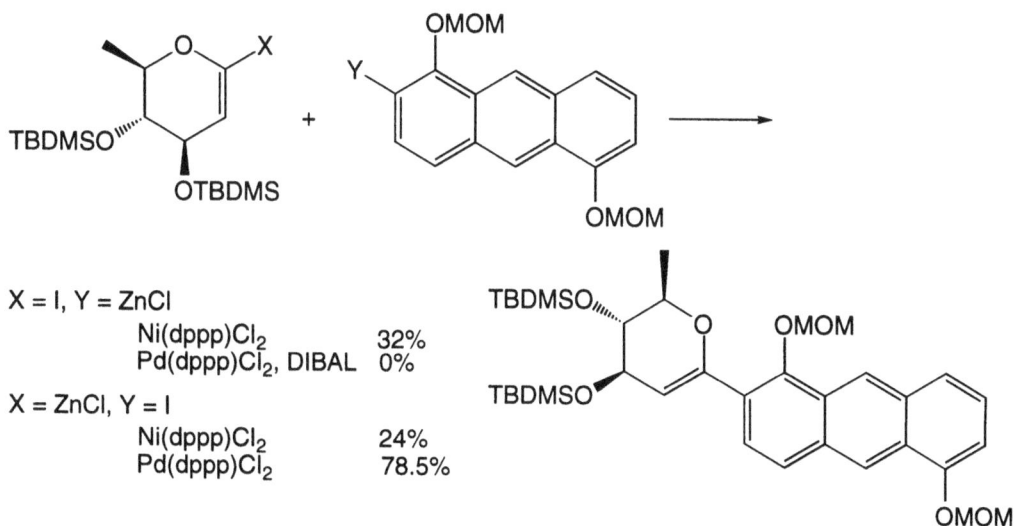

Where C_1 stannylated glycals have been useful in the formation of *C*-glycosides, zinc and iodo substituted glycals have provided complimentary avenues to these compounds. Utilizing either of these classes of glycals, the results of the coupling reactions were shown to be highly susceptible to the transition metal catalyst used.

As shown in Scheme 4.2.3, Tius, *et al.*,[15] compared the results of cross coupling reactions between iodo glycals and aryl zinc reagents to the corresponding coupling reactions between stannyl glycals and aryl iodides. As illustrated, substituted glycals were reacted with substituted aryl compounds to give C_1 aryl substituted glycals. When the starting glycal was substituted with iodine and an aryl zinc reagent was used, a 32% yield of the coupled product was obtained utilizing a nickel catalyst. Use of a palladium catalyst failed to promote this reaction. However, when the glycal was substituted with zinc and an aryl iodide was used, nickel catalysis produced a 24% yield while palladium effected yields greater than 78%. This technology was ultimately used in the preparation of the vineomycinone B2 methyl ester illustrated in Figure 4.2.1.

Figure 4.2.1   Vineomycinone B2 Methyl Ester

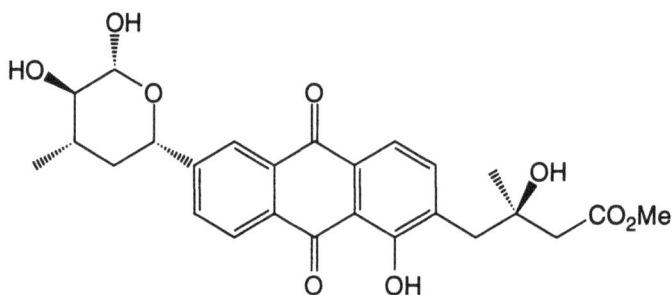

Scheme 4.2.4   Preparation of C_1 Iodo Glycals

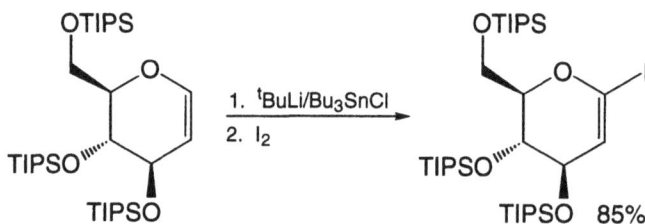

Centering specifically on C_1 iodo glycals, prepared as shown in Scheme 4.2.4, Friesen, *et al.*,[16] initiated coupling reactions with both arylzinc and arylboron compounds. The results, shown in Scheme 4.2.5, illustrate approximately 75% yields for both cases. Thus, the formation of C_1 substituted glycals *en route* to the preparation of C-glycosides is possible through a variety of methods utilizing C_1 metallated and halogenated glycals with a variety of catalysts.

Scheme 4.2.5   *C*-Glycosidation with C$_1$ Iodo Glycals

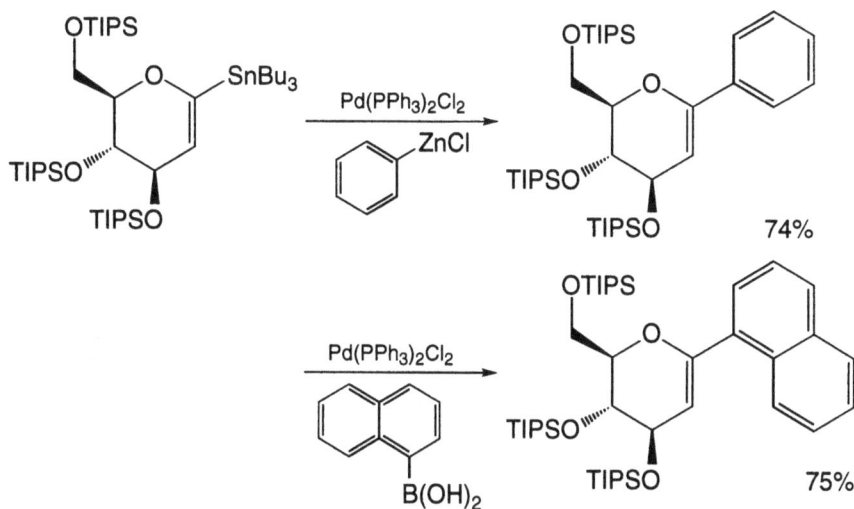

4.3   Coupling of π-Allyl Complexes of Glycals

Until this point, all couplings have involved glycals/glycal derivatives and aryl compounds. However, the generality of this chemistry also applies to the use of 1,3-dicarbonyl compounds and other molecules with suitably electron deficient centers. These reactions proceed with both the neutral and anionic forms of these compounds and generally exhibit moderate to high yields. Unlike the examples shown in section 4.2, these reactions produce *C*-glycosides bearing unsaturation between carbons 2 and 3. Furthermore, the variety of compatible hydroxyl protecting groups include acetate, ketal, and benzyl groups.

Scheme 4.3.1   *C*-Glycosides from
Glycals and 1,3-Diketones

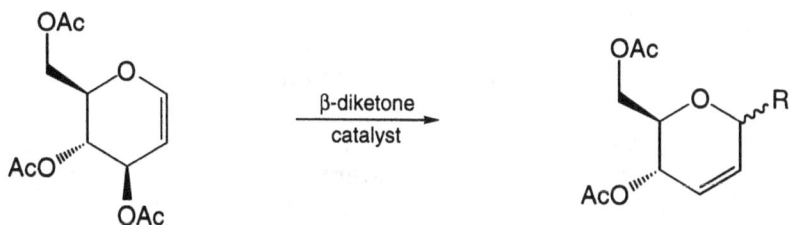

Figure 4.3.1   *C*-Glycosides from
Glycals and 1,3-Diketones

β-Dicarbonyl Compound	*O*-Acetylated Glycal	Catalyst	% Yield	α : β
Acetylacetone	Glycal	Pd(PhCN)$_2$Cl$_2$	83	6 : 1
Acetylacetone	Glycal	BF$_3$	73	5 : 1
Acetylacetone	Galactal	Pd(PhCN)$_2$Cl$_2$	59	α only
Acetylacetone	Galactal	BF$_3$	72	α only
Acetylacetone	Allal	Pd(PhCN)$_2$Cl$_2$	62	5 : 1
Acetylacetone	Allal	BF$_3$	81	5 : 1
Methyl Acetoacetate	Glucal	Pd(PhCN)$_2$Cl$_2$	85	4 : 1
Methyl Acetoacetate	Xylal	Pd(PhCN)$_2$Cl$_2$	65	1 : 4
Ethyl Benzoylacetate	Galactal	Pd(PhCN)$_2$Cl$_2$	65	α only
Ethyl Benzoylacetate	Galactal	BF$_3$	81	α only
Ethyl 2-Oxocyclohexane carboxylate	Glucal	Pd(PhCN)$_2$Cl$_2$	82	---
Ethyl 2-Oxocyclohexane carboxylate	Glucal	BF$_3$	82	---

An early example of the utility of this approach to the formation of *C*-glycosides was presented by Yougai, *et al.*,[17] and compares the use of different Lewis acids and transition metal catalysts. The general reaction reported is shown in Scheme 4.3.1 and utilizes tri-*O*-acetyl-D-glucal as a substrate for coupling with a variety of dicarbonyl compounds. The results of the cited study are summarized in Table 4.3.1.

As the technology developed in this area, the advantages of utilizing glycal activating groups became apparent. Two reports by Brakta, *et al.*,[18,19] involved the initial formation of phenolic glycosides on treatment of glycals with phenol. These species were then coupled, utilizing transition metal catalysts, to a variety of 1,3-dicarbonyl compounds. The general reaction and some results are presented in Scheme 4.3.2 and show that anomeric ratios as high as 100% can be obtained for either anomer and in respectable yields. A notable observation is the retention of the anomeric configuration or conversion from the phenolic glycoside to the corresponding *C*-glycoside.

While the use of phenolic glycosides as activated glycals for use in the preparation of *C*-glycosides was shown to be successful in couplings with 1,3-dicarbonyl compounds, deprotonated 1,3-dicarbonyl compounds showed some utility in comparable reactions with unactivated glycals. As shown in Scheme 4.3.3, RajanBabu, *et al.*,[20] utilized such anionic species in conjunction with transition metal catalysts to effect successful *C*-glycosidations. These reactions proceeded in good yields and were dependent on the nature of the group at C$_3$. Specifically, electron deficient groups such as trifluoroacetoxy groups proved to be the most successful while no reactions were observed utilizing acetates.

Scheme 4.3.2   *C*-Glycosides from Phenolic
                         Glycosides and 1,3-Diketones

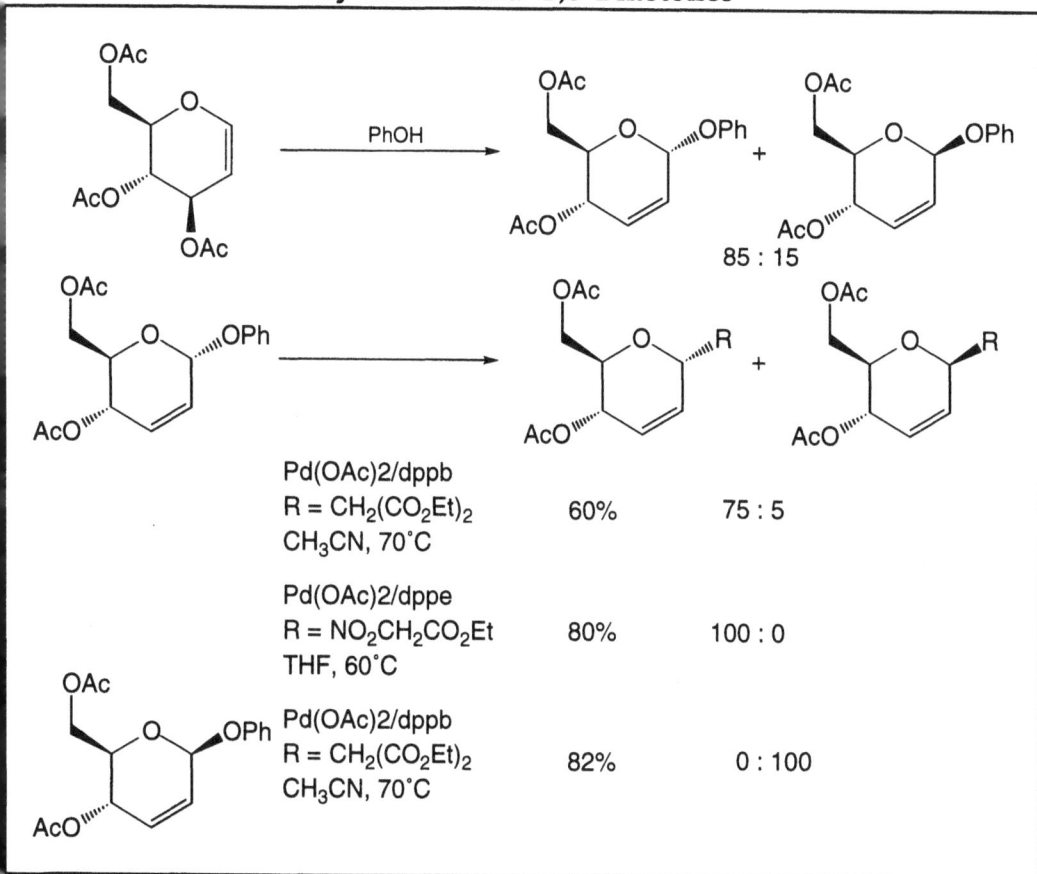

A subsequent study, reported by Dunkerton, *et al.*,[21] showed that the above described chemistry was also applicable to 1-*O*-acetyl-2,3-anhydrosugars. The results, directly analogous to the use of phenolic glycosides shown in Scheme 4.3.2, are illustrated in Scheme 4.3.4. As shown, excellent yields and high anomeric selectivity was observed with a net retention of the anomeric configuration. Although complementary to similar *C*-glycosidations discussed thus far, it should be noted that the applications of this specific reaction type is limited with low yields observed when substituents are present at C$_4$ of the substrate.

Further research into the types of substrates available for transition metal mediated *C*-glycosidations demonstrated the utility of glycosyl carbonates. As shown in Scheme 4.3.5, Engelbrecht, *et al.*,[22] effected the formation of glycosyl carbonates on reaction of anhydrosugars with isobutyl chloroformate and pyridine. Under these conditions, the α glycosyl carbonate was isolated in 75% yield. Subsequent treatment with diethyl malonate in the presence of a palladium catalyst afforded an 81% yield of the *C*-glycoside exhibiting a net retention of stereochemistry at the anomeric center. Similar results were observed with corresponding β glycosyl carbonates.

Scheme 4.3.3   *C*-Glycosides from Activated Glycals and Anionic 1,3-Diketones

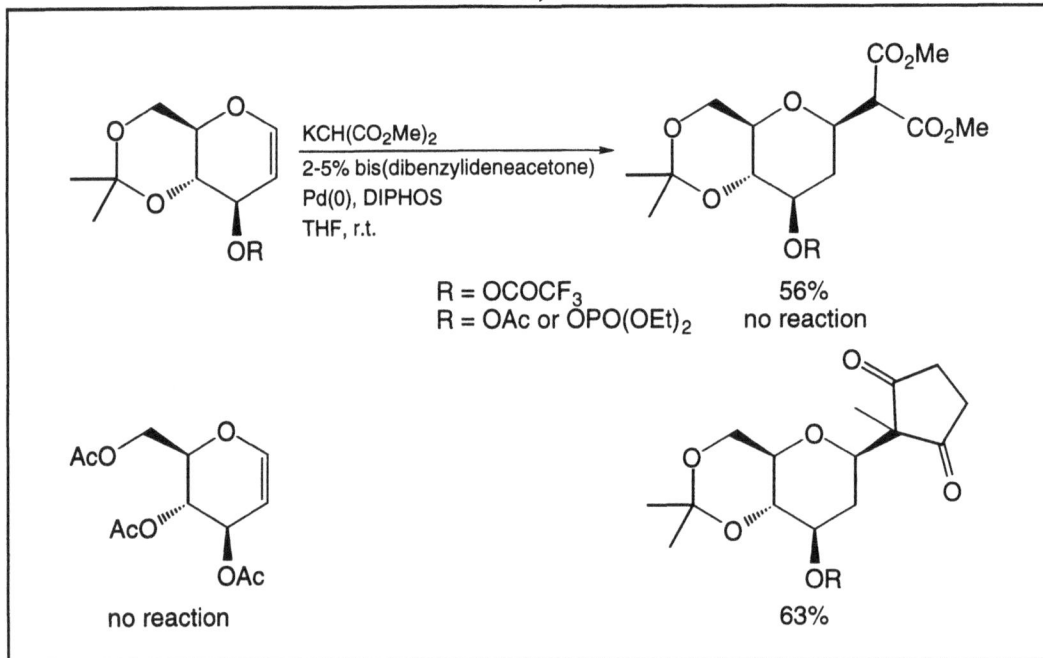

KCH(CO$_2$Me)$_2$

2-5% bis(dibenzylideneacetone)
Pd(0), DIPHOS
THF, r.t.

R = OCOCF$_3$ 56%
R = OAc or OPO(OEt)$_2$ no reaction

no reaction

63%

Scheme 4.3.4   *C*-Glycosides from Anhydrosugars and Anionic Nucloephiles

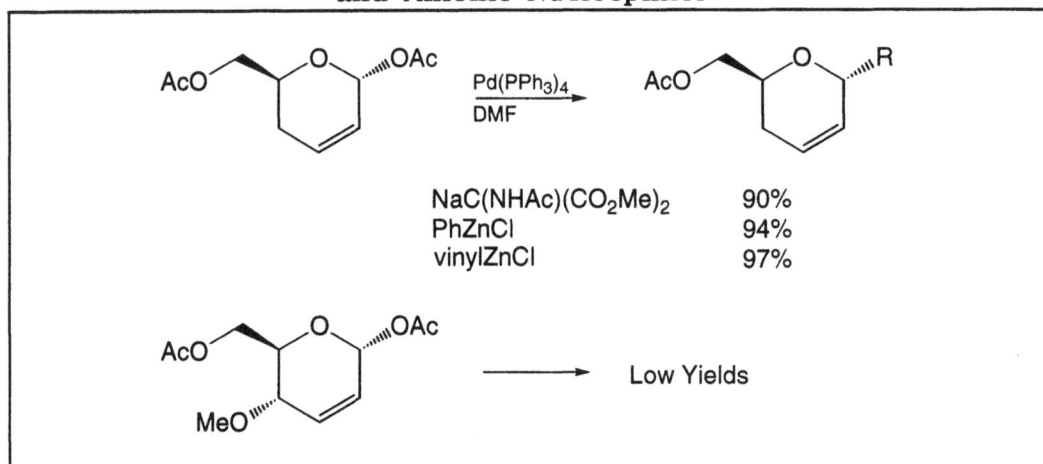

Pd(PPh$_3$)$_4$
DMF

NaC(NHAc)(CO$_2$Me)$_2$ 90%
PhZnCl   94%
vinylZnCl   97%

Low Yields

Scheme 4.3.5 **C-Glycosides from Glycosyl Carbonates**

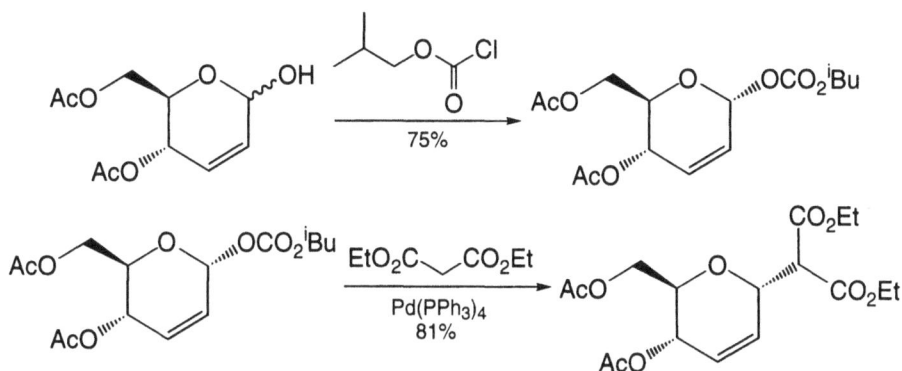

4.4 Mechanistic Considerations

As extensive research has been carried out in the field of palladium mediated *C*-glycosidations, a brief explanation of the mechanistic aspects of these reactions is warranted. Considering the coupling of glycals with pyrimidinyl mercuric acetates, the first step in the reaction involves transmetallation to a pyrimidinyl palladium complex. Subsequent π-complexation to the glycal is followed by formation of a σ-bond adduct between the palladium and one of the pyrimidinyl carbonyls. This process, explained in a review article by Daves[23] and shown in Scheme 4.4.1, proceeds, under specific conditions, to the products observed. Specifically, acidic conditions promote ring opening, heat effects migration of the double bond without acetate elimination, formic acid promotes double bond migration with acetate elimination, and hydrogenation yields saturated *C*-glycosides.

An alternative mechanism applies to the chemistry surrounding the pyranose analogs utilized in Schemes 4.3.2, 4.3.4, and 4.3.5. Addition of palladium to these species forms a π-allylpalladium complex. As shown in Scheme 4.4.2, approach of the palladium is from the side of the sugar opposite that of the glycosidic substituent. This allows the nucleophilic species to approach from the opposite side of the palladium giving the product with a net retention of stereochemistry. Further insight into the mechanistic aspects of this chemistry is available from any of the references cited within this chapter.

Scheme 4.4.1   Mechanisms of
                     Palladium Mediated Couplings

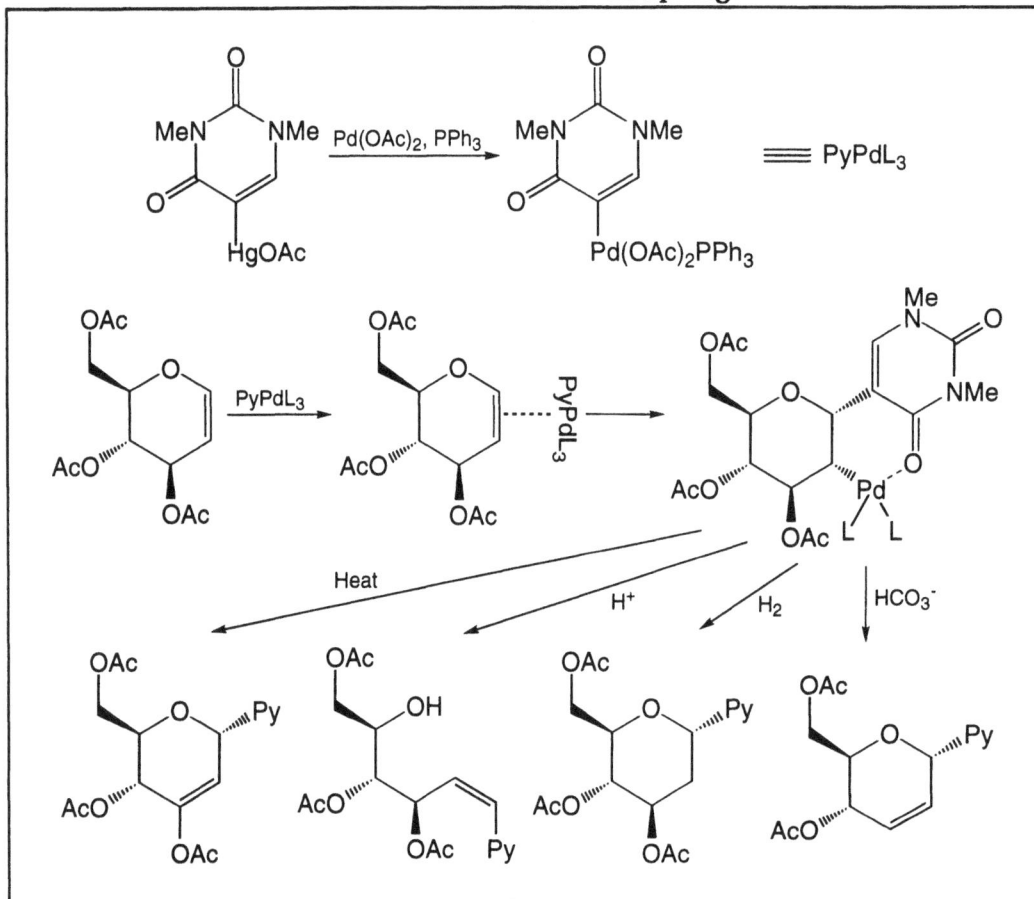

Scheme 4.4.2   Mechanisms of
                     Palladium Mediated Couplings

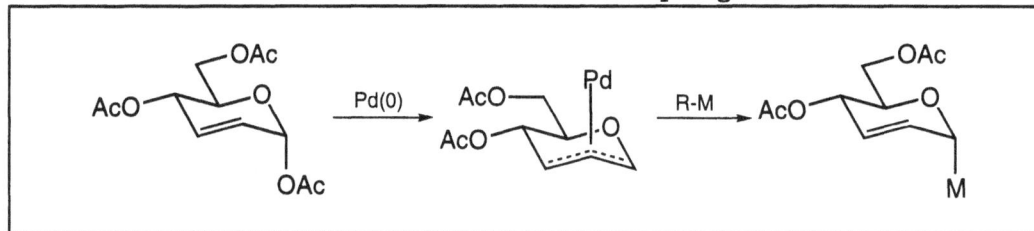

4.5   References

1.   Czernecki, S.; Gruy, E. *Tetrahedron Lett.* **1981**, *22*, 437.
2.   Czernecki, S.; Dechavanne, V. *Can. J. Chem.* **1983**, *61*, 533.
3.   Outten, R. A.; Daves, D. G. Jr. *J. Org. Chem.* **1987**, *52*, 5064.
4.   Outten, R. A.; Daves, D. G. Jr. *J. Org. Chem.* **1989**, *54*, 29.
5.   Cheng, J. C.-Y.; Daves, G. D. Jr. *J. Org. Chem.* **1987**, *52*, 3083.
6.   Ari, I.; Lee, T. D.; Hanna, R.; Daves, G. D. Jr. *Organometallics* **1982**, *1*, 742.
7.   Farr, R. N.; Daves, G. D. Jr. *J. Carbohydrate Chem.* **1990**, *9*, 653.
8.   Kwok, D. I.; Farr, R. N.; Daves, G. D. Jr. *J. Org. Chem.* **1991**, *56*, 3711.
9.   Farr, R. N.; Kwok, D. I.; Daves, G. D. Jr. *J. Org. Chem.* **1992**, *57*, 2093.
10.   Zhang, H. C.; Daves, G. D. Jr. *J. Org. Chem.* **1993**, *58*, 2557.
11.   Zhang, H. C.; Daves, G. D. Jr. *J. Org. Chem.* **1992**, *57*, 4690.
12.   Dubois, E.; Beau, J.-M. *J. Chem. Soc. Chem. Comm.* **1990**, 1191.
13.   Dubois, E.; Beau, J.-M. *Carbohydrate Res.* **1992**, *228*, 103.
14.   Friesen, R. W.; Sturino, F. C. *J. Org. Chem.* **1990**, *55*, 2572.
15.   Tius, M. A.; Gomez, Galeno, J.; Gu, X. Q.; Zaidi, J. H. *J. Am. Chem. Soc.* **1991**, *113*, 5775.
16.   Friesen, R. W.; Loo, R. W. *J. Org. Chem.* **1991**, *56*, 4821.
17.   Yougai, S.; Miwa, T. *J. Chem. Soc. Chem. Comm.* **1982**, 68.
18.   Brakta, M.; Le Borgne, F.; Sinou, D. *J. Carbohydrate Chem.* **1987**, *6*, 307.
19.   Brakta, M.; Lhoste, P.; Sinou, D. *J. Org. Chem.* **1989**, *54*, 1890.
20.   RajanBabu, T. V. *J. Org. Chem.* **1985**, *50*, 3642.
21.   Dunkerton, L. V.; Euske, J. M.; Serino, A. J. *Carbohydrate Res.* **1987**, *171*, 89.
22.   Engelbrecht, G. J.; Holzapfel, C. W. *Heterocycles* **1991**, *32*, 1267.
23.   Daves, D. G. Jr. *Acc. Chem. Res.* **1990**, *23*, 201.

5   Introduction

Thus far, the *C*-glycosidations discussed involve ionic mechanisms. However, neutral free radicals are also useful in the formation of *C*-glycosides. Specifically, the ability to form free radicals at anomeric centers, illustrated in Figure 5.0.1, provides an additional dimension to the developing *C*-glycosidation technologies. Interesting features of free radical chemistry applied to *C*-glycosidations include the ability to couple these species with numerous readily available radical acceptors. The stereochemical course of these reactions generally provides α selectivity. This chemistry is nicely applied to acylated sugars and tolerates many substrate functional groups. Finally, intramolecular radical reactions provide avenues into compounds bearing difficult stereochemical relationships as well as fused ring systems.

Figure 5.0.1   Anomeric Radicals

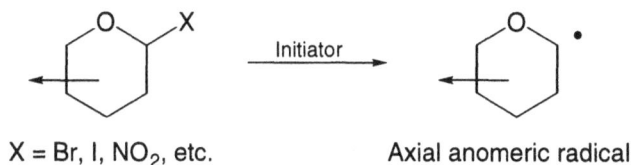

X = Br, I, NO$_2$, etc.   Axial anomeric radical

5.1   Sources of Anomeric Radicals and Stereochemical Consequences

Anomeric free radicals can be generated from sugar substrates bearing a variety of activating groups. Additionally, the conditions required for the generation of these reactive species include both chemical and photolytic methods. Through combining specific reaction conditions and appropriate activating groups, both α and β oriented *C*-glycosides are available. In this section, the generation of anomeric radicals is discussed with respect to the utility of various activating groups. Furthermore, the stereochemical consequences of reaction conditions, as applied to different activating groups, are explored.

5.1.1   Nitroalkyl C-Glycosides as Radical Sources

One source of anomeric radicals and evidence for the preference of axial radical orientations has already been introduced in work reported by Vasella, et al.,[1,2] and illustrated in Scheme 3.3.11. The specific reaction, reviewed in Scheme 5.1.1, involves the cleavage of a nitro group under radical forming conditions. The starting nitroglycoside existed as a mixture of anomers and the product was composed of the β anomer exclusively. This observation suggests the formation of an α oriented radical which accepts a hydrogen radical from tributyltin hydride. Specific delivery of the hydrogen radical to the stable α anomeric radical thus explains the exclusive formation of the observed β-*C*-glycoside.

Scheme 5.1.1   Axially Oriented Anomeric Radicals

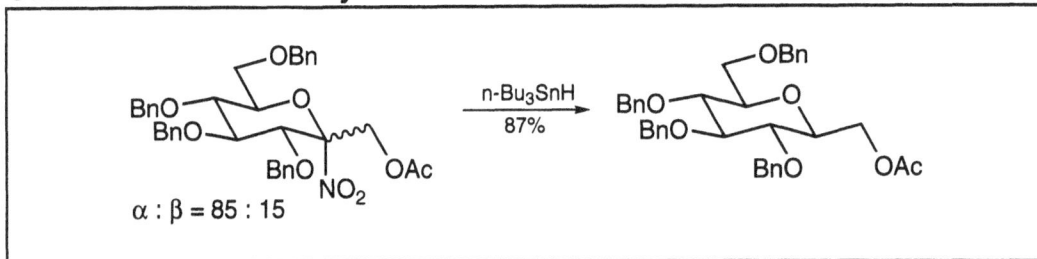

**Scheme 5.1.2   Stereochemical Preference
of Anomeric Radicals**

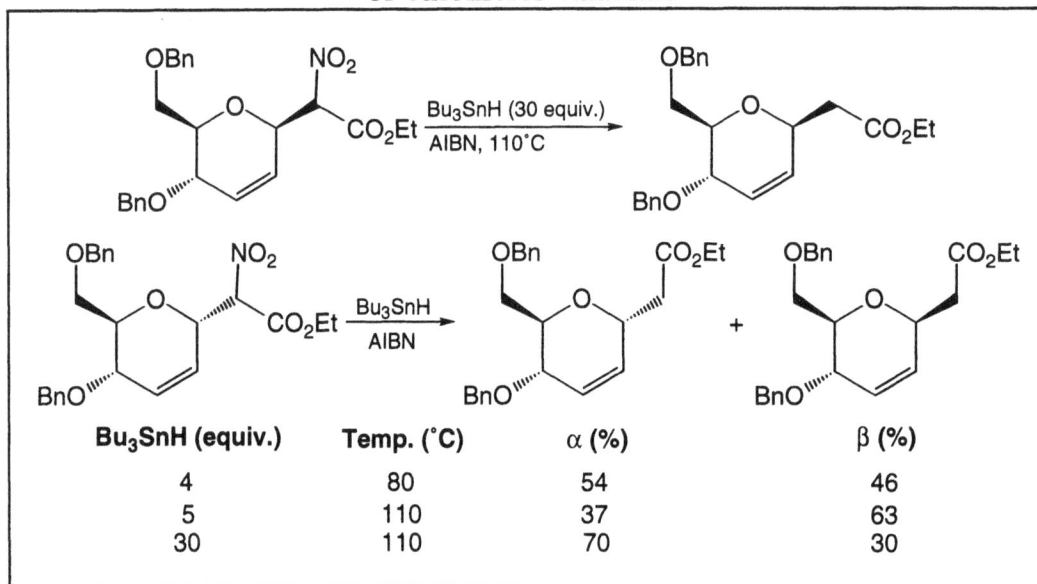

Bu₃SnH (equiv.)	Temp. (°C)	α (%)	β (%)
4	80	54	46
5	110	37	63
30	110	70	30

An excellent exploration into the actual stability of anomeric radicals was reported by Brakta, et al.,[3] and involves the migration of a free radical to the anomeric center of pre-formed C-glycosides. As shown in Scheme 5.1.2, the C-glycosidic nitroacetates were subjected to various radical mediated denitration conditions. Utilizing the β-C-glycoside, treatment with 30 equivalents of tributyltin hydride effected denitration with exclusive isolation of the modified β-C-glycoside. However, when the α-C-glycoside was used, substantial scrambling of the stereochemistry at the anomeric center was observed. The mechanism, shown in Scheme 5.1.3, involves a 1,2-shift of a hydrogen atom thus forming the more stable anomeric radical. Subsequent isomerization from the β to the α anomeric radical thus explains the loss of stereochemical integrity.

Scheme 5.1.3   **Stereochemical Preference**
                            of Anomeric Radicals

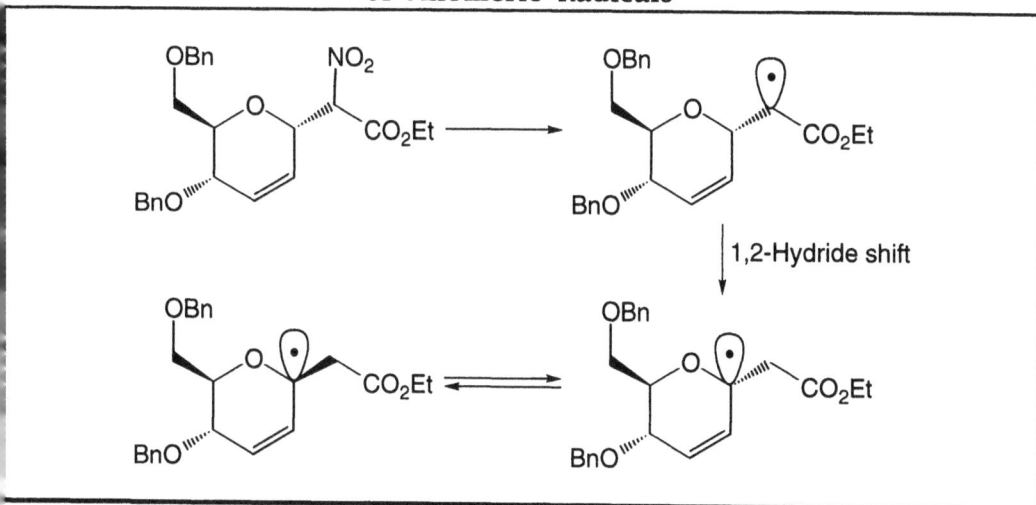

1,2-Hydride shift

5.1.2   Radicals from Activated Sugars

As alluded to in the introduction of this chapter, glycosidic radicals prefer to adopt an α anomeric configuration. However, this preference can be manipulated through the use of a variety of anomeric activating groups and radical initiating conditions. Some glycosidic substrates used in the generation of anomeric free radicals are shown in Figure 5.1.1 and include oxygen-derived glycosides,[4] thioglycosides,[5,6] and selenoglycosides.[7] From these compounds, as well as utilizing either chemical or photolytic radical formations, both α and β C-glycosides have been shown to be accessible. Specific examples utilizing these substrates will be presented later in this chapter.

In addition to the use of substrates similar to those shown in Figure 5.1.1, glycosyl halides have found much utility in the chemistry of glycosidic radicals.

As will be elaborated upon later in this chapter, glycosyl fluorides and bromides tend to produce α anomeric radicals while chlorides and iodides show substantially less selectivity. One particular study, involving the reduction of glycosyl bromides under various conditions, was reported by Somsak, *et al.*,[8] and is shown in Scheme 5.1.4. As shown, this study compared a variety of reducing conditions on two 1-bromo-1-cyano glycosides. In both cases, utilizing tributyltin hydride and a radical initiator, the major products observed resulted from delivery of a hydride with retention of the anomeric configurations. These observations support the preferential formation of α anomeric radicals.

Figure 5.1.1   **Substrates for Anomeric**
                              Radical Formation

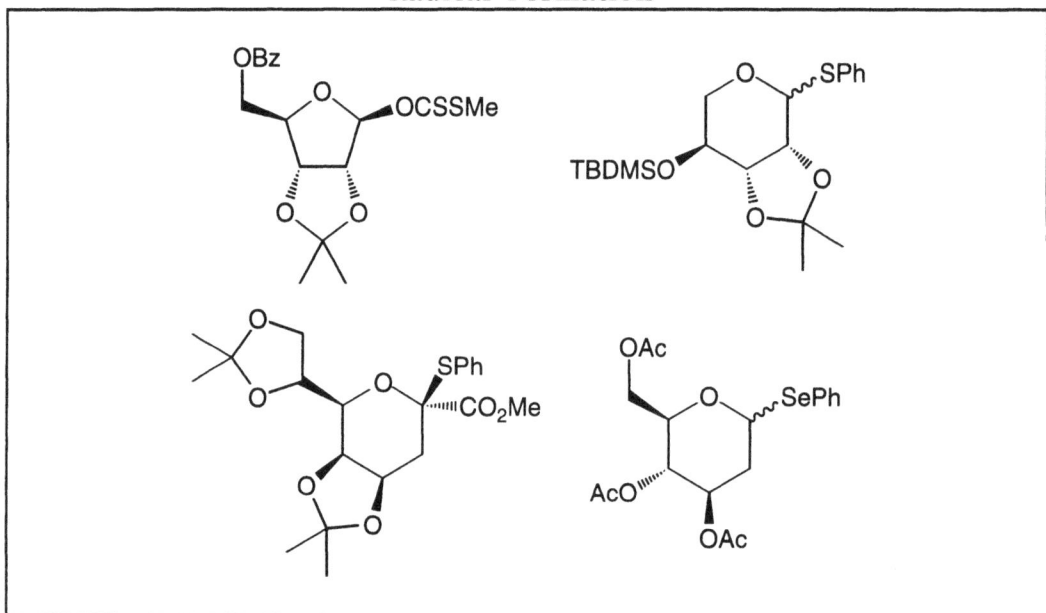

Although the anomeric configuration of glycosidic radicals is relatively stable, groups migrating from C_2 to C_1 of a given sugar migrate in a *cis* fashion. This can, in many cases, yield products contrary to the already stated observations of the stereochemical preference for glycosidic free radicals. As shown in Schemes 5.1.5 and 5.1.6, Giese, *et al.*,[9,10] treated 2,3,4,6-tetra-*O*-acetyl-α-D-glucopyranosyl bromide with tributyltin hydride as the limiting reagent and a radical initiator. Through the mechanism shown, the C_2 acetate migrated to the anomeric position yielding the observed α anomer in 92% yield. Similarly, utilizing 2,3,4,6-tetra-*O*-acetyl-α-D-mannopyranosyl bromide the corresponding β anomer is obtained in 65% yield thus confirming the nature of the migration. Similar observations for the *cis* migration of C_2 acetoxy groups to the anomeric position of sugars were observed for both pyranosyl iodides and furanosyl selenides.

Scheme 5.1.4   Reduction of Glycosyl Bromides

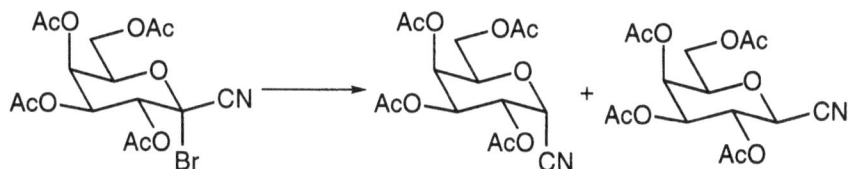

Conditions	Time (min)	α/β	Yield (%)
Zn/AcOH, reflux	15	7/3	64
Zn/^{i}PrOH, reflux	20	6/4	78
Bu$_3$SnH/C$_6$H$_6$, AIBN, reflux	15	4/6	77
NaBH$_4$/DMF, room temp.	10	6/4	64

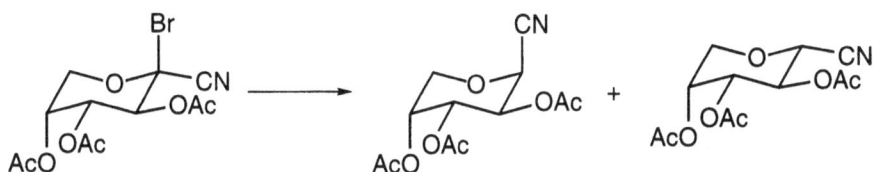

Conditions	Time (min)	α/β	Yield (%)
Zn/AcOH, reflux	15	6/4	64
Zn/^{i}PrOH, reflux	20	6/4	66
Bu$_3$SnH/C$_6$H$_6$, AIBN, reflux	15	2/8	67
NaBH$_4$/DMF, room temp.	10	6/4	67

Scheme 5.1.5   1,2-Migrations of Anomeric Radicals

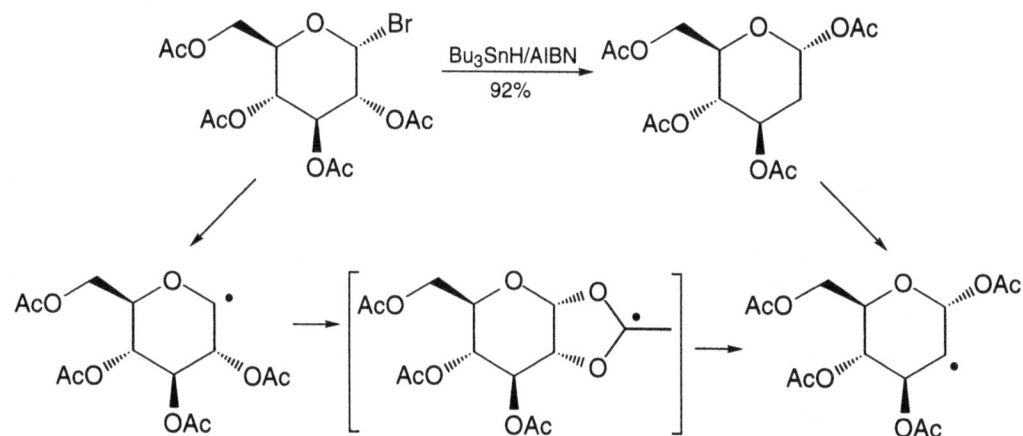

Scheme 5.1.6   1,2-Migrations of Anomeric Radicals

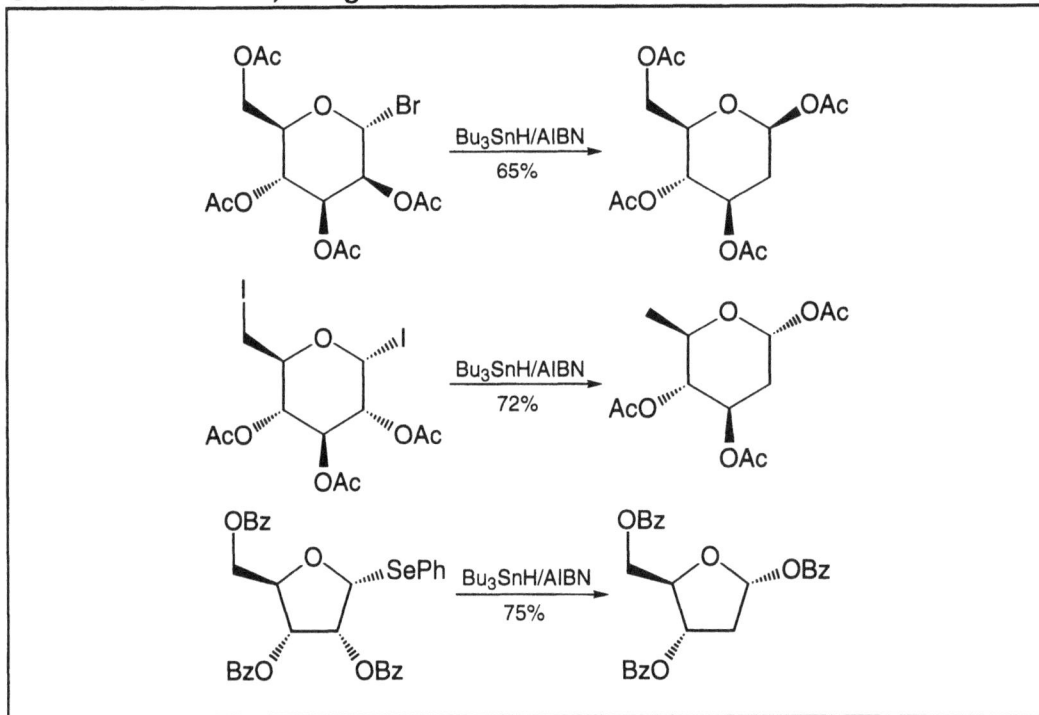

Having introduced the stereochemical course followed by reactions involving glycosidic free radicals, the remainder of this chapter will present examples of the utility of these species in both intermolecular and intramolecular reactions.

5.2   Anomeric Couplings with Radical Acceptors

The intermolecular reactions between glycosidic free radicals and radical acceptors have provided a variety of novel structures inaccessible from other glycosidation pathways. Although the types of radical acceptors useful are quite varied, the different anomeric activating groups used in the generation of free radicals substantially influence the yields and stereochemical outcome of the reactions. The two divisions within this section address the use of glycosyl halides and non-halogenated sugar derivatives in the exploitation of free radical *C*-glycosidations.

5.2.1   Non-Halogenated Radical Sources

Scheme 5.2.1   Radical C-Glycosidations
                             with Nitroglycosides

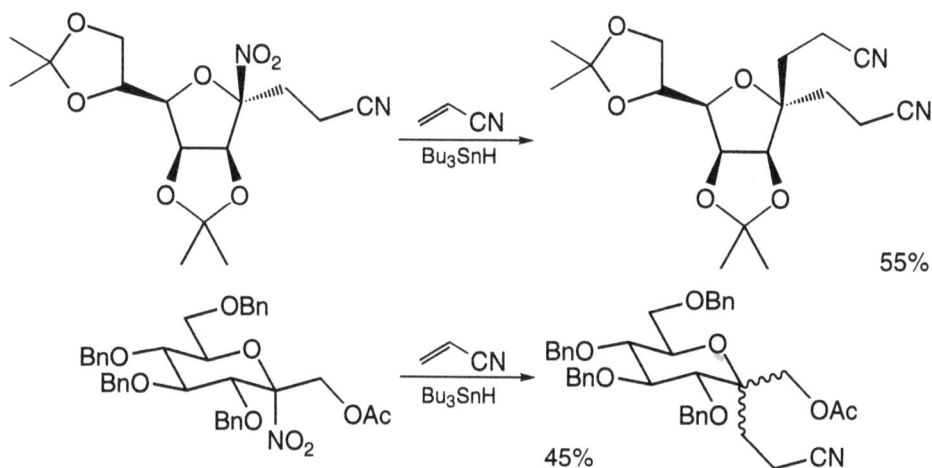

Scheme 5.2.2   Radical C-Glycosidations with
                    Glycosidic Methylthiothiocarbonates

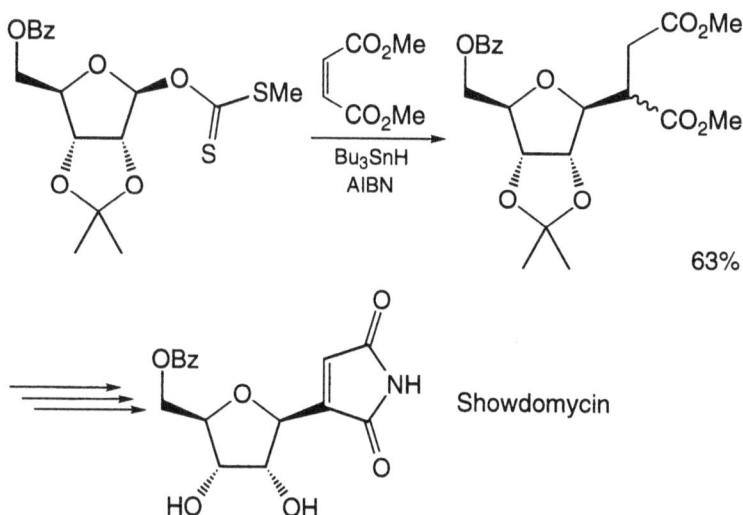

Showdomycin

Understanding the chemistry surrounding the formation and subsequent stereochemical preference of anomeric radicals, the various reactions utilizing these species in the preparation of *C*-glycosides can be discussed. Beginning with nitroglycosides, Depuis, *et al.*,[11] effected reactions with acrylonitrile. As shown in Scheme 5.2.1, the radicals were formed utilizing tributyltin hydride. The observed products were isolated in approximately 50% yields and, in the case of the hexose, exhibited no stereochemical preference.

Scheme 5.2.3   **Radical *C*-Glycosidations**
                         with Phenylthioglycosides

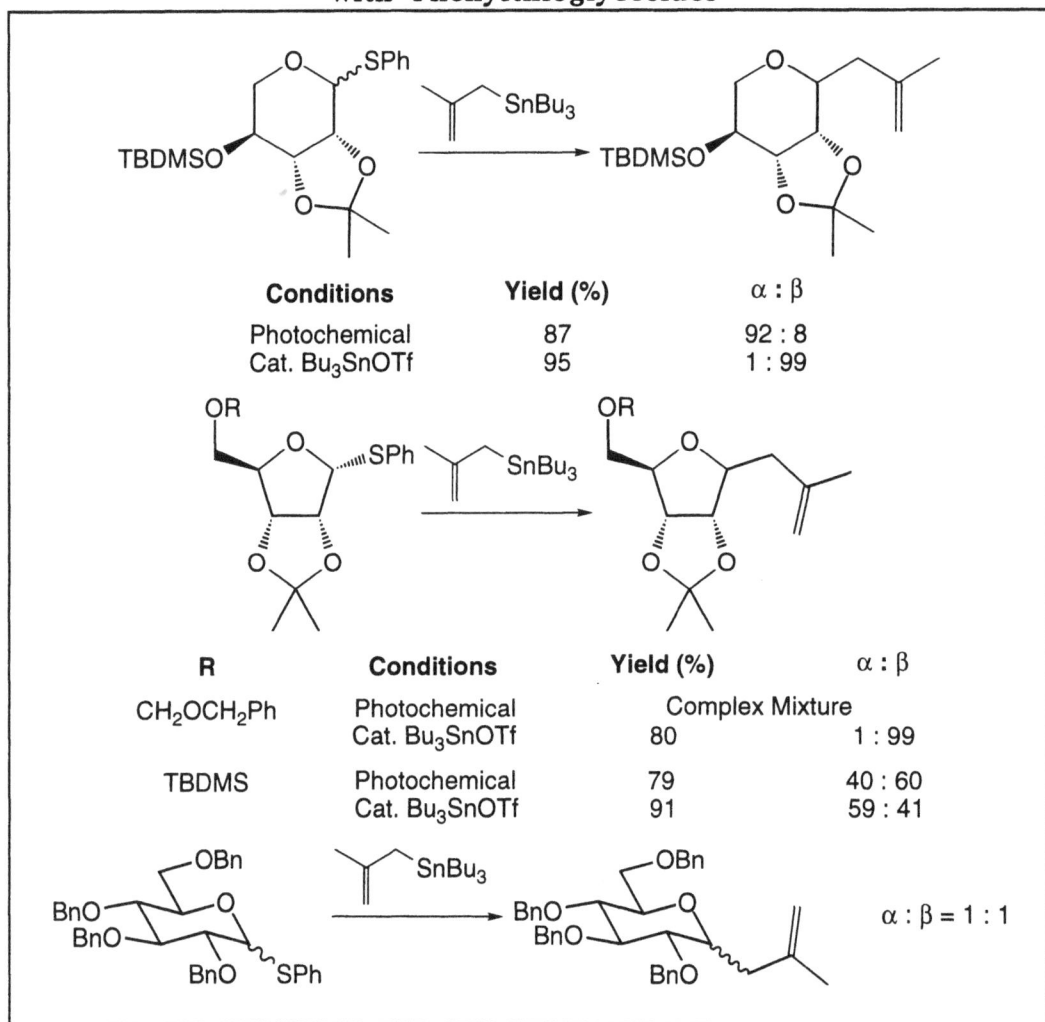

Conditions	Yield (%)	α : β
Photochemical	87	92 : 8
Cat. Bu₃SnOTf	95	1 : 99

R	Conditions	Yield (%)	α : β
CH₂OCH₂Ph	Photochemical	Complex Mixture	
	Cat. Bu₃SnOTf	80	1 : 99
TBDMS	Photochemical	79	40 : 60
	Cat. Bu₃SnOTf	91	59 : 41

$$\alpha : \beta = 1 : 1$$

An example exhibiting a higher yield and substantially better stereoselectivity involves a radical coupling with dimethylmalonate. As shown in Scheme 5.2.2, Araki, *et al.*,[4] generated anomeric radicals from the illustrated glycosidic methylthiothiocarbonate on treatment with tributyltin hydride and a radical initiator. The resulting product, formed in 63% yield, was subsequently utilized in the preparation of showdomycin.

Additional work demonstrating the utility of anomeric radicals in the preparation of *C*-glycosides was reported by Keck, *et al.*,[5] and involved transformations utilizing phenylthioglycosides. As shown in Scheme 5.2.3, *C*-methallylglycosides were prepared from both pyranose and furanose sugars in yields ranging from 79% to 95%. Interestingly, the stereochemical course of the reactions shown were dependent upon the method used for generation of the anomeric radicals. Specifically, with respect to pyranose sugars, photochemical conditions induced predominant formation of α anomers while β anomers were formed utilizing chemical methods. Furthermore, in the case of furanose sugars, these observations were dependent upon the protecting groups used. Finally, the complete lack of stereochemistry observed under all conditions in the reaction involving 2,3,4,6-tetra-*O*-benzyl phenylthioglucopyranose deserves mention.

Scheme 5.2.4   **Radical *C*-Glycosidations
with Phenylthioglycosides**

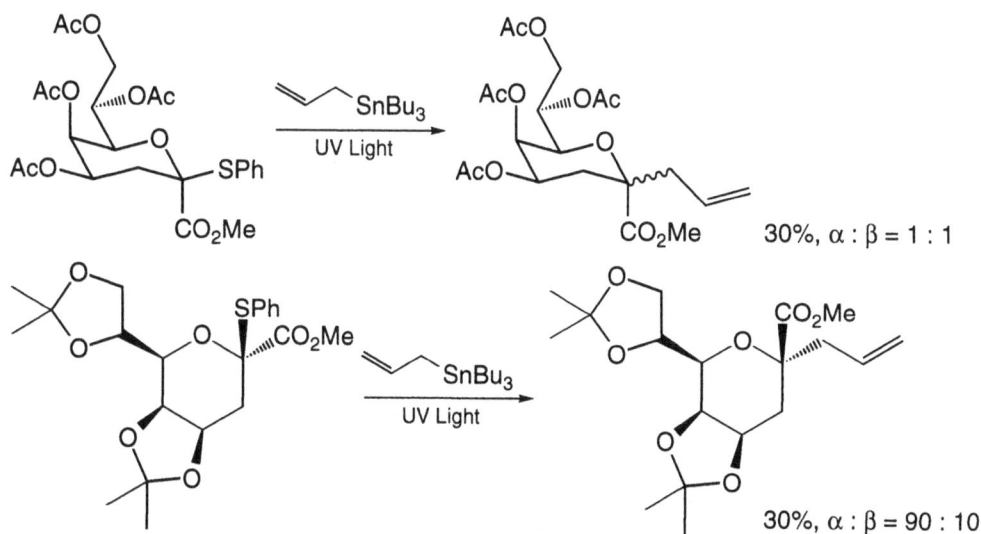

30%, α : β = 1 : 1

30%, α : β = 90 : 10

An additional example illustrating the utility of thiophenyl glycosides in free radical mediated *C*-glycosidations was reported by Waglund, *et al.*[6] As shown in Scheme 5.2.4, 1-methylcarboxy-1-phenylthioglycosides were treated with allyltributyltin and irradiated with ultraviolet light. The resulting *C*-allyl derivatives were formed in approximately 30% yields with the stereochemical course of the reactions highly dependent upon the nature of the protecting groups used. For example, the peracetylated sugar was transformed to a 1 : 1 anomeric mixture of C-glycosides while the acetonide protected sugar exhibited predominant formation of the α oriented C-allylglycoside.

5.2.2   Glycosyl Halides as Radical Sources

With all the anomeric activating groups useful in the generation of sugar-derived free radicals, glycosyl halides have been most exploited. In the remainder of this section, various reactions involving glycosyl halides will be discussed. Additionally, the stereochemical consequences of these reactions will be analyzed.

Scheme 5.2.5   **Radical C-Glycosidations with Glycosylfluorides**

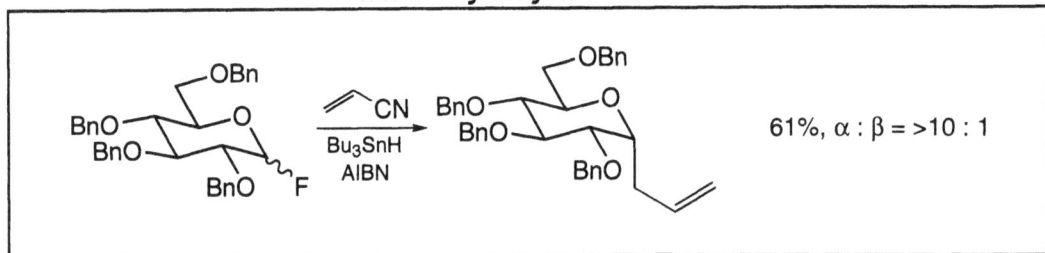

Glycosyl fluorides, already addressed as substrates for cyanations and allylations, are excellent substrates for radical mediated C-glycosidations. As shown in Scheme 5.2.5, Nicolau, et al.,[12] demonstrated such reactions in the coupling of 2,3,4,6-tetra-O-benzyl glucopyranosyl fluoride with acrylonitrile. The reaction proceeded in 61% yield with an anomeric ratio of greater than 10 : 1 favoring the α isomer.

Scheme 5.2.6   **Radical C-Glycosidations with Glycosylchlorides**

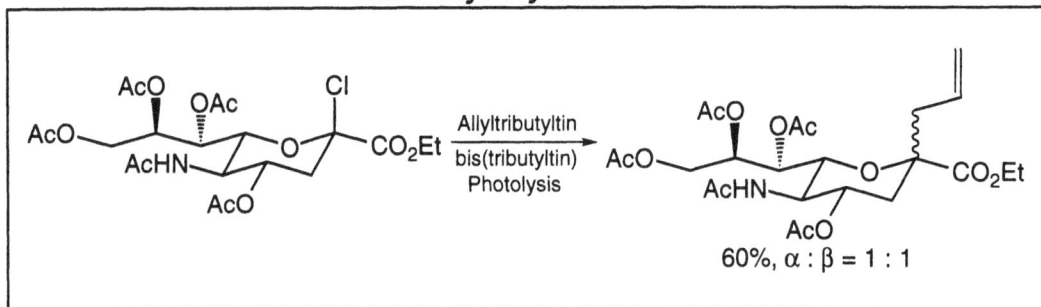

Complimentary to the use of glycosylfluorides is the ability to convert glycosylchlorides to *C*-glycosides under radical conditions. As shown in Scheme 5.2.6, Bednarski, *et al.*,[13] effected such reactions utilizing allyltributyltin and photolytic conditions. Although the modified sialic acid was isolated in 60% yield, no stereochemical induction was observed.

Additional work involving glycosylchlorides as substrates for radical mediated *C*-glycosidations have shown interesting stereochemical consequences. As shown in Scheme 5.2.7, Paulsen, *et al.*,[14] effected the conversion of 1-methylcarboxy-1-chloroglycosides to the corresponding 1-methylcarboxy-1-allylglycosides in excellent yield. Unfortunately, the stereochemical course of the reaction involving 2-deoxy sialic acid favored the α-allyl isomer in a ratio of 1.8 : 1. However, utilizing similar conditions, the analogous sialic acid glycosylbromide provided similar 2-hydroxy compounds favoring the β-allyl isomer in a ratio of 3 : 1. Although the yield was lower than that observed for the 2-deoxy sialic acid derivative, these results demonstrate the versatility available from the use of a variety of sugar derivatives in the formation of *C*-glycosides.

Scheme 5.2.7   Radical *C*-Glycosidations with Glycosylchlorides/bromides

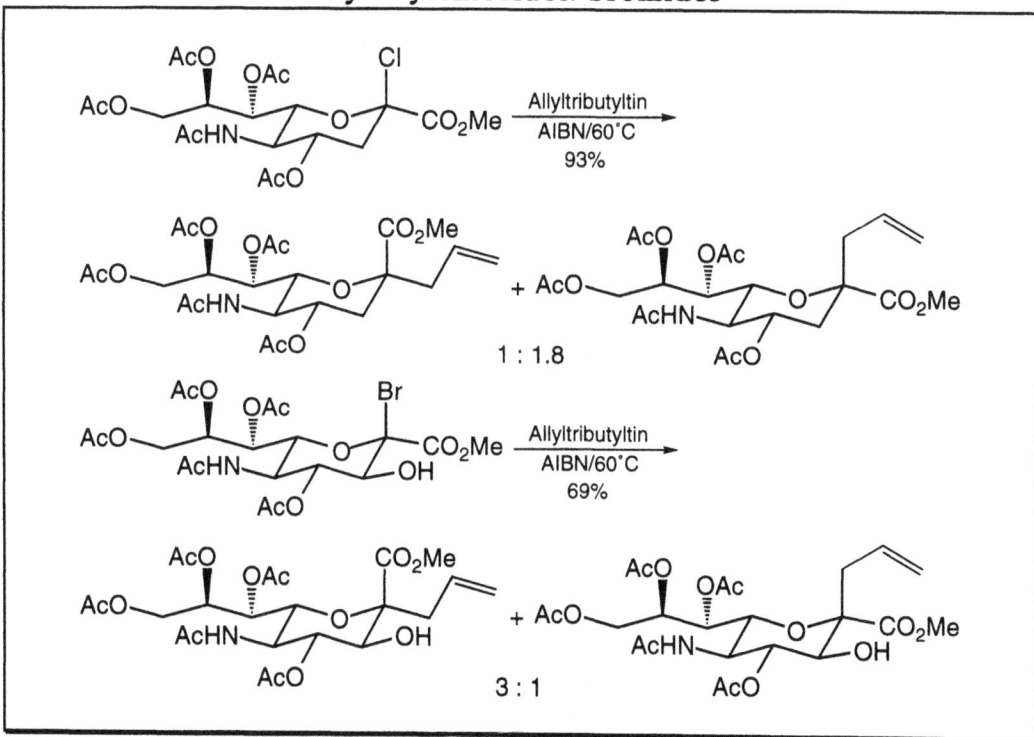

Acrylonitrile has already been established as an excellent radical acceptor for the formation of *C*-glycosides and further examples utilizing this reagent will be presented shortly. However, before continuing with the chemistry surrounding this reagent, the applicability of the electronically similar α,β-unsaturated carbonyl compounds deserves mention. As shown in Scheme 5.2.8, Giese, *et al.*,[15] utilized the lactone, shown, as an acceptor for radicals derived from 2,3,4,6-tetra-*O*-acetyl-D-glucopyranosyl bromide. The anomeric radicals were generated utilizing tributyltin hydride and a radical initiator. Following treatment of the radicals with the acceptor, the product was isolated in 22% yield with an anomeric ratio of 10 : 1 favoring the α anomer. Unfortunately, the final hydrogen transfer adjacent to the carbonyl proceeds with little stereoselectivity exhibiting an isomeric ratio of approximately 60 : 40. In spite of the low yield for the illustrated reaction, the utility of carbonyl-derived radical acceptors has been demonstrated and additional examples will be examined later in this section.

Scheme 5.2.8   **Radical *C*-Glycosidations**
                        with Glycosylbromides

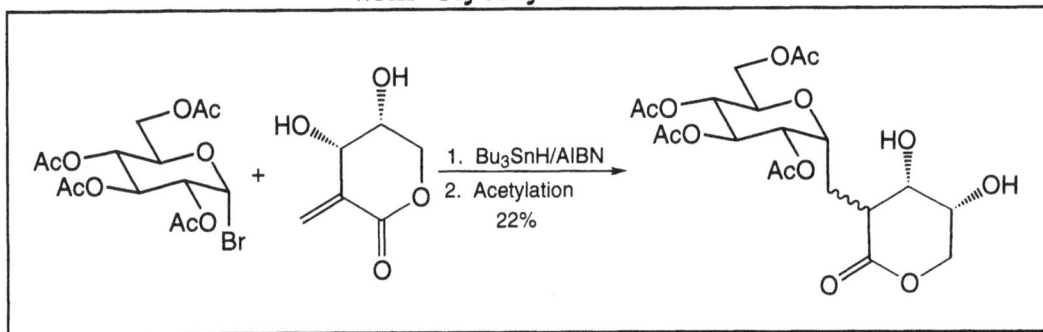

Scheme 5.2.9   **Radical *C*-Glycosidations**
                        with Glycosylbromides

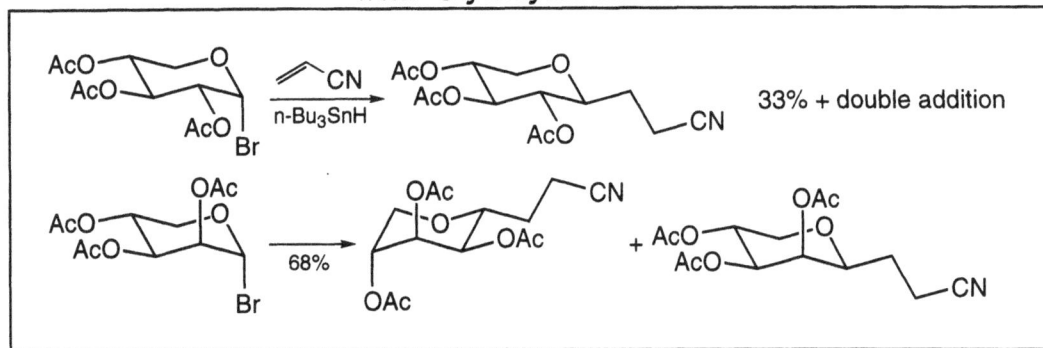

Reactions involving the coupling of acrylonitrile with anomeric radicals derived from glycosyl bromides have been useful for the preparation of *C*-

cyanoalkyl glycosides. As shown in Scheme 5.2.9, Giese, *et al.*,[16] generated glycosidic radicals utilizing chemical methods. Subsequent treatment with acrylonitrile effected the formation of the desired products. In the case of 2,3,4-tri-*O*-acetyl-D-xylopyranosyl bromide, the reaction produced the β anomer in 33% yield with substantial double addition observed. However, when 2,3,4-tri-*O*-acetyl-D-lyxopyranosyl bromide was used, a 68% yield comprising a mixture of anomers was obtained.

The relatively low yields and poor anomeric selectivity observed in the reactions with pentose-derived glycosyl bromides does not translate to higher sugars. As shown in Scheme 5.2.10, Giese, *et al.*,[16] demonstrated reactions with acrylonitrile and other radical acceptors. The results observed with 2,3,4,6-tetra-*O*-acetyl-D-glucopyranosyl bromide include isolation of a 75% yield of the expected *C*-glycoside comprising a mixture of anomers in a ratio of 93 : 7 favoring the α isomer. Additionally, utilizing fumaronitrile as the radical acceptor, high stereoselectivity was observed in light of the substantially lower isolated yield. Finally, utilizing 2,3,4,6-tetra-*O*-acetyl-D-mannopyranosyl bromide, a 68% yield was obtained with predominant approach of the radical acceptor to the α position.

Scheme 5.2.10   **Radical *C*-Glycosidations**
                          with Glycosylbromides

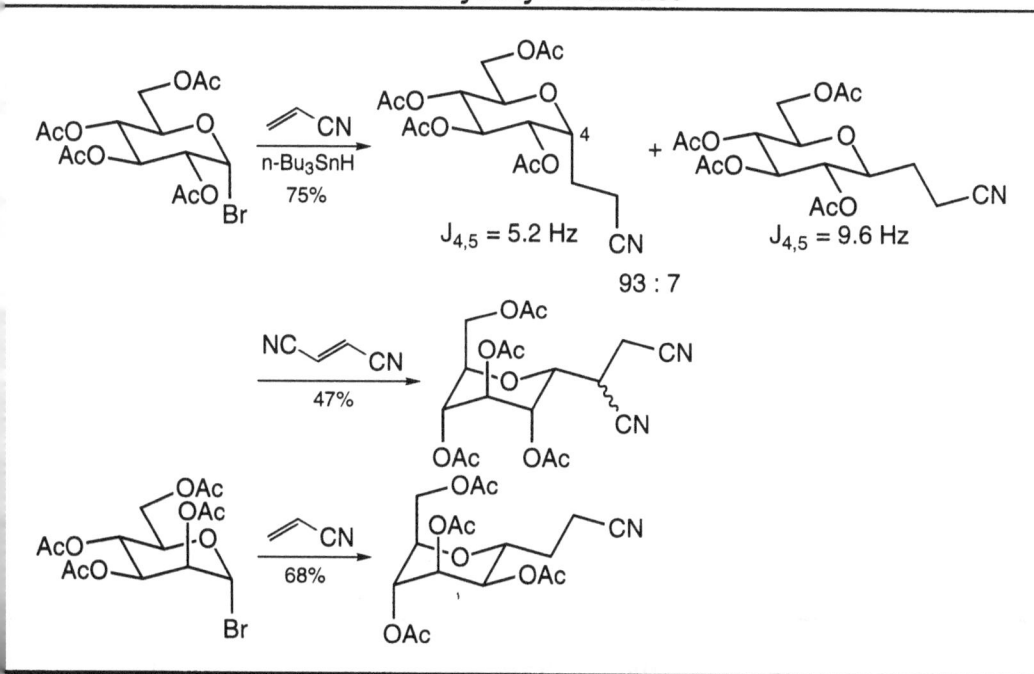

In a novel example of the conversion of xylose derivatives to more complex sugars, Blattner, *et al.*,[17] treated 1,2,3,4-tetra-*O*-acetyl-5-bromo-β-D-xylopyranose with tributyltin hydride followed by acrylonitrile. As shown in Scheme 5.2.11, the resulting product was converted to the glycosylbromide under standard conditions. This product was then subjected to radical C-glycosidation conditions similar to those previously described thus providing a 60% overall yield of the C_1, C_5 modified xylopyranose.

Scheme 5.2.11   **Radical C-Glycosidations**
                          with Glycosylbromides

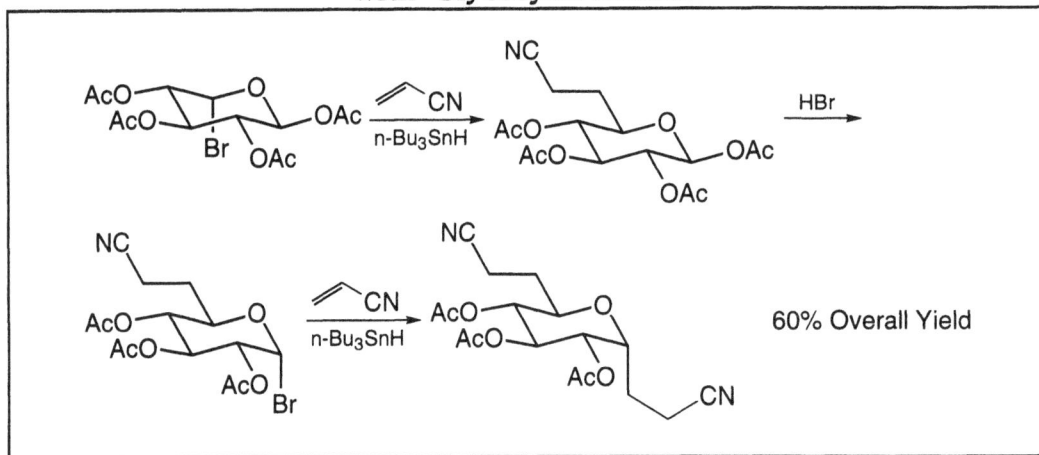

Figure 5.2.1   **Stereoview of Vitamin B_{12}**

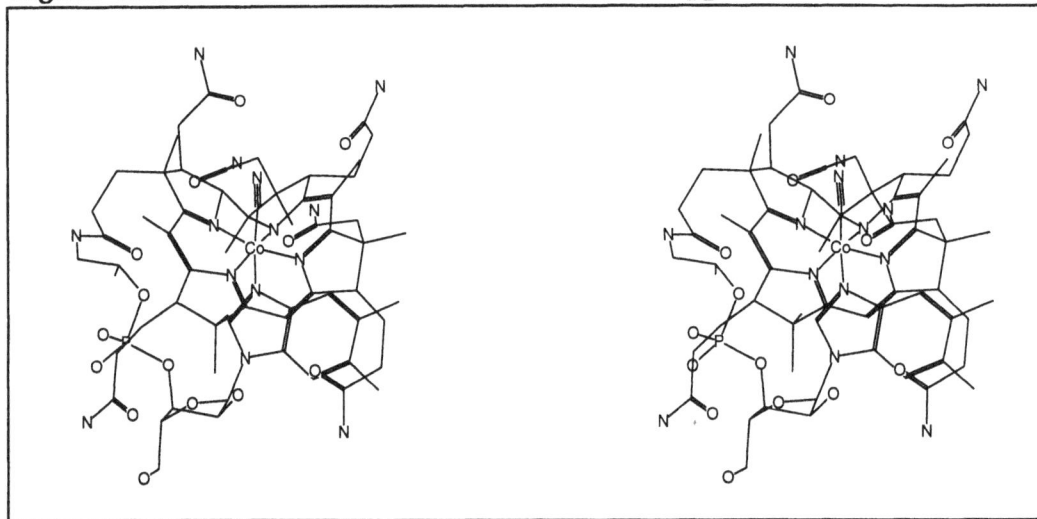

Chemical methods for the generation of glycosidic free radicals are not limited to tin hydrides or irradiation. Catalysts such as vitamin B_{12}, shown in Figure 5.2.1, have also been used. As shown in Scheme 5.2.12, Abrect, et al.,[18]

utilized vitamin B_{12} in the generation of free radicals from 2,3,4,6-tetra-*O*-acetyl-α-D-glucopyranosyl bromide. Subsequent treatment with acrylonitrile or methyl vinyl ketone provided the desired α anomeric *C*-glycosides in yields greater than 50%.

Scheme 5.2.12   **Radical *C*-Glycosidations**
                          with Glycosylbromides

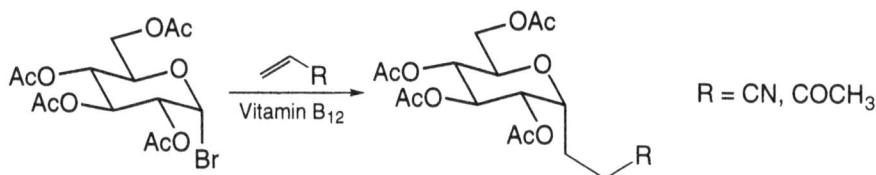

R = CN, COCH$_3$

Scheme 5.2.13   **Radical *C*-Glycosidations**
                          with Glycosylbromides

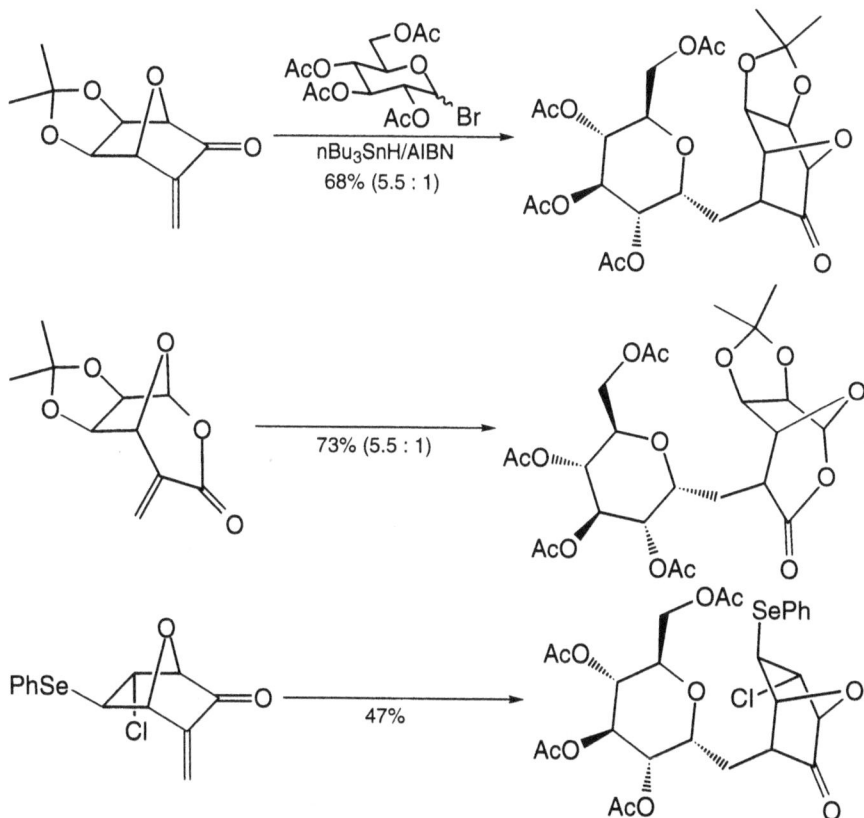

The use of methyl vinyl ketone in Scheme 5.2.12 and the previously described conjugated lactone used in Scheme 5.2.8 illustrate the utility of various radical acceptors complimentary to the commonly used acrylonitrile. In the next several examples, the use of conjugated carbonyl compounds will be addressed in the general technology surrounding the chemistry of anomeric free radicals.

Extensive exploration into the applicability of conjugated carbonyl compounds for the preparation of *C*-glycosides was carried out by Bimwala, *et al.*,[19,20] and involves both conjugated ketones and lactones. As shown in Scheme 5.2.13, the anomeric radicals were generated utilizing tin hydride reagents and the couplings, all exhibiting a preference for α anomeric selectivity, proceeded in moderate to good yields. Further elaboration of the products shown provided strategies to novel *C*-linked disaccharides and will be further discussed in Chapter 8.

Figure 5.2.2   **Radical Acceptor and Final *C*-Disaccharide**

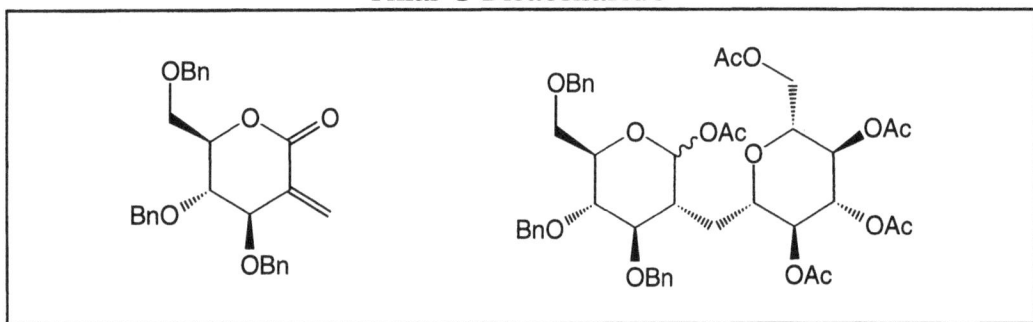

Scheme 5.2.14   **Radical *C*-Glycosidations with Glycosylbromides**

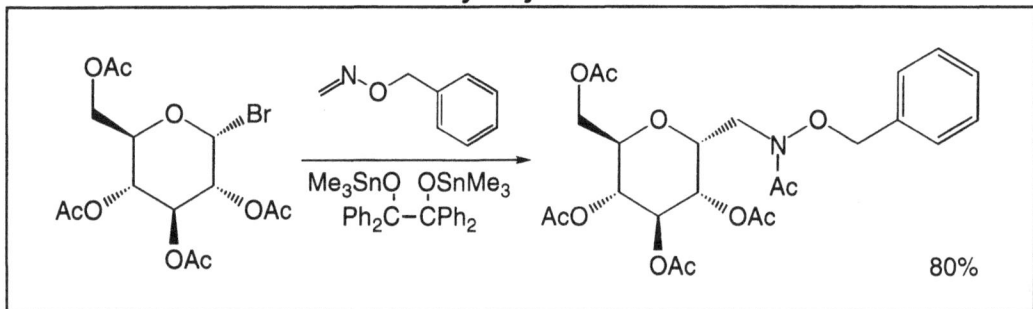

Earlier work in the area of anomeric radicals demonstrated the feasibility of directly joining two sugar units to form *C*-disaccharides. The work, described by Giese, *et al.*,[21] involved use of the modified sugar shown in Figure 5.2.2. The reported coupling, to be elaborated upon in Chapter 8, proceeded in 70% yield with high selectivity for the α anomer. Through these observations, the ability to prepare higher *C*-glycosides has been demonstrated and applied to a variety of interesting and useful targets.

While tin reagents have provided ample methods for the generation of anomeric radicals, the variety of structures useful as radical acceptors is extremely diverse. In addition to the conjugated nitriles and conjugated carbonyl compounds, oximes have also been used. As shown in Scheme 5.2.14, Hart, *et al.*,[22] effected a *C*-glycosidation utilizing an *O*-benzyl oxime and a tin reagent. The reaction proceeded in 80% yield with α anomeric selectivity.

Scheme 5.2.15   **Radical *C*-Glycosidations with Glycosylbromides/iodides**

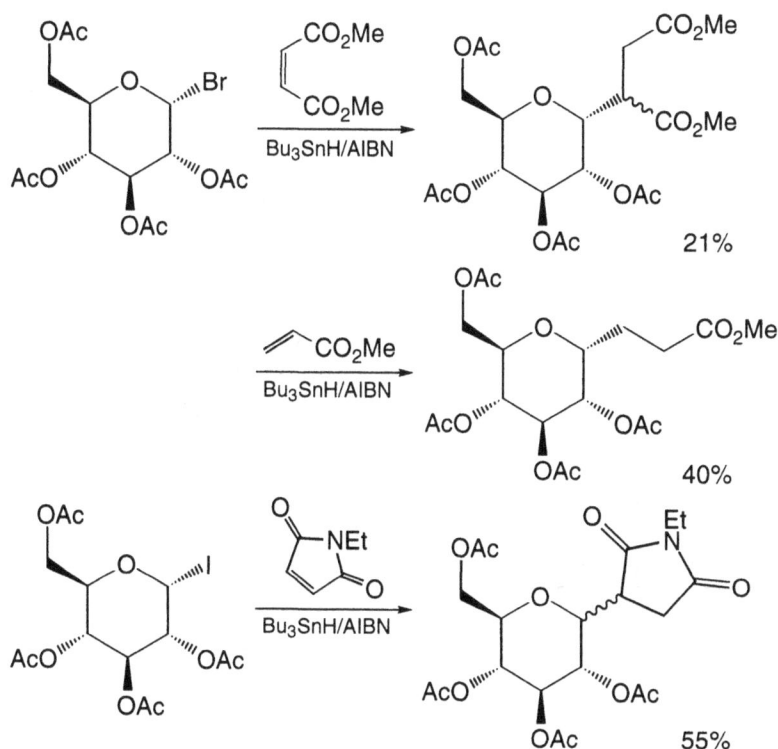

In closing this section, the use of iodides in radical mediated *C*-glycosidations is addressed. As illustrated throughout this section, radicals generated from bromides generally provide products bearing the α configuration in wide margins with respect to their yields. As shown in Scheme 5.2.15, these observations were confirmed by Araki, *et al.*[23] The radicals were generated utilizing chemical methods and the radical acceptors were conjugated ketones. However, when the corresponding glycosyliodide was used in conjunction with a photochemical radical generation, the observed yield was significantly higher with no stereochemical preference at the anomeric center.

The use of free radicals in the formation of C-glycosides has been fruitful and a wide variety of novel and versatile structures are now available. This section has served to introduce this technology with respect to intermolecular chemical reactions. Further examples illustrating the utility of this chemistry will be touched upon in later discussions surrounding the syntheses of C-di and trisaccharides.

5.3 Intramolecular Radical Reactions

Through the use of free radicals, C-glycosidic structures otherwise inaccessible are easily formed. However, the variability in yields and the frequently observed inconsistencies with respect to the degree of stereochemical induction regarding intermolecular radical reactions have revealed limitations in the general applicability of this technology. In order to alleviate some of these problems, the use of the intramolecular delivery of free radical acceptors was developed. An early example of intramolecular radical reaction is shown in Scheme 5.3.1. As illustrated, Giese, *et al.*,[24] stereospecifically formed bicyclic dideoxy sugars. The isolated yield was 53% and the stereochemical approach of the allyl group was entirely dictated by the stereochemistry at C_3. Although, as shown in Figure 5.3.1, several intermediate conformations regarding the radical species are possible, the actual intermediate radical species is postulated to be that bearing the boat conformation. The two other illustrated conformations are ruled out, as illustrated in Scheme 5.3.2, due to the inability of the 6-deoxy-6-allyl compound to cyclize under the same conditions used in Scheme 5.3.1.

Scheme 5.3.1 Intramolecular Radical C-Glycosidations

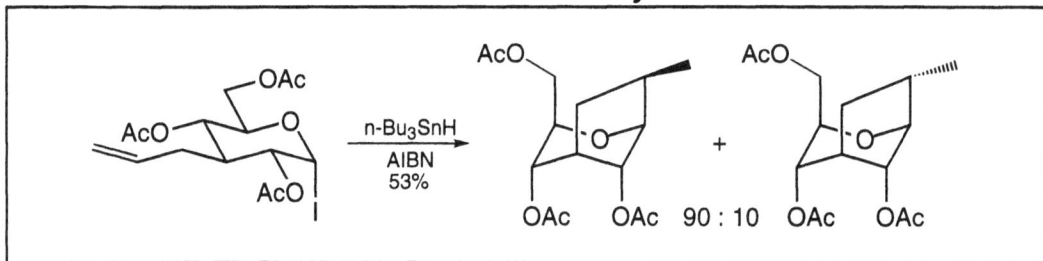

Complimentary to the direct modification of the sugar ring as a means of attaching radical receptors is the selective blocking of the sugar hydroxyl groups. As shown in Scheme 5.3.3, Mesmaekerm, *et al.*,[25] substituted the 2-hydroxyl group of glucose derivatives with allyl and trimethylsilyl propargyl groups. Subsequent treatment with tributyltin hydride and a radical initiator effected yields in excess on 90% in both cases. The stereochemistry of the cyclizations proceeded as expected with approach of the radical acceptors forming cis-fused ring systems.   Finally, unlike the previous examples presented in this section, the radicals generated in Scheme 5.3.3 were generated from phenylselenoglycosides.

Figure 5.3.1   **Proposed Conformations**
of Anomeric Radicals

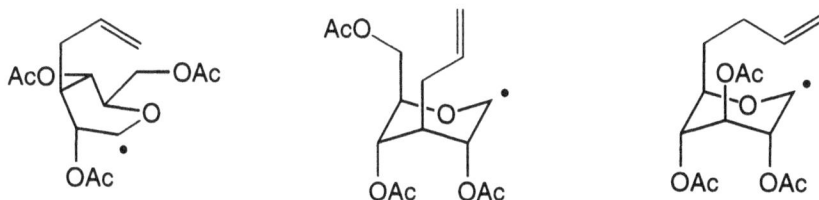

Scheme 5.3.2   **Attempted Radical Cyclization**
of 6-Deoxy-6-Allyl Sugars

A further extension of the modification of sugar hydroxy groups is the actual tethering of free radical acceptors to these groups via cleavable linkers. Convenient groups used to make these tethers are silyl groups. As shown in Scheme 5.3.4, Stork, et al.,[26] successfully introduced styrene units to the anomeric position of various sugars. This chemistry was applicable to both pyranose and furanose sugars. As with the examples presented in Scheme 5.3.3, the free radicals were generated from phenylselenoglycosides. Additionally, as expected, the stereochemical approaches of the radical acceptors were dictated by the stereochemistry of the hydroxyl group bearing the tethered substituent. Following cyclization, the bicyclic system was cleaved on removal of the silicon via treatment with potassium fluoride. In general, the products were formed in high yields with high stereoselectivity at the anomeric positions. Furthermore, the products contained one unprotected hydroxyl group making accessible further functional changes on the sugar ring.

**Scheme 5.3.3   Attempted Radical Cyclization
of 6-Deoxy-6-Allyl Sugars**

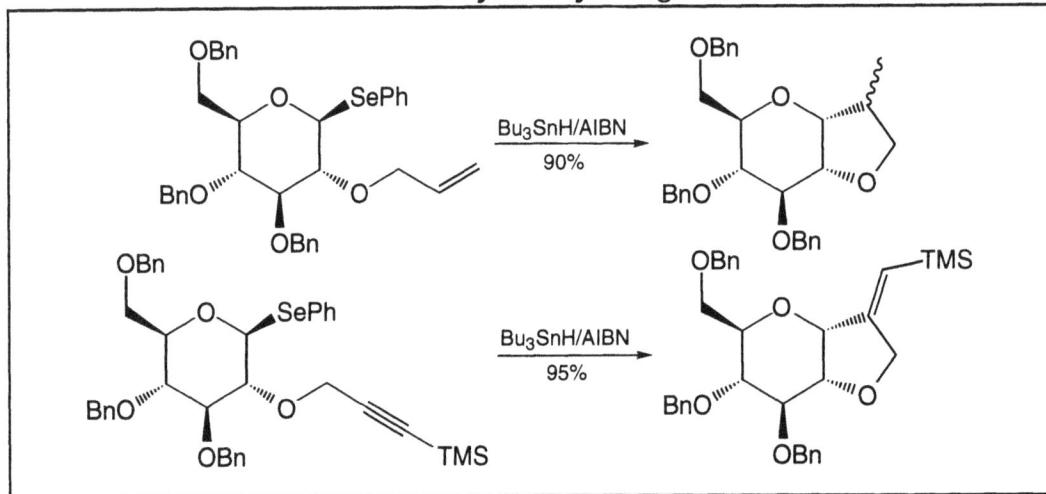

**Scheme 5.3.4   Intramolecular *C*-Glycosidations
with Tethered Radical Acceptors**

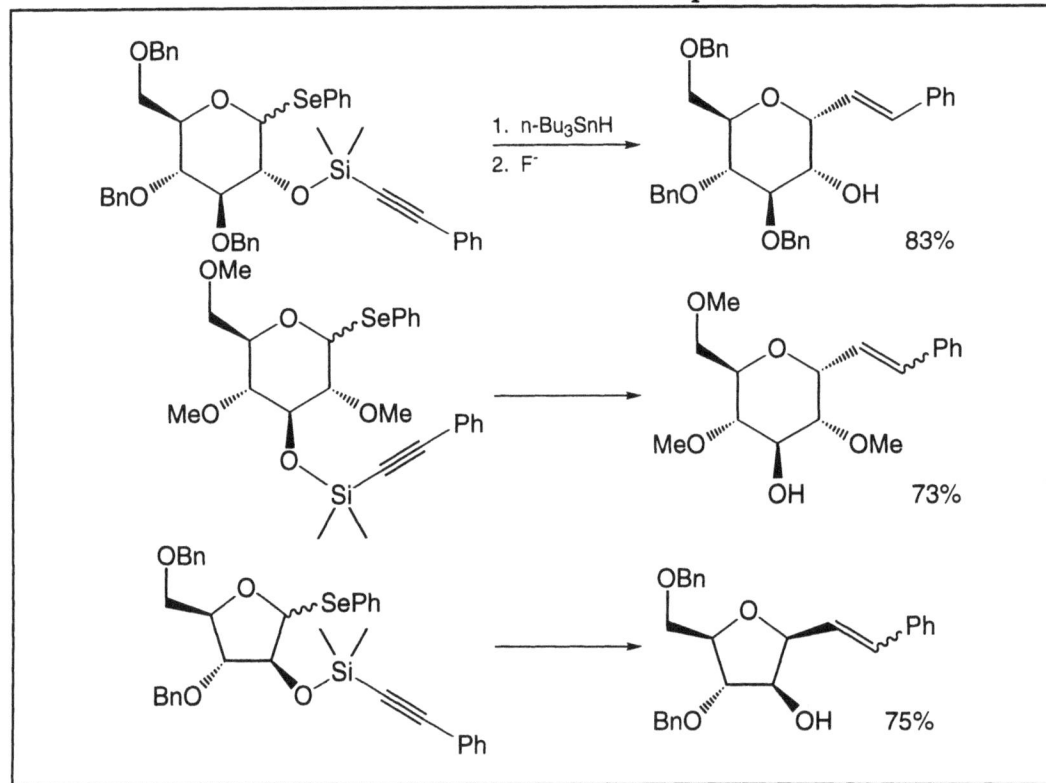

As a natural extension to the utility of silicon tethered radical acceptors for the preparation of *C*-glycosides, Xin, *et al.*,[27] applied this technology to the preparation of a *C*-maltose derivative. This example, centering around the preparation of a *C*-disaccharide, will be discussed in detail in Chapter 8.

In one final example of the utility of intramolecular delivery of free radical acceptors to the preparation of *C*-glycosides, Lee, *et al.*,[28] utilized radical chemistry to actually form the sugar ring from appropriate enol ethers. As shown in Scheme 5.3.5, the enol ethers were prepared on coupling ethyl propargylacetate with various alcohols. In all cases, the radicals were generated from tin reagents and radical initiators. The yields observed exceeded 96% and the chemistry was adapted to reactions in which radicals were generated from acetylenes. This, and other applications are contained within the cited report.

Scheme 5.3.5   Formation of *C*-Glycosides *via* Radical Cyclizations of Acyclic Precursors

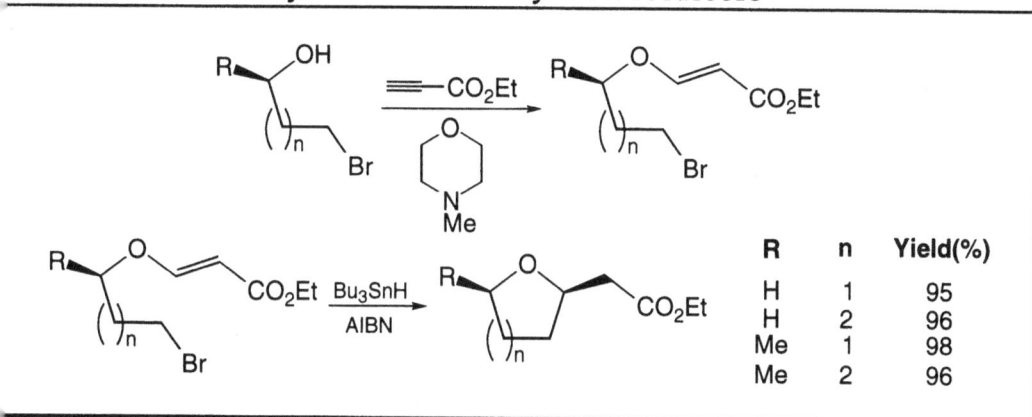

R	n	Yield(%)
H	1	95
H	2	96
Me	1	98
Me	2	96

This chapter has endeavored to present approaches to the preparation of *C*-glycosides utilizing the chemistry of free radicals. Through the use of intermolecular and intramolecular reactions, a variety of novel structures are available. Further insight into the variations available within this technology, reviewed by Descotes,[29] include discussions of the kinetics of radical systems and topics complimenting the specific realm of *C*-glycosides.

5.4   References

1.   Acbischer, B.; Vasella, A. *Helv. Chim. Acta* **1983**, *66*, 2210.
2.   Baumberger, F.; Vasella, A. *Helv. Chim. Acta* **1983**, *66*, 2210.
3.   Brakta, M.; Lhoste, P.; Sinou, D.; Banoub, J. *Carbohydrate Res.* **1990**, *203*, 148.

4. Araki, Y.; Endo, T.; Tanji, M.; Nagasawa, J. *Tetrahedron Lett.* **1988**, *29*, 351.
5. Keck, G. E.; Enholm, E. J.; Kachensky, D. F. *Tetrahedron Lett.* **1984**, *25*, 1867.
6. Waglund, T.; Claesson, A. *Acta Chem. Scand.* **1992**, *46*, 73.
7. Giese, B.; Ruckert, B.; Groninger, K. S.; Muhn, R.; Lindner, H. J. *Liebig's Ann. Chem.* **1988**, 997.
8. Somsak, L.; Batta, G.; Farkas, I. *Tetrahedron Lett.* **1986**, *27*, 5877.
9. Giese, B.; Gilges, S.; Groninger, K. S.; Lamberth, C.; Witzel, T. *Liebig's Ann. Chem.* **1988**, 615.
10. Giese, B.; Groninger, K. S. *Org. Syn.* **1990**, *69*, 66.
11. Dupuis, J.; Giese, B.; Hartung, J.; Leising, M.; Korth, H.-G.; Sustmann, R. *J. Am. Chem. Soc.* **1985**, *107*, 4332.
12. Nicolau, K. C.; Dolle, R. E.; Chucholowski, A.; Randall, J. L. *J. Chem. Soc. Chem. Comm.* **1984**, 1153.
13. Nagy, J. O.; Bednarski, M. D. *Tetrahedron Lett.* **1991**, *32*, 3953.
14. Paulsen, H.; Matschulat, P. *Liebig's Ann. Chem.* **1991**, 487.
15. Giese, B.; Witzel, T. *Angew. Chem. Int. Ed. Engl.* **1986**, *25*, 450.
16. Giese, B.; Dupuis, J.; Leising, M.; Nix. M.; Lindner, H. J. *Carbohydrate Res.* **1987**, *171*, 329.
17. Blattner, R.; Ferrier, R. J.; Renner, R. *J. Chem. Soc. Chem. Comm.* **1987**, 1007.
18. Abrecht, S.; Scheffold, R. *Chimia* **1985**, *39*, 211.
19. Bimwala, R. M.; Vogel, P. *Tetrahedron Lett.* **1991**, *32*, 1429.
20. Bimwala, R. M.; Vogel, P. *J. Org. Chem.* **1992**, *57*, 2076.
21. Giese, B.; Hoch, M.; Lamberth, C.; Schmidt, R. R. *Tetrahedron Lett.* **1988**, *29*, 1375.
22. Hart, D. J.; Seely, F. L. *J. Am. Chem. Soc.* **1988**, *110*, 1631.
23. Araki, Y.; Endo, T.; Tanji, M.; Nagasawa, J. *Tetrahedron Lett.* **1987**, *28*, 5853.
24. Groninger, K. S.; Jager, K. F.; Giese, B. *Liebigs Ann. Chem.* **1987**, 731.
25. Mesmaekerm, A. D.; Hoffmann, P.; Ernst, B.; Hug, P.; Winkler, T. *Tetrahedron Lett.* **1989**, *30*, 6307.
26. Stork, G.; Suh, H. S.; Kim, G. *J. Am. Chem. Soc.* **1991**, *113*, 7054.
27. Xin, Y. C.; Mallet, J. M.; Sinay, P. *J. Chem. Soc. Chem. Comm.* **1993**, 864.
28. Lee, E.; Tae, J. S.; Lee, C.; Park, C. M. *Tetrahedron Lett.* **1993**, *30*, 4831.
29. Descotes, G. *J. Carbohydrate Chem.* **1988**, *7*, 1.

6   Introduction

As alluded to in section 5.3, the ability to direct reactions to the anomeric centers of modified sugars provides novel routes to C-glycosides. In Chapter 5, these reactions involved the intramolecular delivery of radical acceptors to anomeric free radicals. However, as the chemistry of C-glycosides developed, the use of a variety of cycloadditions and structural rearrangements became useful.   Some substrates for these reactions, both intermolecular and intramolecular, are shown in Figure 6.0.1. By examining these structures, a variety of specific reaction types can be envisioned and recognized as applicable to the preparation of C-glycosides.   Among these are Claisen rearrangements, Wittig rearrangements, and carbenoid displacements. In this chapter, these and other reactions will discussed through examples of their use in the formation of novel C-glycosidic structures.

Figure 6.0.1   **Rearrangement and Cycloaddition Substrates**

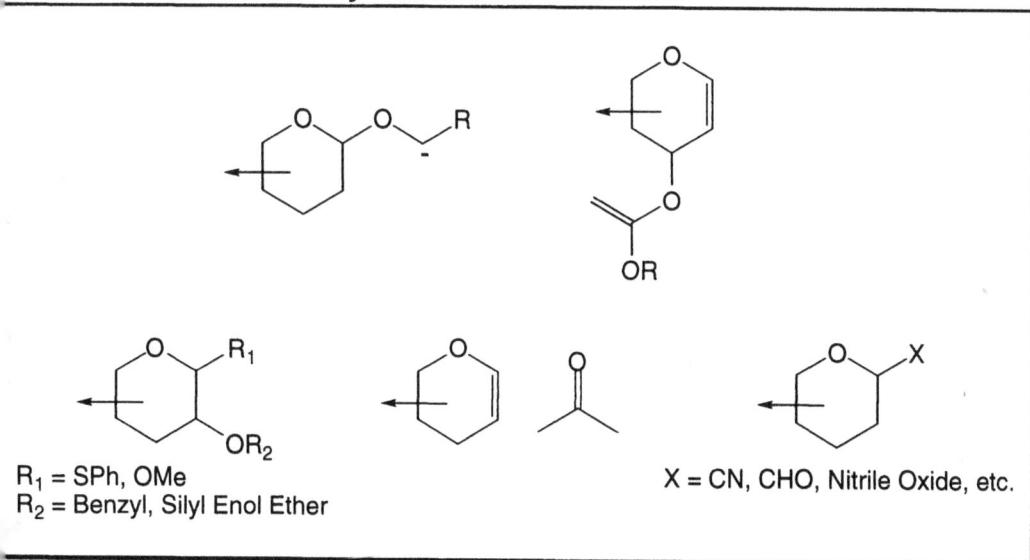

R_1 = SPh, OMe
R_2 = Benzyl, Silyl Enol Ether

X = CN, CHO, Nitrile Oxide, etc.

6.1   Rearrangements by Substituent Cleavage and Recombination

Many rearrangements involve the cleavage of substituents from a given substrate followed by recombination of the cleaved group to yield a new product.   A classical example of this type of reaction is the Wittig rearrangement. In the following paragraphs, this reaction will be presented in the context of its applicability to the preparation of *C*-glycosides. Furthermore, it will be compared to mechanistically similar rearrangements associated with carbenoids.

6.1.1   *Wittig Rearrangements*

Scheme 6.1.1   Wittig Rearrangements

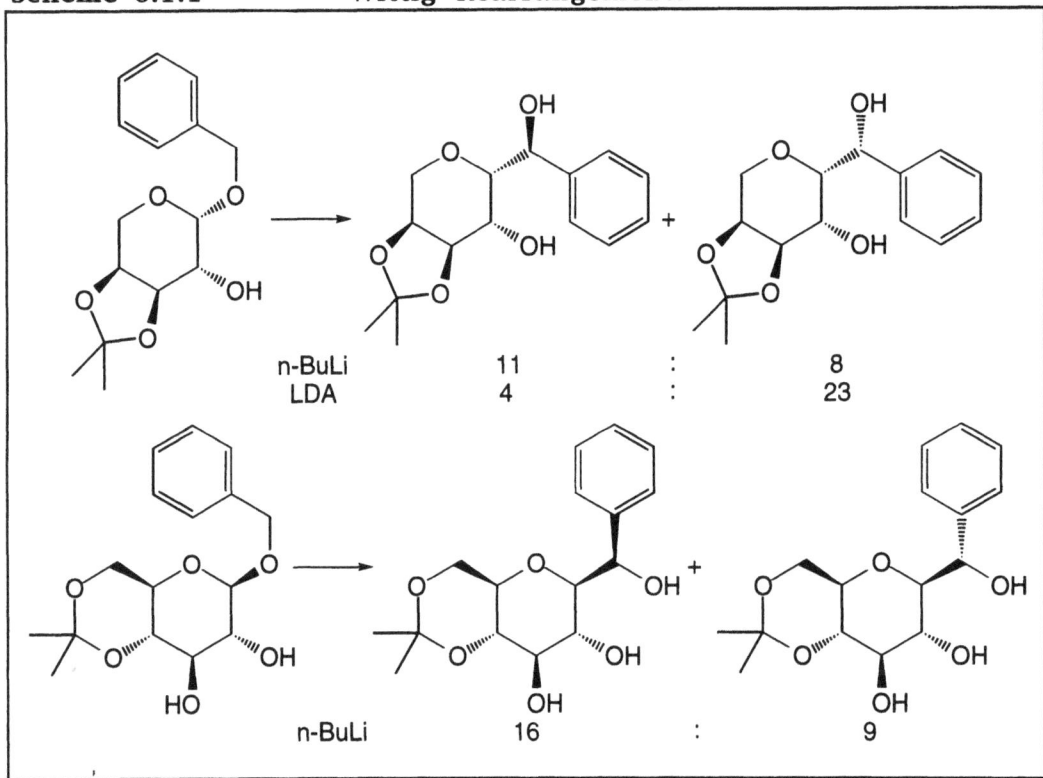

The Wittig rearrangement,[1] useful in the transformation of ethers to alcohols, has found utility in the formation of a variety of *C*-hydroxyalkyl glycosides. The actual mechanism of this reaction is believed to involve a radical cleavage followed by recombination to the final product.[2,3] As shown in Scheme 6.1.1, Grindley, *et al.*,[4] applied this methodology to the preparation of *C*-glycosides from their corresponding *O*-benzyl glycosides. Although the reactions proceeded with retention of stereochemistry at the anomeric center and inversion of the chair conformation, the configuration at the newly formed stereogenic center could not be unambiguously assigned. Regardless of this complication, the product ratio was found to be highly dependent upon the base used to initiate the reaction. For example, when *n*-butyllithium was used, the *O*-benzyl arabinoside was transformed to the corresponding *C*-glycosides with a ratio of 11 : 8. However, when LDA was used, the ratio changed to 4 : 23 favoring the other isomer.

6.1.2   *Carbenoid Rearrangements*

Scheme 6.1.2   **Carbenoid Rearrangements**

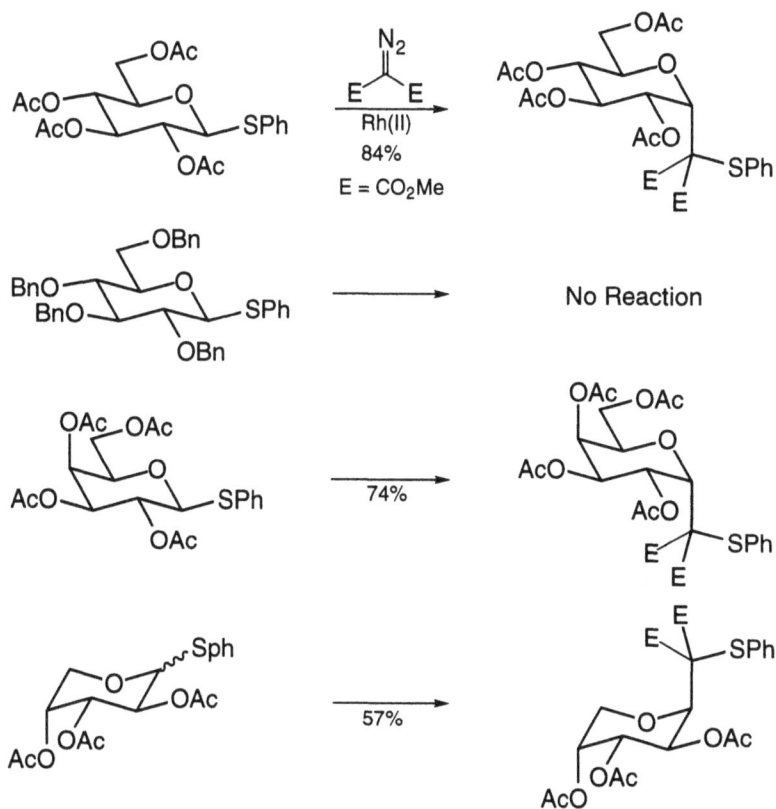

Where the Wittig rearrangement provides products formed with retention of the configuration at the anomeric center, the use of carbenoids in the formation of *C*-glycosides provides a complimentary approach allowing the preparation of products with inverted anomeric configurations. As shown in Scheme 6.1.2, Kametani, *et al.*,[5] observed the formation of a variety of *C*-glycosides in yields as high as 84%. Interestingly, these reactions proceeded well utilizing peracetylated sugars while no reaction was observed with the corresponding perbenzylated sugars. This particular observation is rationalized through the proposed reaction mechanism shown in Scheme 6.1.3. As illustrated, the carbene adds to the thiophenyl group forming an ylide-type species. Addition of the resulting stabilized anion to the neighboring acetate group is followed by cleavage of the thiophenyl group thus forming an oxonium ion. As the acetate group is regenerated, the carbene-derived substituent is stereospecifically transferred to the oxonium ion. Considering this mechanism, it can be inferred that if a thiophenyl glycoside is utilized with the thiophenyl group on the same face as the 2-acetoxy group retention of the anomeric configuration will be observed.

Though not Wittig rearrangements, the illustrated carbenoid rearrangements are mechanistically similar by virtue of the cleavage and recombination of the reactive structural components. One distinct difference between these reactions is the intramolecular nature of the carbenoid rearrangements as compared to the corresponding intermolecular mechanism observed in Wittig rearrangements. However, the similarities and the complimentarity apparent in carbenoid reactions warrants their inclusion in this section. As a final note, the utility of carbenoid reactions was subsequently illustrated in the synthesis of showdomycin[5] complimentary to the preparation illustrated in Scheme 5.2.2.[6]

Scheme 6.1.3   **Mechanism of**
Carbenoid Rearrangements

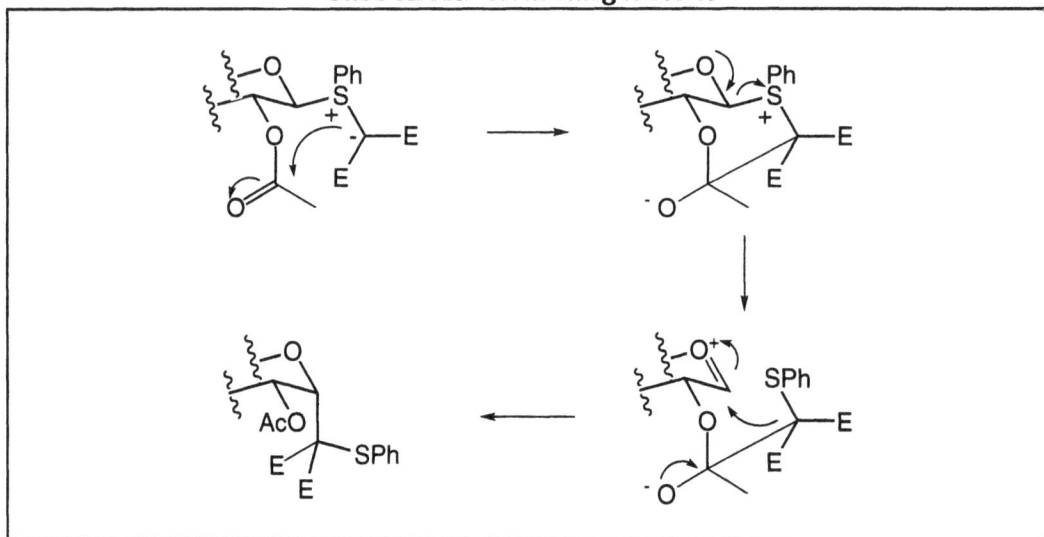

6.2 Electrocyclic Rearrangements Involving Glycals

The rearrangement of modified glycals provides novel strategies to the preparation of C-glycosides. Such reactions involve the sigmatropic mechanisms associated with both Cope and Claisen reactions. In this section, these and related reactions are presented.

6.2.1   Sigmatropic Rearrangements

Sigmatropic rearrangements of glycal derivatives are recognized as useful strategies for the preparation of C-glycosides. In fact, this methodology was used as early as 1984 when Tulshian, et al.,[7] observed the formation of hydroxymethylglycosides on treatment of galactal analogs with iodomethyltributyltin. As shown in Scheme 6.2.1, the structure of the hydroxymethyl compound was confirmed via an alternate synthesis beginning with a glucal analog.

Scheme 6.2.1   1,4-Sigmatropic Rearrangements

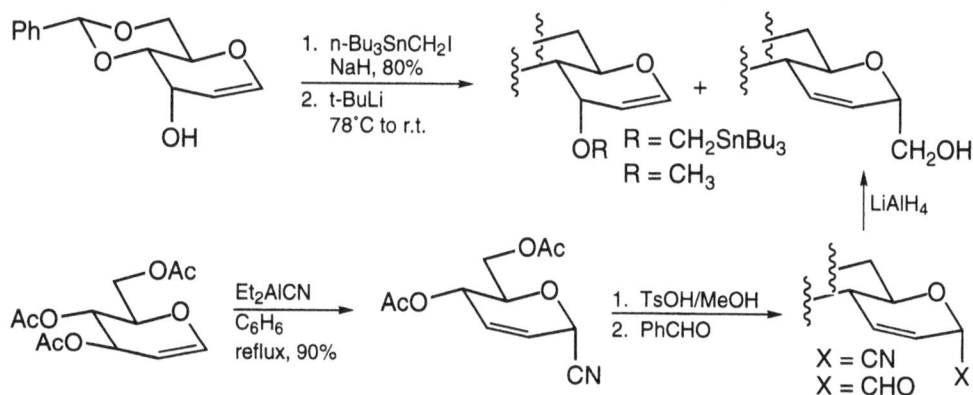

In the same report, the utility of sigmatropic rearrangements was expanded to include orthoester Claisen rearrangements and the related Cope rearrangements. The results, shown in Scheme 6.2.2, illustrate the ease of formation of C-glycosides bearing a variety of ester, aldehyde, and amide substituents. In all examples, the starting galactal was protected as the 4,6-benzylidine acetal. The reagents utilized to effect various reactions include triethyl orthoacetate, ethyl vinyl ether, and 1,1-dimethoxy-1-dimethylaminoethane.

From triethyl orthoacetate, two esters were formed. Though not C-glycosides, the chemistry presented in previous chapters provide ample strategies to the conversion of these compounds to other products. From ethyl vinyl ether, the actual isolated product was the C-glycosidic aldehyde formed in 75% yield. Finally, from 1,1-dimethoxy-1-dimethylaminoethane, the C-glycosidic amide was formed in 85% yield. As illustrated, these products were all easily interconverted as well as derivatized to a variety of other useful products.

Scheme 6.2.2   3,3-Sigmatropic Rearrangements

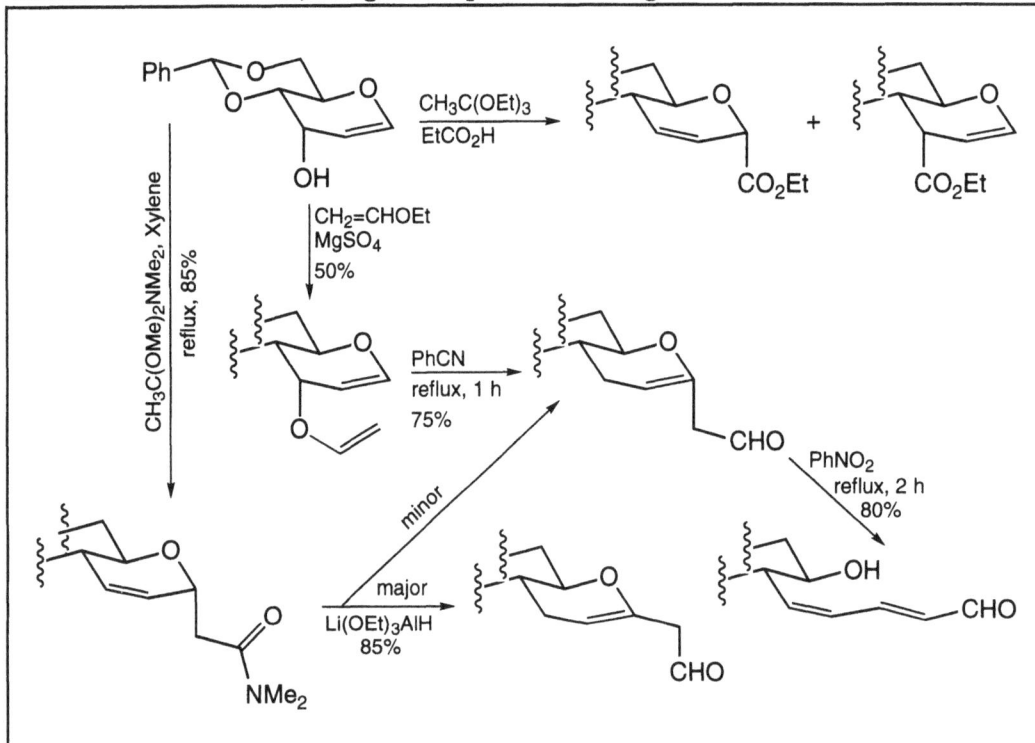

6.2.2   Claisen Rearrangements

Drawing specifically on Ireland's ester enolate Claisen rearrangements,[8-10] Curran, et al.,[11] demonstrated the generality of this methodology in the conversion of a variety of acetoxyglycals to their corresponding methylcarboxymethylglycosides. Not shown, the acetoxyglycals were treated with LDA followed by tert-butyldimethylchlorosilane thus producing near quantitative yields of the ketenesilyl acetals illustrated in Scheme 6.2.3. As shown, the ketenesilyl acetals were converted to the C-glycosides on heating. Notably, all reactions proceeded in yields exceeding 55% with less than 5% of the double Claisen products observed. The stereochemical outcome of these reactions appeared to be influenced by the stereochemical configurations at C_3 coupled with vinyligous anomeric effects.[12]

Scheme 6.2.3   Claisen Rearrangements

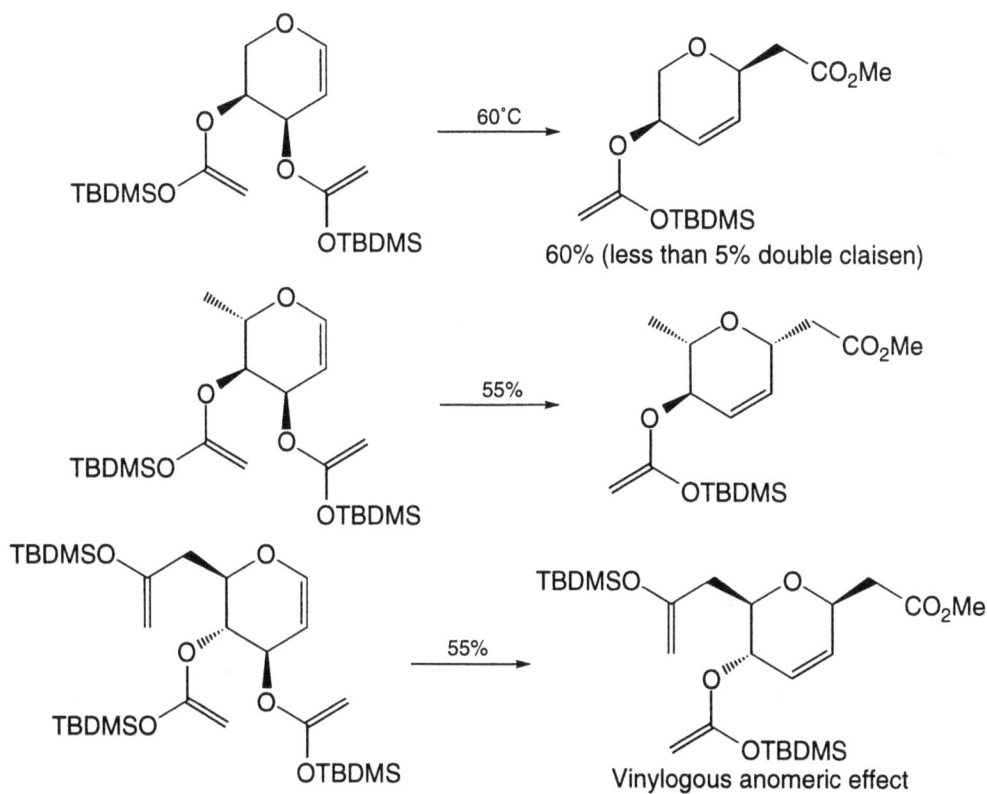

60°C

60% (less than 5% double claisen)

55%

55%

Vinylogous anomeric effect

A final example of the utility of sigmatropic rearrangements in the preparation of *C*-glycosides is shown in Scheme 6.2.4. As illustrated, Colombo, *et al.*,[13] effected Claisen-type rearrangements utilizing heterocycles as the migrating groups. As with the previous examples mentioned in this chapter, this chemistry proceeded with the stereochemistry at the anomeric position dependent upon the stereochemistry at C_3 of the glycal. The high anomeric stereospecificity of this chemistry coupled with the high yields demonstrates the tremendous versatility available within the application of sigmatropic rearrangements for the preparation of *C*-glycosides.

Scheme 6.2.4   Claisen-Type Rearrangements

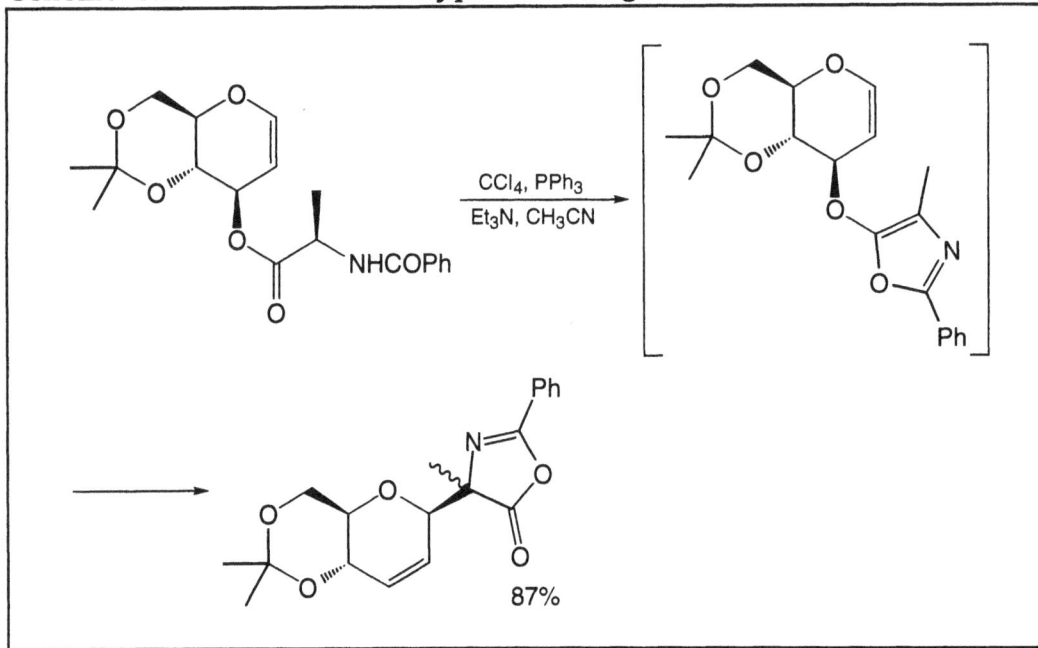

87%

6.3   Rearrangements from the 2-Hydroxyl Group

Where Wittig rearrangements have provided *C*-glycosides *via* rearrangements from anomeric hydroxyl groups and Claisen rearrangements have provided *C*-glycosides *via* migrations from 3-hydroxyl groups, the delivery of substituents from 2-hydroxyl groups have yielded novel *C*-glycosidic structures. Two examples will be illustrated in this section.

The first example, shown in Scheme 6.3.1, was reported by Craig, *et al.*,[14] and previously illustrated in Chapter 2. The reiteration of this example is to show the types of strained fused-ring systems available from the delivery of nucleophilic species to adjacent centers.

The second example, shown in Scheme 6.3.2, was reported by Martin, *et al.*,[15] and augments the discussions surrounding Schemes 2.4.16 through 2.4.19. The examples illustrated fit more within the profile of rearrangements than the example in Scheme 6.3.1 in that following delivery of the benzyl groups to the anomeric centers, cleavage of the benzyl ether can be accomplished utilizing hydrogenation techniques. The net result is a two step rearrangement of a 2-benzylated sugar to a true *C*-glycoside. While direct rearrangements from 2-hydroxyl groups to anomeric centers have not been well documented, the examples presented in this section provide strategies complimentary to the true rearrangements presented thus far.

Scheme 6.3.1   **Rearrangements from 2-Hydroxyl Groups**

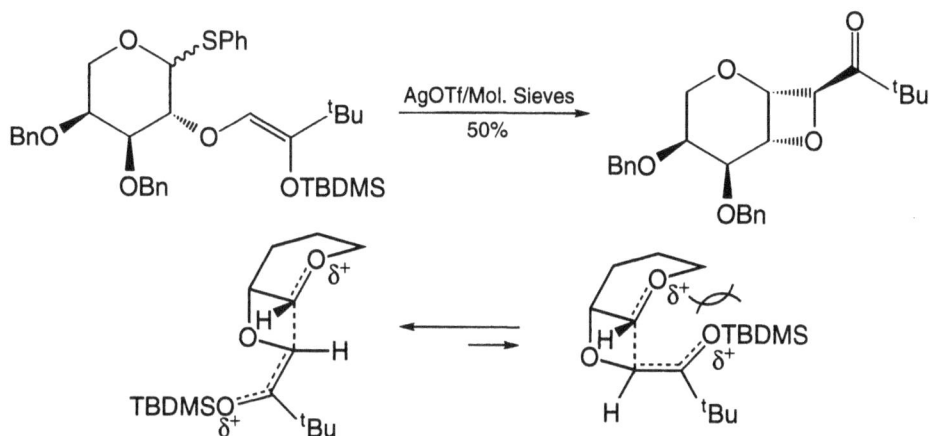

Scheme 6.3.2   **Rearrangements from 2-Hydroxyl Groups**

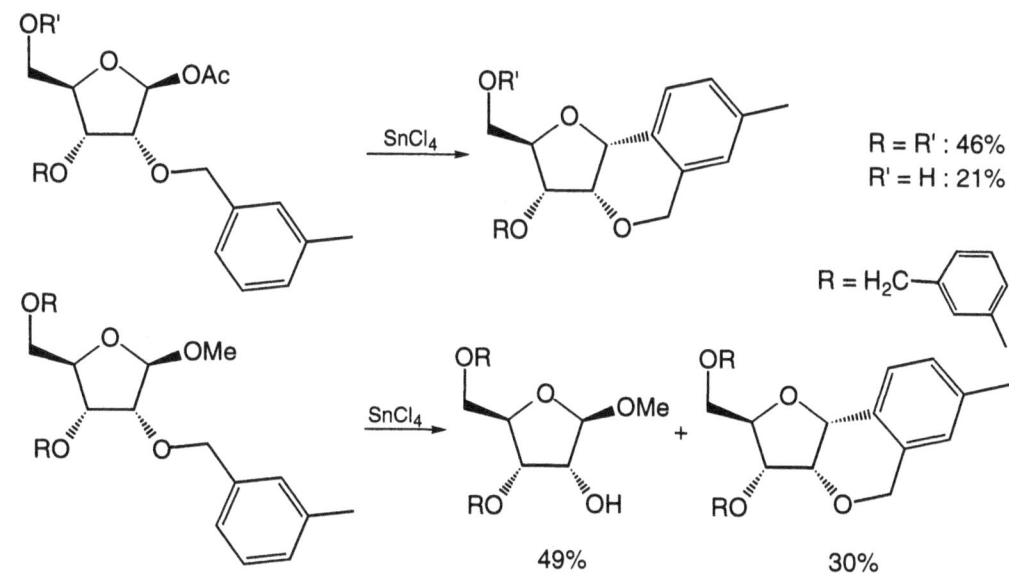

R = R' : 46%
R' = H : 21%

R = H₂C—

49%   30%

6.4 Bimolecular Cycloadditions

Complimentary to the intramolecular rearrangements discussed thus far are the Lewis acid mediated intermolecular reactions of the types discussed in Chapter 2. As relevant examples are extensive in the literature and many have already been presented, the reader is referred to the following schemes: 2.3.16; 2.3.17; 2.3.18; 2.3.38; 2.3.39; 2.3.40; 2.3.41. Through the reexamination of these examples, the mechanistic course of these reactions are realized through comparisons with ene and Prins-type mechanisms. Furthermore, although the reactions differ from the examples in this chapter involving intramolecular mechanisms, high stereoselectivity is observed with a propensity for the formation of α anomeric *C*-glycosides.

6.5 Manipulations of *C*-Glycosides

Thus far, this book has addressed a wide variety of methods useful for the formation of *C*-glycosides bearing different functional groups at the anomeric center. In context with the discussions set forth in this chapter, these functional groups are subject to further modification by a variety of cycloaddition and rearrangement reactions. Though these reactions are being illustrated on pre-formed *C*-glycosides, it is suitable to close this chapter with examples illustrating the ease of several transformations available for the conversion of simple *C*-glycosides to more complex entities. The examples presented in this section will include transformations involving glycosidic aldehydes and glycosidic nitrile oxides.

Scheme 6.5.1   **Nitrile Transformations**

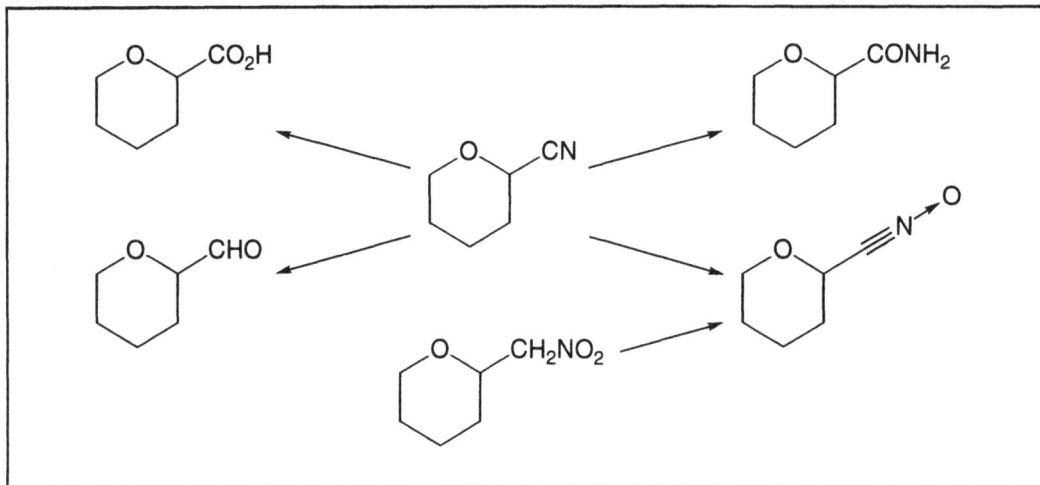

Within this section, we can recognize *C*-glycosidic nitriles, available as set forth in Chapter 2, as important substrates for further modification. As shown in Scheme 6.5.1, these compounds are easily transformed to a variety of different groups including aldehydes[16] and nitrile oxides.[17] Additionally, nitrile oxides are also available from *C*-nitromethylglycosides[18] prepared as discussed in Chapter 2. The discussions set forth in this chapter's final section center around Knoevenagel condensations and 1,3-dipolar cycloadditions.

6.5.1   *Knoevenagel Condensations*

Scheme 6.5.2   Knoevenagel Condensations

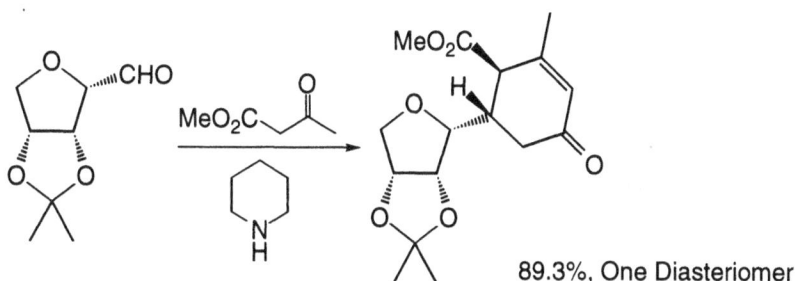

89.3%, One Diasteriomer

Scheme 6.5.3   Knoevenagel Mechanism

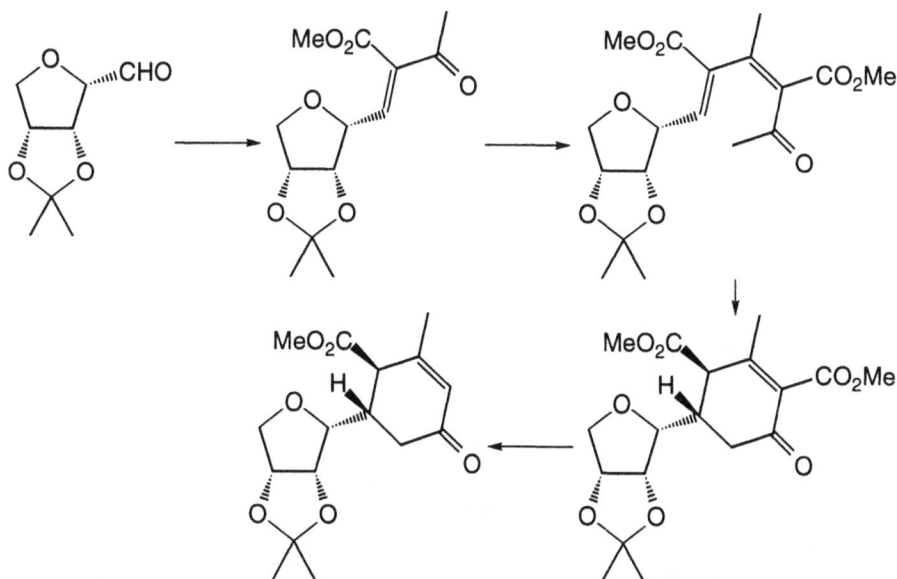

Beginning with *C*-glycosidic aldehydes, the Knoevenagel condensation has provided interesting double ring systems suitable for further elaboration. As shown in Scheme 6.5.2, Herrera, *et al.*,[19] treated the arabinose derivative, shown, with methyl acetoacetate to give the illustrated product in approximately 90% yield as a single diasteriomer. The mechanism for this reaction is shown in Scheme 6.5.3 and involves the initial condensation of the aldehyde with methyl acetoacetate. Following the first condensation, a second molecule of methyl acetoacetate condenses with the ketone and cyclizes *via* a Michael-type addition. Spontaneous decarboxylation liberates the observed product.

6.5.2   *1,3-Dipolar Cycloadditions*

With respect to 1,3-dipolar cycloadditions, several approaches have been utilized in the preparation of *C*-glycosides. As shown in Scheme 6.5.4, Kozikowski, *et al.*,[18] effected the reaction of the ribose-derived nitrile oxide with the olefin, shown. The reaction provided a 70% yield of the isoxazoline glycoside with no deleterious effects on the configuration at the anomeric center.

Scheme 6.5.4   1,3-Dipolar Cycloadditions

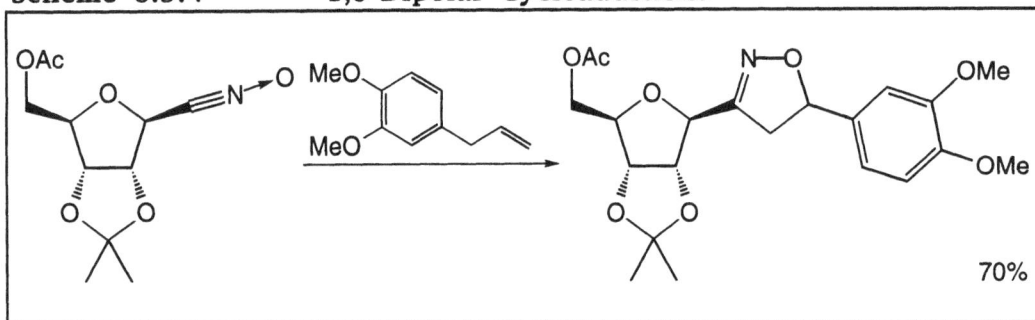

In addition to aliphatic olefins, cyclic olefins have also been utilized in the preparation of *C*-glycosides *via* 1,3-dipolar cycloadditions. As shown in Scheme 6.5.5, Dawson, *et al.*,[20] utilized a 2,3-anhydro glucose analog in the reaction with the sugar-derived nitrile oxide. Again, the observed product was an isoxazoline ring separating two additional ring systems. Unlike the previous example, the oxazoline was fused to the ring previously bearing the double bond. Consistent with the observations of the previous example, this reaction proceeded in approximately 66% yield.

In a final example demonstrating the various types of olefins useful in 1,3-dipolar cycloadditions, RajanBabu, *et al.*,[21] utilized sugar derivatives bearing exoanomeric methylenes. The specific reaction, previously shown in Scheme 2.13.2, is specifically illustrated in Scheme 6.5.6 in order to compare the nature of the observed product with the products of the previous two examples. As shown, the reaction proceeded in 78% yield with the isoxazoline forming a spiro ring system with the methylene bearing glucose derivative.   Thus, the incorporation of 1,3-dipolar cycloadditions into the chemistry surrounding the formation and modification of C-glycosides provides novel heterocyclic structures suitable for isolation or further elaboration.

Scheme 6.5.5   **1,3-Dipolar Cycloadditions**

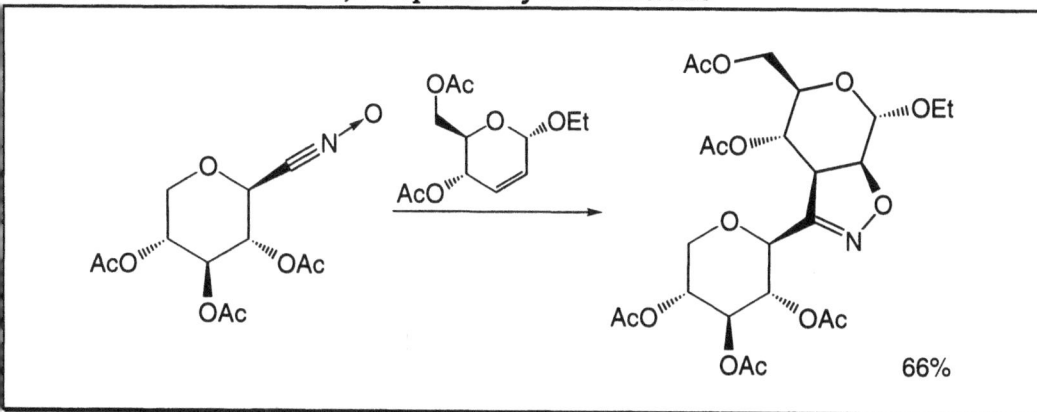

Scheme 6.5.6   **1,3-Dipolar Cycloadditions**

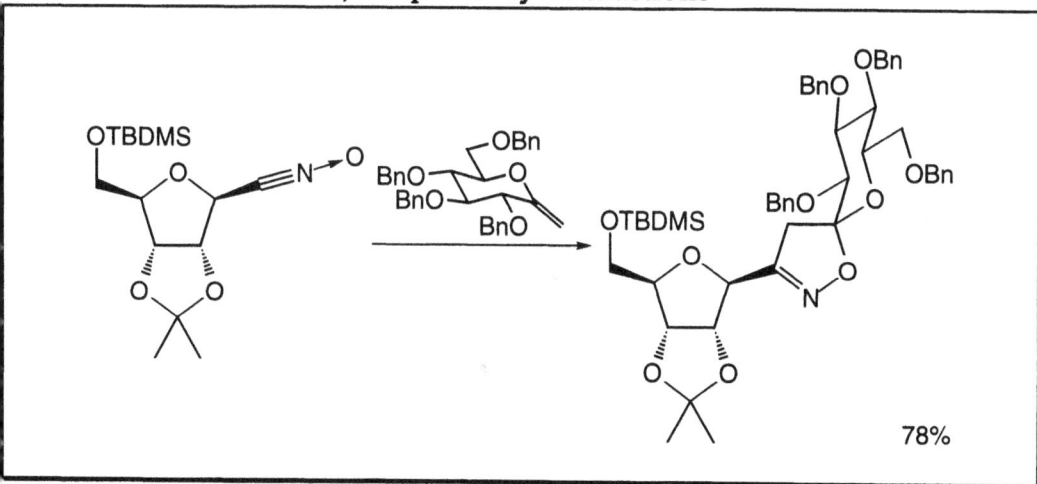

This chapter has not only centered around the formation of C-glycosides *via* rearrangements and cycloadditions, it has also introduced methods used solely for the modification of previously formed C-glycosides. Furthermore, substantial reviews of examples from previous chapters have been presented. In light of this, it is important to note that as discussions surrounding the chemistry of C-glycosides advance, the interdependence of the various reactions becomes more notable. This will become more prominent in the final two chapters where ring cyclizations and methods for the preparation of *C*-di and trisaccharides are discussed.

6.6 References

1.   Schollkof, U. *Angew. Chem. Int. Ed. Eng.* **1970**, *9*, 763.
2.   Lansbury, P.; Pattison, V. A.; Sidler, J. D.; Bieber, J. B. *J. Am. Chem. Soc.* **1966**, *88*, 78.
3.   Hebert, E.; Welvart, Z.; Ghelfenstein, M.; Swarc, H. *Tetrahedron Lett.* **1983**, *24*, 1381.
4.   Grindley, T. B.; Wickramage, C. *J. Carbohydrate Chem.*, **1988**, *7*, 661.
5.   Kametani, T.; Kawamura, K.; Honda, T. *J. Am. Chem. Soc.*, **1987**, *109*, 3010.
6.   Araki, Y.; Endo, T.; Tanji, M.; Nagasawa, J. *Tetrahedron Lett.* **1988**, *29*, 351.
7.   Tulshian, D. B.; Fraser-Reid, B. *J. Org. Chem.*, **1984**, *49*, 518.
8.   Ireland, R. E.; Thaisrivongs, S.; Varner, N. R.; Wilcox, C. S. *J. Org. Chem.* **1980**, *45*, 48.
9.   Ireland, R. E.; Mueller, R. H.; Willard, A. K. *J. Am. Chem. Soc.* **1976**, *98*, 2868.
10.   Ireland, R. E.; Wilcox, C. S. *Tetrahedron Lett.* **1977**, 2839.
11.   Curran, D. P.; Suh, Y. G. *Carbohydrate Res.*, **1987**, *171*, 161.
12.   Denmark, S. E.; Dappen, M. S. *J. Org. Chem.* **1984**, *49*, 798.
13.   Colombo, L.; Casiraghi, G.; Pittalis, A. *J. Org. Chem.* **1991**, *56*, 3897.
14.   Craig, D.; Munasinghem, V. R. N. *J. Chem. Soc. Chem. Comm.* **1993**, 901.
15.   Martin, O. R. *Carbohydrate Res.* **1987**, *171*, 211.
16.   Rabinovitz in Rappoport, Z. (ed), "The Chemistry of the Cyano Group," Wiley, New York, 1970, 307-340.
17.   Torssell, K. B. G. (ed), "Nitrile Oxides, Nitrones, and Nitronates in Organic Synthesis," VCH, New York, 1988.
18.   Kozikowski, A. P.; Cheng, X. M. *J. Chem. Soc. Chem. Comm.* **1987**, 680.
19.   Herrera, F. J. L.; Fernandez, M. V.; Segura, R. G. *Carbohydrate Res.* **1984**, *127*, 217.
20.   Dawson, I. M.; Johnson, T.; Paton, R. M.; Rennie, R. A. C. *J. Chem. Soc. Chem. Comm.* **1988**, 1339.
21.   RajanBabu, T. V.; Reddy, G. S. *J. Org. Chem.* **1986**, *51*, 5458.

7   Introduction

Each of the chapters thus far have dealt with methods used in the conversion of natural sugars, and derivatives thereof, to C-glycosides. Deviating from this trend, this chapter addresses the formation of C-glycosides from non-carbohydrate substrates. Specifically, the formation of the sugar ring is the focus of the chemistry discussed herein. Several methods for the execution of this approach are presented in the following schemes and include cyclizations of alcohols to olefins, alcohols participating in S_N2 displacements, and cycloaddition reactions. The first of these examples, illustrated in Scheme 7.0.1, involves olefination of the ring-opened sugar followed by cyclization to the previously formed double bond. This addition is usually facilitated by the use of halogens or by converting the double bond to an epoxide thus creating more electrophilic species. The second approach mentioned, shown in Scheme 7.0.2, compliments the first through the cyclization of hydroxy halides or hydroxy epoxides.

Scheme 7.0.1   Olefination of Ring-Opened Sugars

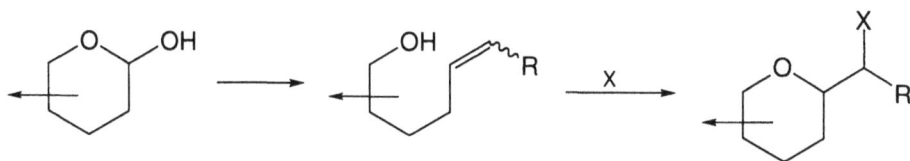

Scheme 7.0.2   Hydroxy Halide/Epoxide Cyclizations

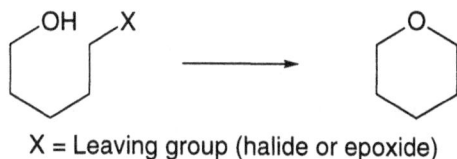

X = Leaving group (halide or epoxide)

With respect to the final approach mentioned above, several types of cycloadditions are applicable to the preparation of *C*-glycosides. One particular example, the hetero Diels-Alder reaction, is illustrated in Scheme 7.0.3. Through this method, some versatility is added through the resulting site of unsaturation inadvertently formed. Through expansion of, and the introduction of chemistry complimentary to, the methodology presented in these introductory schemes, this chapter presents specific examples in which the formation of the sugar ring is the key step involved in the preparation of *C*-glycosides.

Scheme 7.0.3   Hetero Diels-Alder Reactions

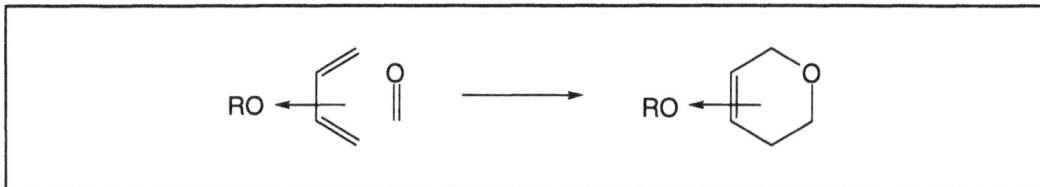

RO   ⟶   RO

7.1   Wittig Reactions of Lactols Followed by Ring Closures

The Wittig reaction is a valuable method for the formation of olefins from aldehydes and ketones. In the case of carbohydrates, the aldehyde is masked in the form of a hemiacetal at the reducing end of a sugar or polysaccharide. However, due to the equilibrium between the hemiacetal and ring opened form of sugars, the Wittig reaction can drive the equilibrium entirely to the ring opened form with the final product being a newly formed olefin. Once prepared, these hydroxyolefins can be cyclized under a variety of conditions allowing the formation of *C*-glycosides.

7.1.1   *Base Mediated Cyclizations*

As Michael-type reactions are among the most useful of base-mediated cyclizations, the first examples presented in this chapter center around this chemistry. Thus, utilizing Wittig methodology, Vyplel, *et al.*,[1] prepared the hydroxyolefin, shown in Scheme 7.1.1, from the starting protected glucosamine. treatment of this intermediate species with DBU effected cyclization of the 5-hydroxy group to the unsaturated ester in a Michael fashion. The resulting product was ultimately used in the preparation of *C*-glycosidic analogs of lipid A and lipid X initially presented in the introductory chapter. Notably, the cyclization provided the α anomer in 43% yield with an additional 7% accounted for in the isolation of the β anomer.

Scheme 7.1.1   **Wittig Reactions Followed**
by Base Cyclizations

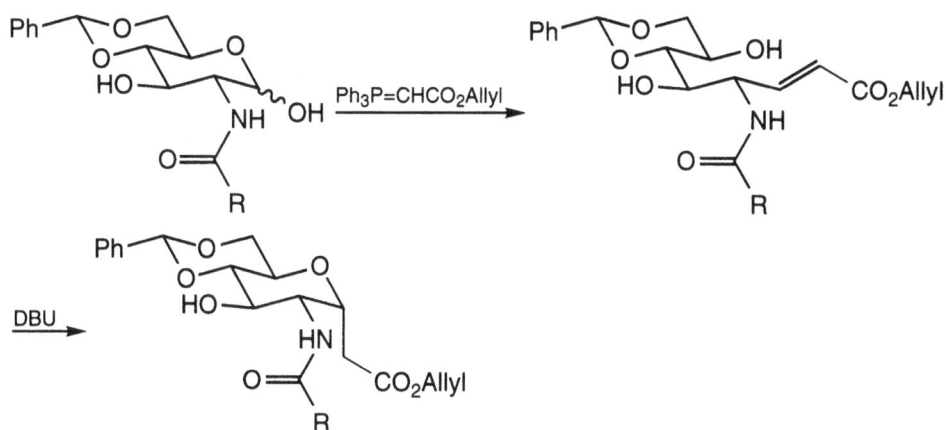

7.1.2   *Halogen Mediated Cyclizations*

Scheme 7.1.2   **Wittig Reactions Followed**
by Bromo Cyclizations

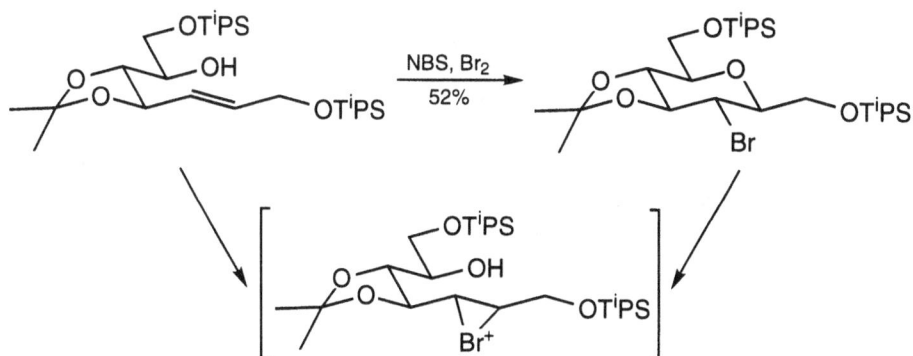

Unlike base mediated cyclizations, halocyclizations do not depend on the presence of Michael acceptors. As shown in Scheme 7.1.2, Armstrong, *et al.*,[2] treated a hydroxyolefin (derived from a Wittig reaction applied to a protected arabinose) with *N*-bromosuccinimide and catalytic bromine to effect the illustrated cyclization. The reaction proceeded through an intermediate bromonium ion providing a 52% yield of the β anomer. This observation compliments the preference for α anomers observed in base mediated cyclizations. Furthermore, the isolation of a bromine substituted *C*-glycoside provides opportunities for the preparation of new and novel structures.

Scheme 7.1.3   **Wittig Reactions Followed by Iodocyclizations**

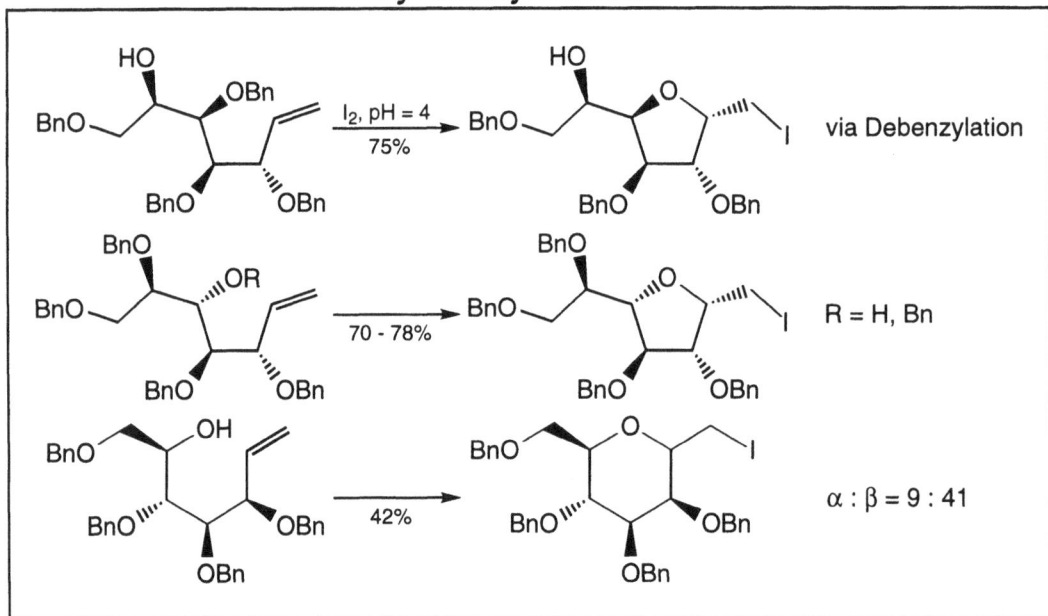

The use of bromine mediated cyclization reactions is nicely complimented by iodocyclizations. The mechanism of these reactions proceeds through a similar iodonium ion and the products all bear potential iodide leaving groups. However, as illustrated by the products shown in Scheme 7.1.3, the iodides are all primary and aliphatic suggesting a preference for approach of the hydroxy group from the opposite side as that observed in the case of bromocyclizations. The examples presented in Scheme 7.1.3 were reported by Nicotra, *et al.*,[3] and demonstrate the utility of these reactions as emphasized by yields as high as 78%. Furthermore, where choices were available, the reactions proceeded with a preference for the formation of five membered rings.

7.1.3 *Metal Mediated Cyclizations*

Methodologies that compliment the halogen mediated cyclizations are found within the chemistry surrounding mercury and selenium. Beginning with mercury, oxymercurations provide access to cyclic ethers *via* the coordination of a mercury reagent to olefins. The resulting products, as those isolated from halogen mediated cyclizations, generally bear mercury substituents which are easily modified or replaced under a variety of conditions. As shown in Scheme

7.1.4, Pougny, *et al.*,[4] utilized an oxymercuration in the cyclization of a hydroxyolefin resulting from a Wittig reaction applied to 2,3,4,6-tetra-*O*-benzyl-D-glucopyranose.   Subsequent cleavage of the mercury was accomplished under reductive conditions yielding a *C*-methylglycoside. Furthermore, oxidative conditions yielded the corresponding hydroxymethyl derivative.

Scheme 7.1.4   **Wittig Reactions Followed**
                          by Oxymercurations

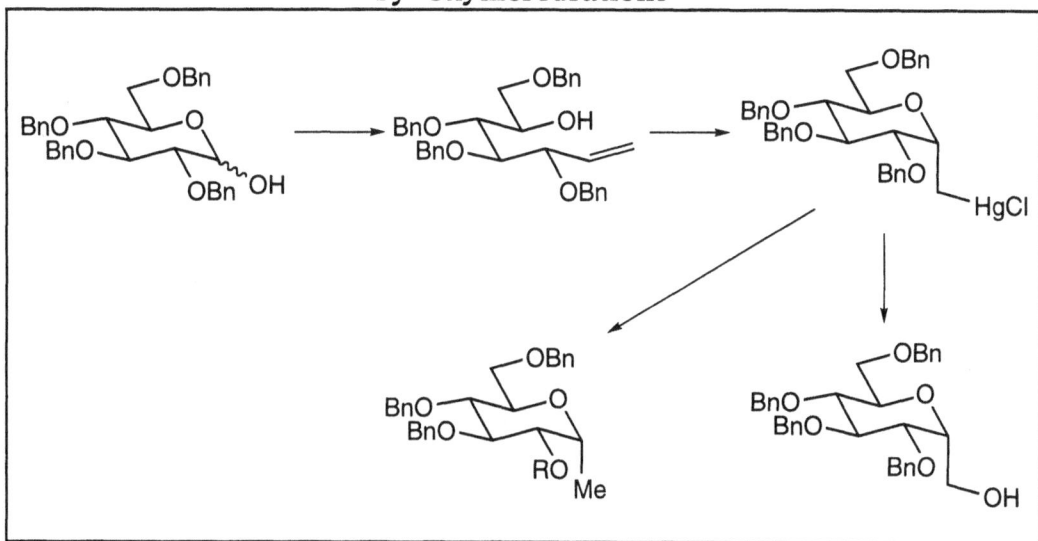

Scheme 7.1.5   **Wittig Reactions Followed**
                          by Oxymercurations

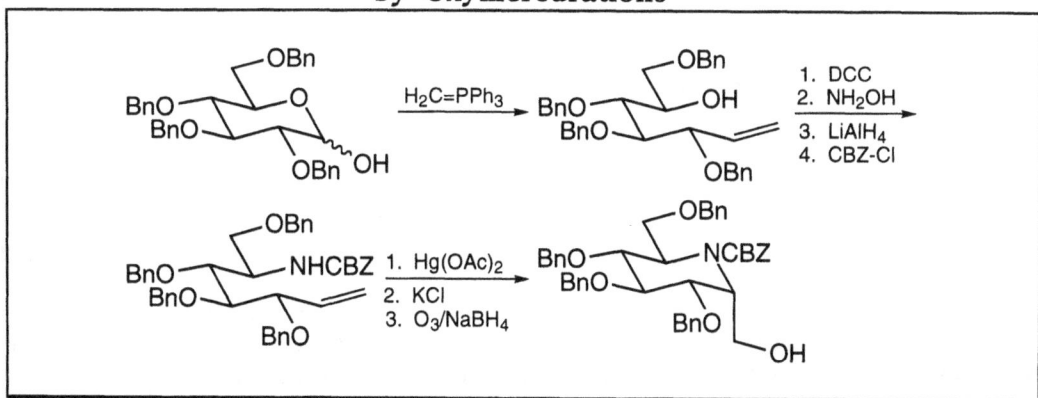

A reaction sequence similar to that just described was subsequently applied to the preparation of piperidinyl sugars. As shown in Scheme 7.1.5, Liu, *et al.*,[5]   treated   2,3,4,6-tetra-*O*-benzyl-D-glucopyranose   with methylenetriphenylphosphorane and the resulting hydroxyolefin was converted to the *N*-CBZ amino derivative in four steps. Subsequent cyclization was accomplished on treatment with mercuric acetate followed by potassium chloride. Finally, ozonolysis liberated the final alcohol. This product was ultimately utilized in the preparation of α-glucosidase inhibitors.

In a final example specifically devoted to the use of mercury reagents following Wittig olefinations, Qiao, *et al.*,[6] utilized mercuric trifluoroacetate to effect ring closure. This example, shown in Scheme 7.1.6, was utilized in the synthesis of the *C*-phosphonate disaccharide, shown in Figure 7.1.1, as a potential inhibitor of peptidoglycan polymerization by transglycosylase.

Scheme 7.1.6   **Wittig Reactions Followed by Oxymercurations**

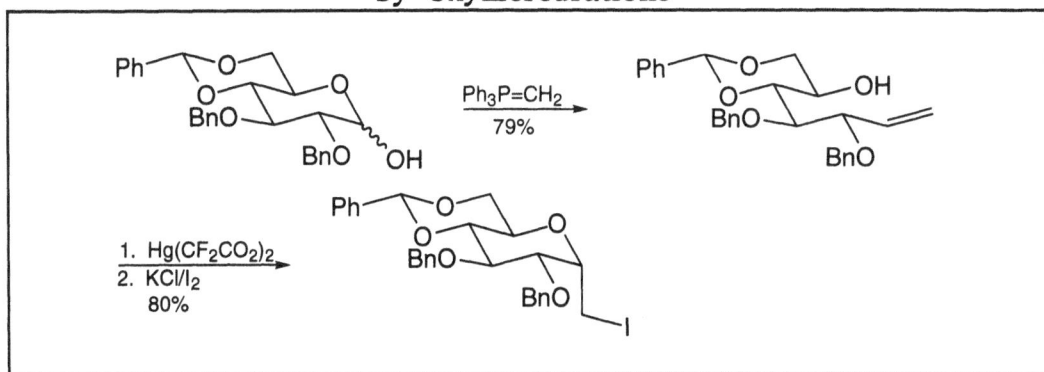

Figure 7.1.1   **Potential Peptidoglycan Polymerization Inhibitor**

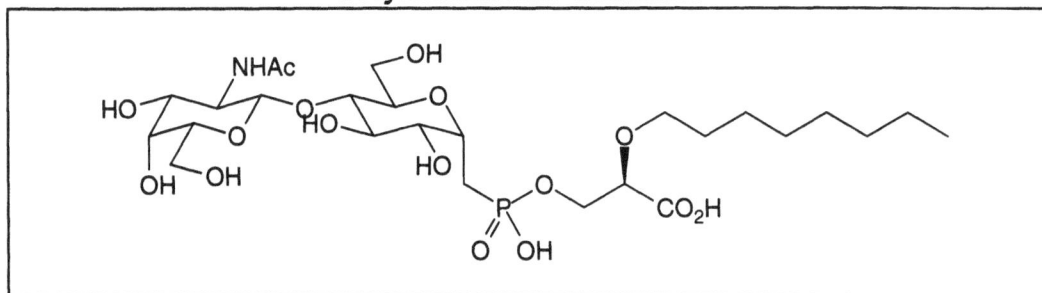

Mercury mediated cyclizations are not unique in the use of metals to initiate the formation of pyranose and furanose *C*-glycosides. As shown in Scheme 7.1.7, Lancelin, *et al.*,[7] effected similar reactions utilizing selenium reagents. As illustrated, the starting hydroxyolefin, prepared utilizing Wittig methodology, was treated with *N*-phenylselenophthalimide and camphor sulfonic acid (CSA). The resulting product mixture exhibited a 60% yield of the pyranosyl phenylselenide and a 30% yield of the furanosyl *C*-glycoside. Further reactions included the conversion of the pyranosyl phenylselenide to the olefin on treatment with sodium periodate and sodium bicarbonate. This final reaction proceeded in 90% yield and provided a product suitable for further manipulations as set forth in section 2.13.

Scheme 7.1.7   **Selenium Mediated**
Cyclization of Wittig Products

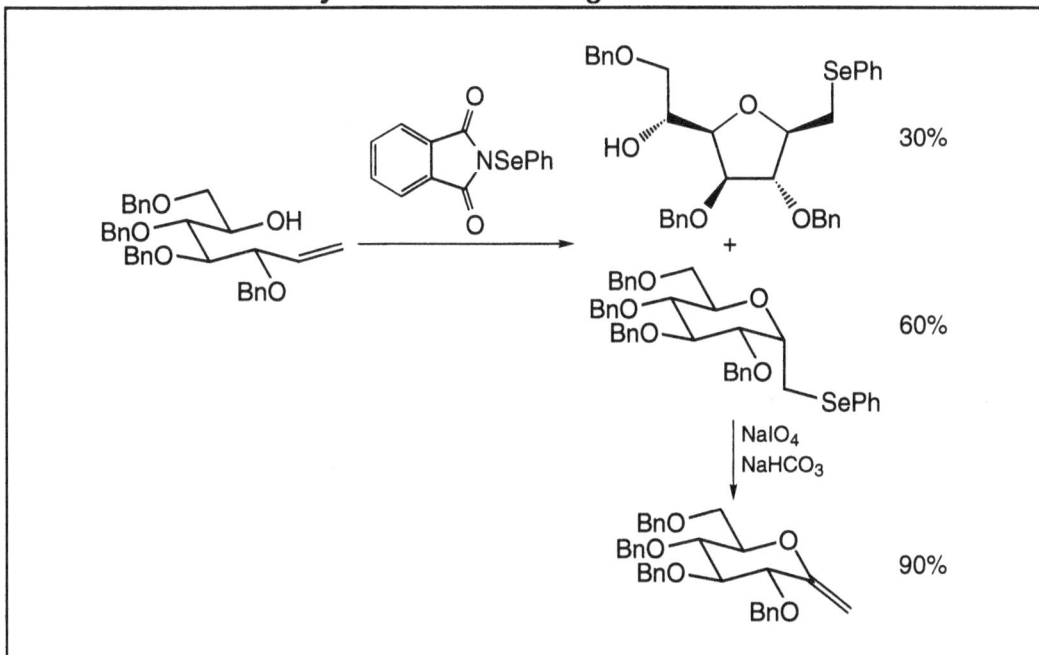

7.1.4   Comparison of Cyclization Methods

With the variety of methods used, in conjunction with Wittig chemistry, for the formation of C-glycosides, it is helpful to compare the results of one method to another. Several reports have accomplished this. As shown in Scheme 7.1.8, Freeman, et al.,[8] effected the cyclization of a Wittig-derived hydroxy olefin to the furanosyl C-glycoside shown. The reagents utilized in this report encompass both halogen and metal mediated cyclizations and provided the desired product in yields as high as 65%.

A later comparison of the methods available for the cyclization of hydroxyolefins was reported by the same group and includes a study of the stereochemical consequences of the reactions.[9] As illustrated in Scheme 7.1.9, the cyclization proceeded in yields as high as 87% and generally exhibited preferences for the β anomeric C-glycosides. However, when phenyl selenochloride was used as the cyclization reagent, the α anomer was favored in a ratio of 60 : 40.

Scheme 7.1.8　　Hydroxyolefin Cyclization Methods

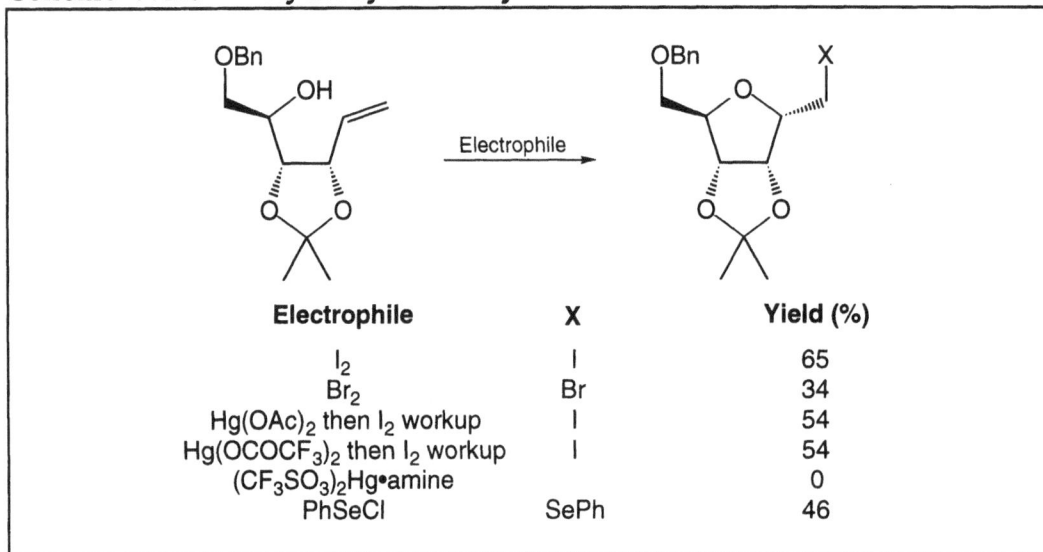

Electrophile	X	Yield (%)
I_2	I	65
Br_2	Br	34
$Hg(OAc)_2$ then I_2 workup	I	54
$Hg(OCOCF_3)_2$ then I_2 workup	I	54
$(CF_3SO_3)_2Hg \cdot amine$		0
PhSeCl	SePh	46

Scheme 7.1.9　　Hydroxyolefin Cyclization Methods

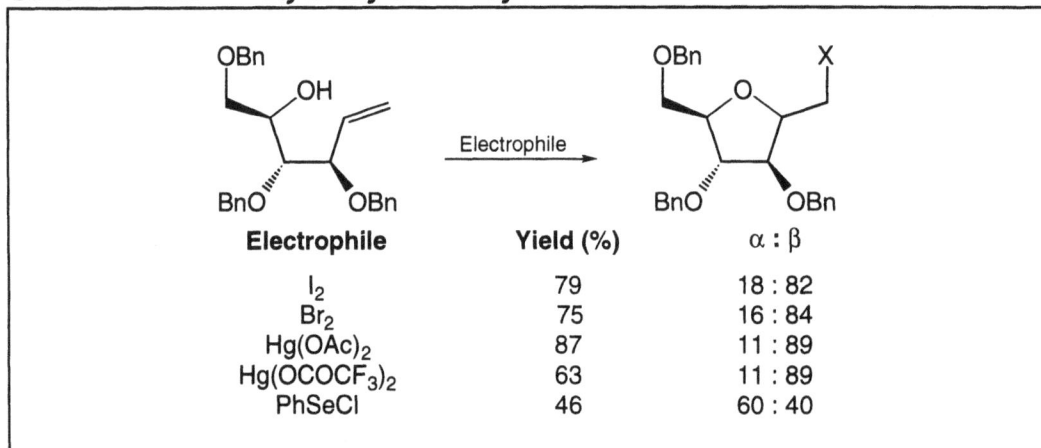

Electrophile	Yield (%)	α : β
I_2	79	18 : 82
Br_2	75	16 : 84
$Hg(OAc)_2$	87	11 : 89
$Hg(OCOCF_3)_2$	63	11 : 89
PhSeCl	46	60 : 40

Additional reports examining the differences between the various methods for hydroxyolefin cyclizations are available and will not be discussed further. Instead, this chapter will continue with discussions centered around alternate methods used in the formation of hydroxyolefinic species as intermediates to the formation of C-glycosides.

7.2   Addition of Grignard and Organozinc Reagents to Lactols

Wittig reactions are only one means of converting lactols to hydroxy olefins. Complimentary olefinations are available utilizing Grignard and organozinc reagents. Where Wittig reactions rely on the olefination of a reactive aldehyde, the use of vinylzinc and the corresponding magnesium reagents provide simple additions to carbonyl groups and effectively provide unprotected hydroxyl groups suitable for selective manipulations. Finally, the reagents ultimately utilized to effect ring closure of the resulting hydroxyolefins lie within those described in the previous section.

As shown in Scheme 7.2.1, Boschetti, *et al.*,[10] effected olefinations on a variety of furanoses utilizing divinyl zinc. In general, the initial additions proceeded in high yields and stereoselectivity with the vinyl group being delivered from the same face as the 2-benzyloxy groups. In this report, the final cyclizations were effected utilizing mercury reagents and provided yields ranging from 51% to 71%.

Scheme 7.2.1   **Divinyl Zinc Additions**
                            and Mercury Cyclizations

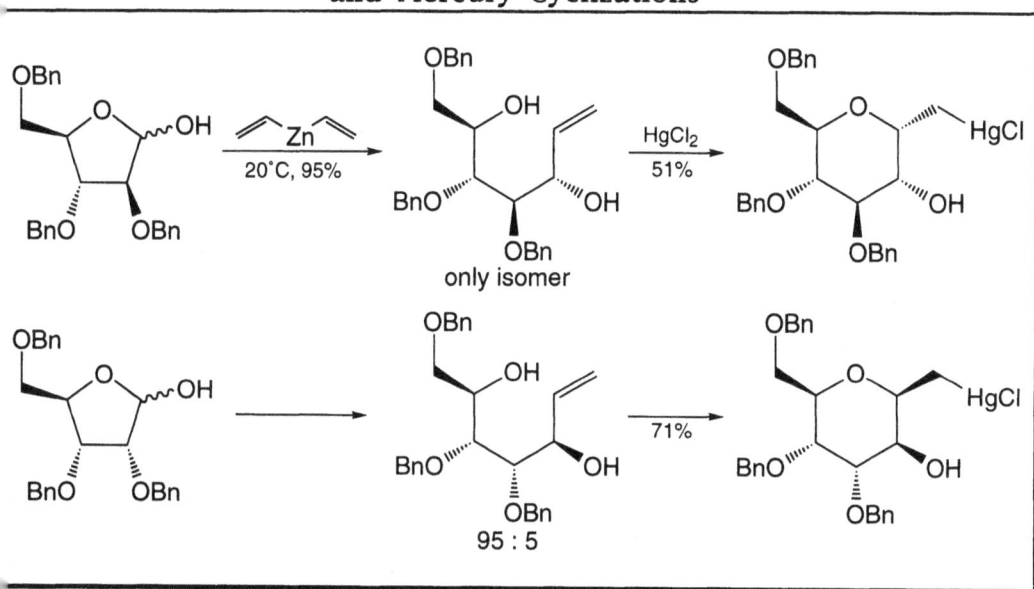

A similar reaction sequence, reported by Lay, *et al.*,[11] is shown in Scheme 7.2.2. In this study, Grignard reagents were used to convert furanose *N*-glycosides to 2-amino-2-deoxy pyranose *C*-glycosides. The reactions proceeded with a preference for the formation of α anomers. A particular advantage to this chemistry is that conventional *C*-glycosidations do not work well for glucosamine analogs. Similar results were independently reported by Carcano, *et al.*,[12] thus confirming these observations.

Scheme 7.2.2   Addition-Oxymercuration
                            with Aminoglycosides

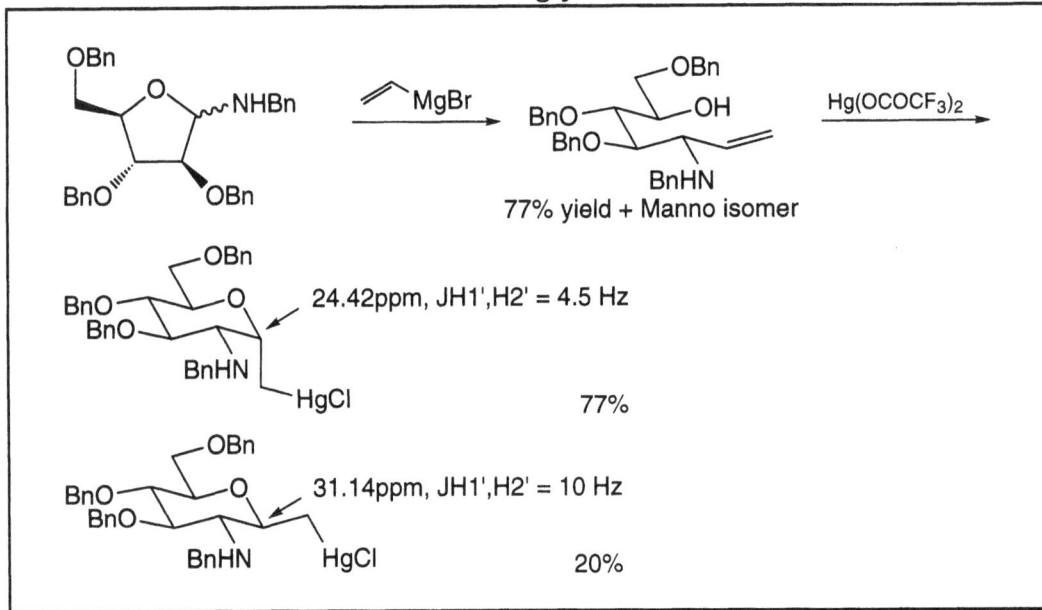

7.3 Cyclization of Suitably Substituted Polyols

The chemistry presented in the previous two sections centered around the formation of hydroxyolefins followed by cyclization to the desired C-glycosides. The latter steps in the examples shown essentially rely upon the cyclization of protected polyols. In this section, cyclization methods including ether formations, ketal formations, and halide displacements are presented.

7.3.1   Cyclizations via Ether Formations

Among the simplest of the cyclizations utilized in the formation of C-glycosides involves the acid mediated formation of ethers. Such reactions can be viewed as a dehydration between two alcohol units with the driving force for the reaction being the elimination of water. The following paragraphs present this reaction in the context of the cyclization of polyol units as a means of preparing C-glycosidic products.

As shown in Scheme 7.3.1, Frick, et al.,[13] formed the perbenzylated polyol on treatment of the lithiated flavonoid with the ring opened form of penta-O-benzylglucose. As illustrated, catalytic hydrogenation in acetic acid produced a C-furanoside after three hours and a C-pyranoside after two days. Additionally, the furanoside form was easily converted to the pyranoside on treatment with hydrochloric acid in dioxane.

Scheme 7.3.1   **Acid Mediated Cyclizations**

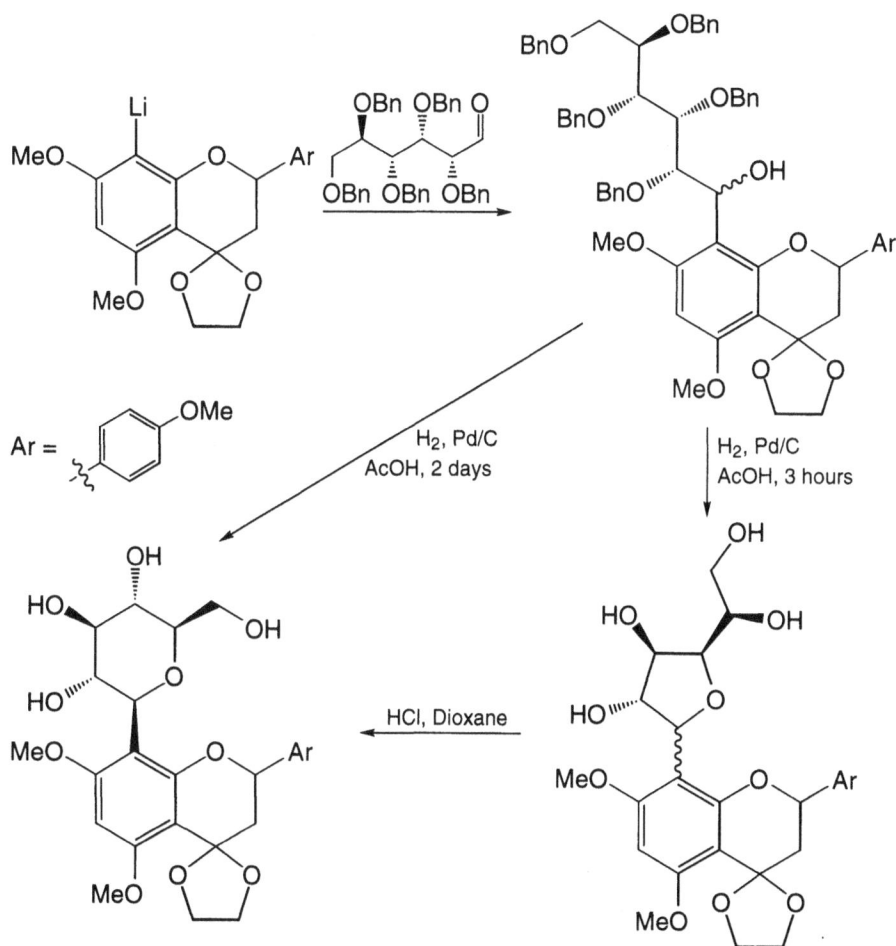

Similar acid mediated cyclizations have been reported on a variety of substrates. For example, as shown in Scheme 7.3.2, Casiraghi, *et al.*,[14] treated 2,3,4,6-tetra-*O*-benzyl-D-glucopyranose with a pyrrole-derived Grignard reagent and a titanium based Lewis acid. As shown, the addition proceeded in 64% yield with the formation of a single diasteriomer. Subsequent exposure to an acidic DOWEX resin provided a 50% yield of the β-*C*-glycoside and an additional 21% isolated as the corresponding α anomer.

Scheme 7.3.2   Acid Mediated Cyclizations

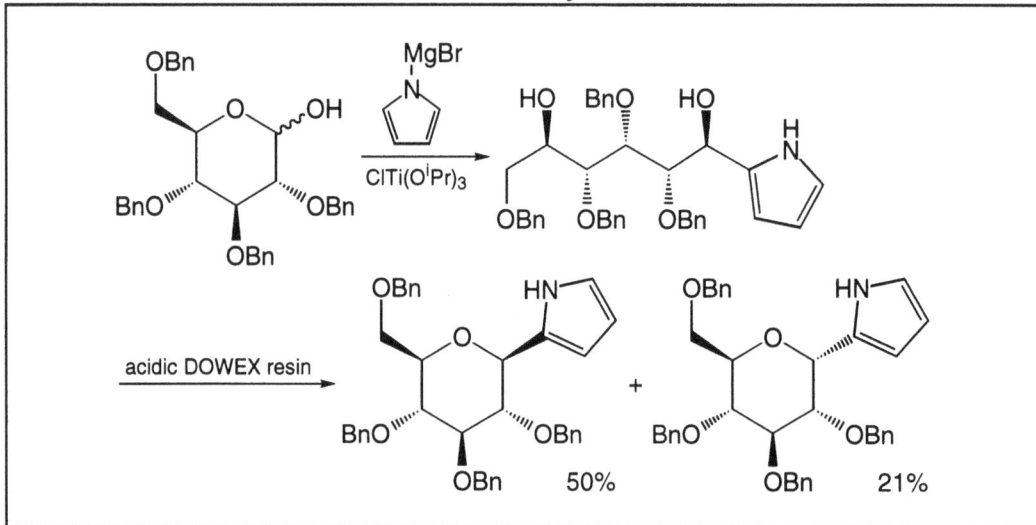

7.3.2   Cyclizations via Ketal Formations

Scheme 7.3.3   Cyclizations via Ketal Formations

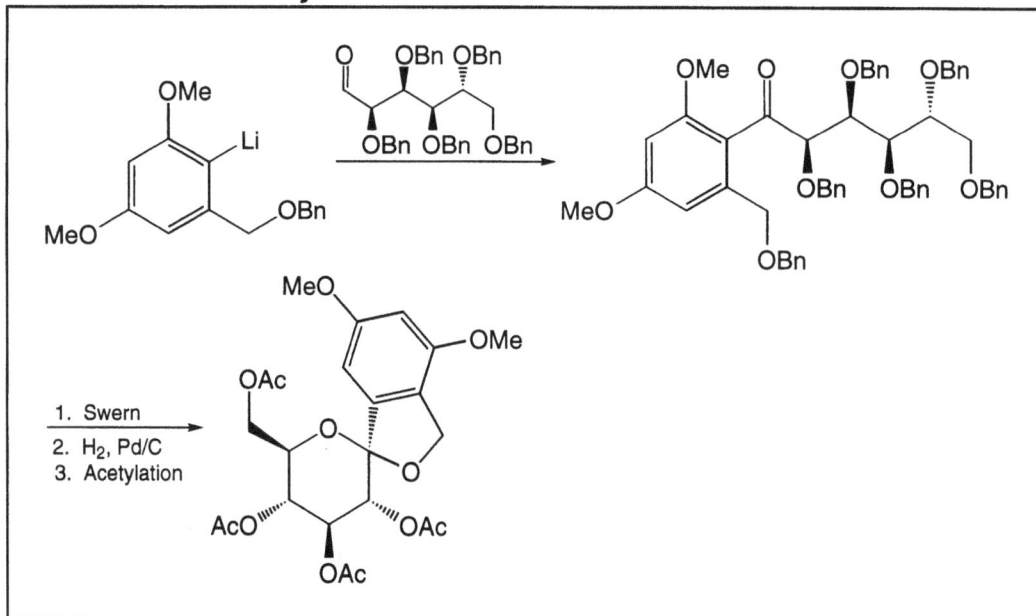

Ketals are capable of being formed under acidic conditions. Consequently, this technology compliments the acid mediated cyclizations set forth in the previous two examples. As applied to the preparation of *C*-glycosides, the general approach is similar to the previous examples in that a nucleophile is added to a sugar derivative. As shown in Scheme 7.3.3, Schmidt, *et al.*,[15] utilized a perbenzylated open-chain glucose. The resulting alcohol was then oxidized to the corresponding ketone and the benzyl groups were removed under catalytic conditions. On acetylation, the resulting polyol cyclized with formation of a single diasteriomer of the papulacandin spiroketal. It should be noted that the ketals formed utilizing this technology are anomerically alkylated *C*-glycosides bearing resemblances in chemistry and appearance to both *O* and *C*-glycosides.

Scheme 7.3.4   Cyclizations *via* Ketal Formations

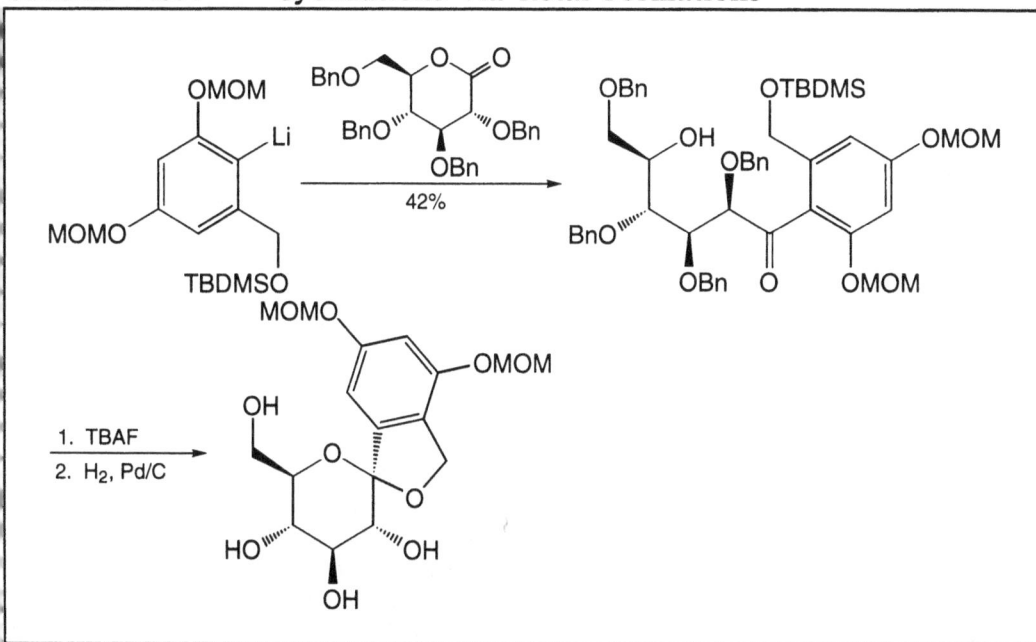

The papulacandins are a class of antifungal antibiotics isolated from *Papularia sphaerosperma.*[16] As the antifungal properties of these compounds are interesting, substantial effort has been directed towards their preparation. Utilizing a similar approach of spiroketal formation, Rosenblum, *et al.*,[17] achieved a somewhat shorter preparation of the above described product. As shown in Scheme 7.3.4, perbenzylated gluconolactone was treated with an aryllithium compound producing a 42% yield of the desired hydroxyketone. Subsequent removal of the silyl protecting group followed by hydrogenation and yielded a polyol which spontaneously cyclized to a spiroketal. This final ketal formation afforded the key intermediate in the illustrated papulacandin preparation.

7.3.3   Cyclizations via Halide Displacements

Though technically ether formations, the displacement of halides by hydroxyl groups differs from the examples presented earlier in this section in that these reactions are base mediated. As such, *C*-glycosidations under these conditions compliment the previously described cyclization strategies in that they allow reactions with acid sensitive substrates. As shown in Scheme 7.3.5, Schmid, *et al.*,[18] utilized this approach in the preparation of *C*-furanosides. Specifically, the illustrated triol was protected with silyl groups. Subsequent treatment of the ketone with allylmagnesium chloride effected formation of a hydroxide which readily displaced the terminal chloride. The final *C*-glycoside was isolated as a 2.7 : 1 mixture of diastereomers.

Scheme 7.3.5   Cyclizations *via* Halide Displacements

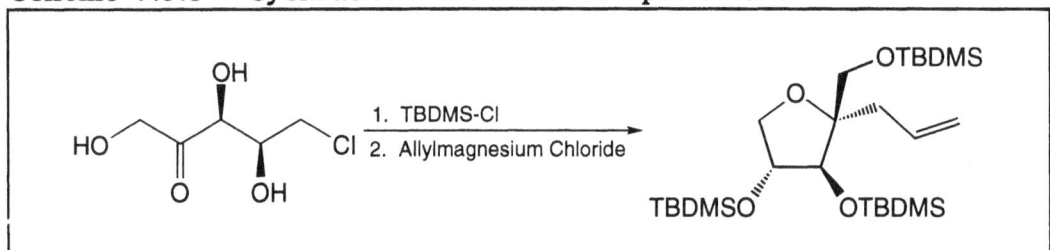

7.3.4   Ring Cyclizations: A Case Study

Perhaps the most extensive studies into the cyclizations of protected polyols as methods for the preparation of *C*-glycosides was presented by Kishi, *et al.*[19] This study involves the exploration of three routes to similar end products from similar starting materials. In the first of these strategies, shown in Scheme 7.3.6, a glucose derivative was converted to a hemiketal by Swern oxidation of the free alcohol followed by removal of the acetonide. Following spontaneous cyclization, the primary alcohol was protected. Utilizing reductive conditions to remove the anomeric hydroxyl group, the desired β-*C*-glycoside was obtained.

In the second approach, shown in Scheme 7.3.7, the same glucose derivative used in Scheme 7.3.6 was converted to the *p*-nitrobenzoate in the following manner. The acetonide was removed under hydrolytic conditions and the resulting diol was cleaved oxidatively. Spontaneous cyclization afforded the hemiacetal which was subsequently protected as the corresponding *p*-nitrobenzyl ester. This *O*-PNB glycoside was then converted to a *C*-allenylglycoside utilizing propargyltrimethylsilane and a Lewis acid. The allene was then cleaved on treatment with ozone. Reduction of the resulting aldehyde and protection of the resulting alcohol gave a product identical to that formed in Scheme 7.3.6 with the exception of the anomeric stereochemistry.

Scheme 7.3.6   **Kishi Polyols -- Route A**

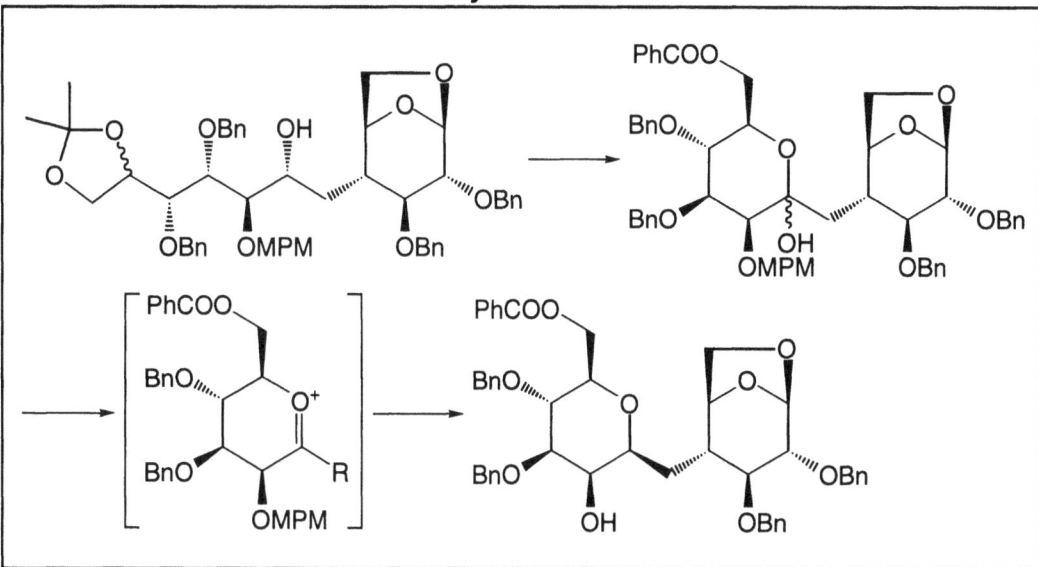

Scheme 7.3.7   **Kishi Polyols -- Route B**

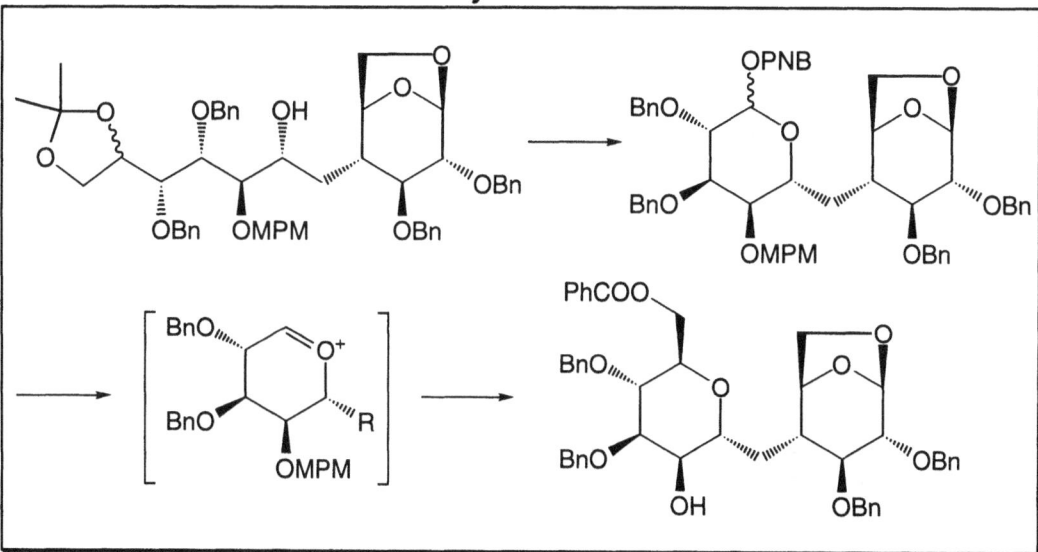

In the third approach, shown in Scheme 7.3.8, a glucose derivative was converted to an epoxide *via* hydrolysis of the acetonide followed by tosylation of the terminal hydroxyl group and treatment with sodium hydride. Subsequent removal of the PMB group with ceric ammonium nitrate and treatment with *p*-toluenesulfonic acid effected cyclization to the pyranoside.

Protection as previously described afforded a product identical to that obtained in Scheme 7.3.7 with the exception of an acetylated hydroxyl group.

Scheme 7.3.8   Kishi Polyols -- Route C

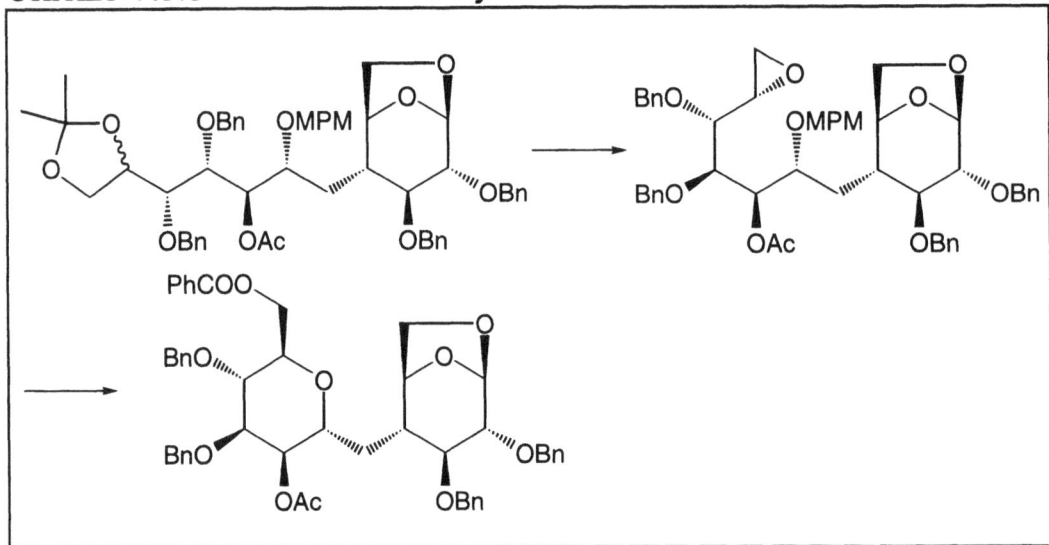

The work reported by Kishi, *et al.*, provided landmark insight into the preparation of *C*-disaccharides. As the examples presented in this section are representative of *C*-disaccharides, they were intentionally vague and will be discussed in much greater detail in Chapter 8.

7.4 Rearrangements

Chapter 6 addressed the subject of the formation of *C*-glycosides *via* rearrangements. In the examples presented therein, the rearrangements were executed on substrates that already possessed suitable ring configurations for the desired *C*-glycosidic products. The rearrangements presented in this section differ from those in Chapter 6 in that the core ring required for the *C*-glycosides are either formed or substantially modified. The rearrangement types presented include electrocyclic reactions as well as ring contractions.

7.4.1   *Electrocyclic Rearrangements*

The preparation of *C*-glycosides, *via* formation of the sugar ring, can be achieved through the rearrangement of suitable substrates. For example, as shown in Scheme 7.4.1, Burke, *et al.*,[20] utilized the enolate Claisen rearrangement. In the cited study, a series of 6-alkenyl-1,4-dioxane-2-ones was converted to dihydropyrans bearing *C*-methylcarboxyl glycosidic units. All reactions were stereospecific and the yields and substitution patterns are shown in Table 7.4.1.

Scheme 7.4.1 Enolate Claisen Rearrangement

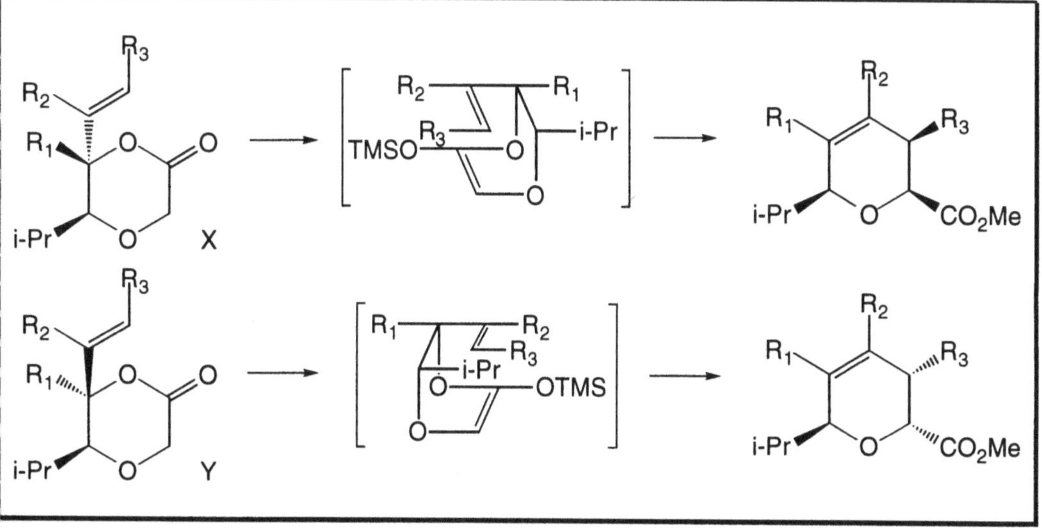

Table 7.4.1 Enolate Claisen Rearrangement Results

	Dihydropyran Product			
Substrate	R_1	R_2	R_3	Yield (%)
X	H	H	H	67
X	H	Me	H	75
X	H	H	$SiMe_3$	52
X	H	H	Me	70
X	Me	H	H	90
X	Me	Me	H	80
Y	H	H	H	69
Y	H	Me	H	78
Y	H	H	$SiMe_3$	61
Y	H	H	Me	81
Y	Me	H	H	91

7.4.2 Ring Contractions

In addition to Claisen-type rearrangements, other methods are applicable to the formation of *C*-glycosides. As shown in Scheme 7.4.2, Fleet, *et al.*,[21] utilized ring contraction techniques in the transformation of *O*-methyl-pyranosyl glycosides to *C*-furanosyl glycosides. The reactions proceeded in 51%

to 91% yields, depending upon the anomeric configuration of the starting galactosides, and produces approximately 2 : 1 ratios of diastereomeric products. A potential mechanism is also presented in Scheme 7.4.2 and involves the displacement of the triflate by the pyranose ring oxygen followed by opening of the three membered intermediate by an azide group. In this respect, the mechanism resembles that belonging to the classical Favorskii rearrangement.

Scheme 7.4.2   **Ring Contractions**

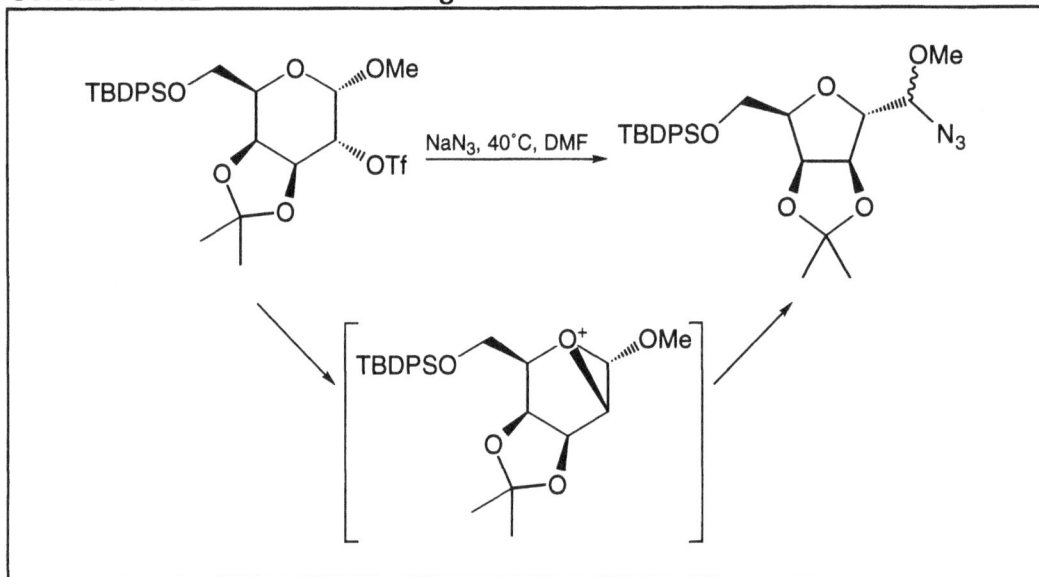

Scheme 7.4.3   **Ring Contractions**

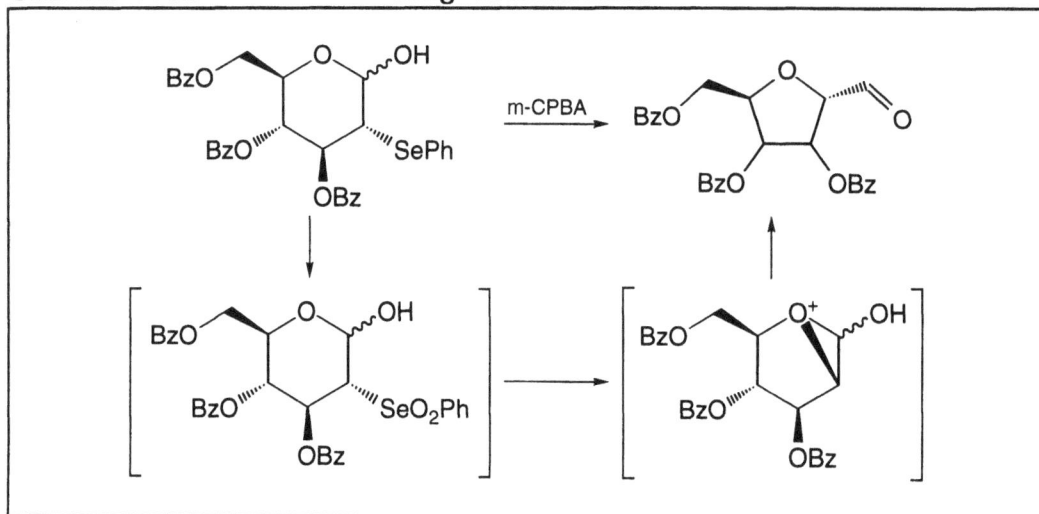

Additional types of rearrangements have been used in the preparation of C-glycosides. One particular example, reported by Kaye, *et al.*,[22] involves an oxidative pathway and was ultimately used in the preparation of showdomycin analogs. The reaction, shown in Scheme 7.4.3, proceeds in a similar fashion to that in the previous example with the ring oxygen displacing the selenium leaving group. However, a substantial difference is noted by the observation that once oxidation of the phenylselenide is complete, no additional reagents need be added in order to induce reaction.

7.4.3   Other Rearrangements

Scheme 7.4.4   Benzylidine Acetal Rearrangements

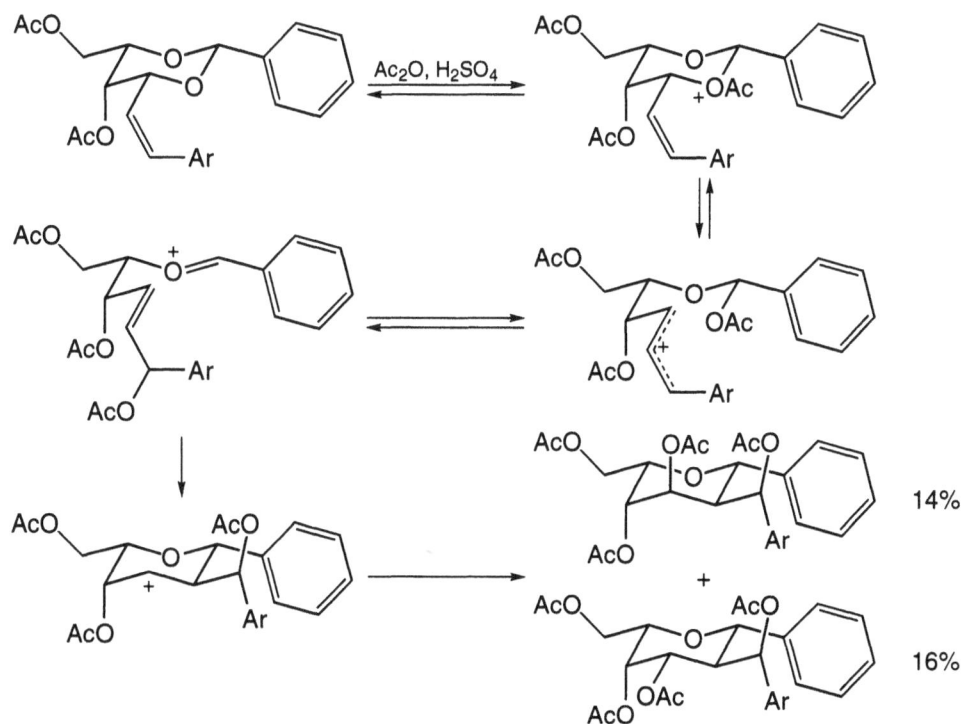

In a final example of the utility of rearrangements in the formation of C-glycosides, Kuszmann, *et al.*,[23] reported reactions involving benzylidine acetals. A specific reaction, illustrated in Scheme 7.4.4, yielded two diastereomeric C-glycosides in a combined yield of approximately 30%. The mechanism associated with this reaction involves initial acetylation of one of the ring oxygens followed by cleavage of the ring with formation of a conjugated carbocation. Migration of an acetate to the carbocation yields a species bearing an olefin on one side and an oxonium ion on the other. Ring closure followed by quenching of the resulting carbocation by an acetate group gives the observed products.

7.5 Cycloadditions

Perhaps the most useful method for the formation of substituted ring systems from acyclic substrates is the Diels-Alder reaction. As will be explored in this section, the Diels-Alder reaction fits well within the theme of this chapter through its use in the direct formation of ring systems capable of being converted to sugars. As shown in Scheme 7.5.1, Katagiri, *et al.*,[24] incorporated this methodology in transformations involving furan derivatives. As illustrated, the Diels-Alder reaction between 3,4-dibenzyloxyfuran and dimethyl (acetoxymethylene)malonate afforded a 57% yield of the bicyclic adduct. This product was then subjected to catalytic hydrogenation followed by basic reductive conditions. The resulting cascade of reductions and rearrangements produced the dimethylmalonyl-*C*-lyxoside in an 85% isolated yield.

Scheme 7.5.1   *C*-Glycosides *via* Diels-Alder Reactions

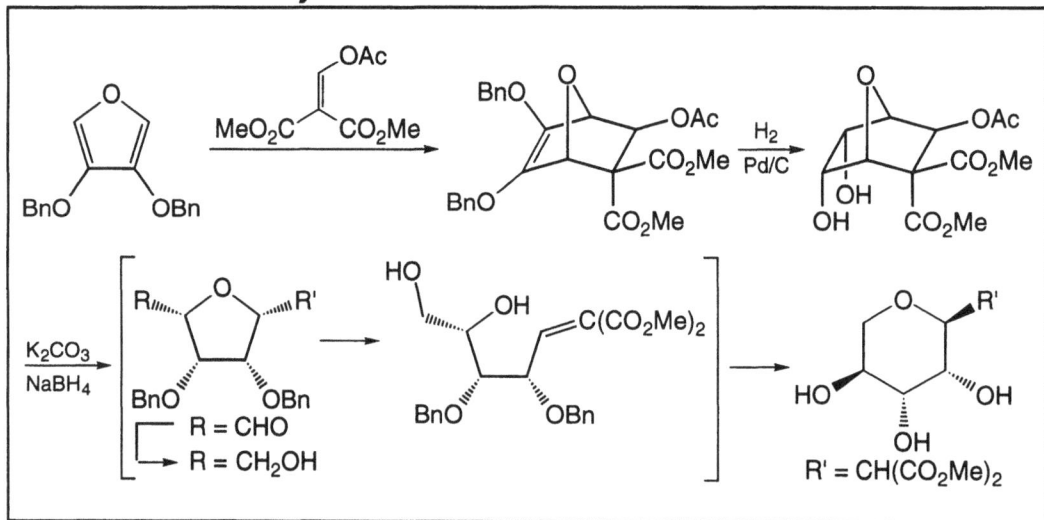

Where, as illustrated in the above described example, the Diels-Alder reaction provides avenues into the formation of *C*-glycosides from furan-derived dienes, the hetero Diels-Alder reaction allows for the direct formation of sugar rings from carbonyl groups. As shown in Scheme 7.5.2, Schmidt, *et al.*,[25] effected reactions between conjugated carbonyl compounds and olefins. The illustrated reaction proceeded in 81% yield giving an adduct which, after further manipulations, was converted to a *C*-aryl glycoside.

Scheme 7.5.2   *C-Glycosides via*
                        Hetero Diels-Alder Reactions

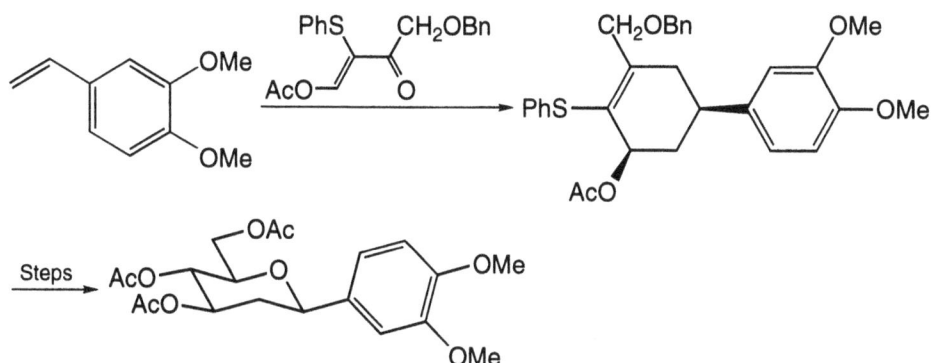

Scheme 7.5.3   *C-Glycosides via*
                        Hetero Diels-Alder Reactions

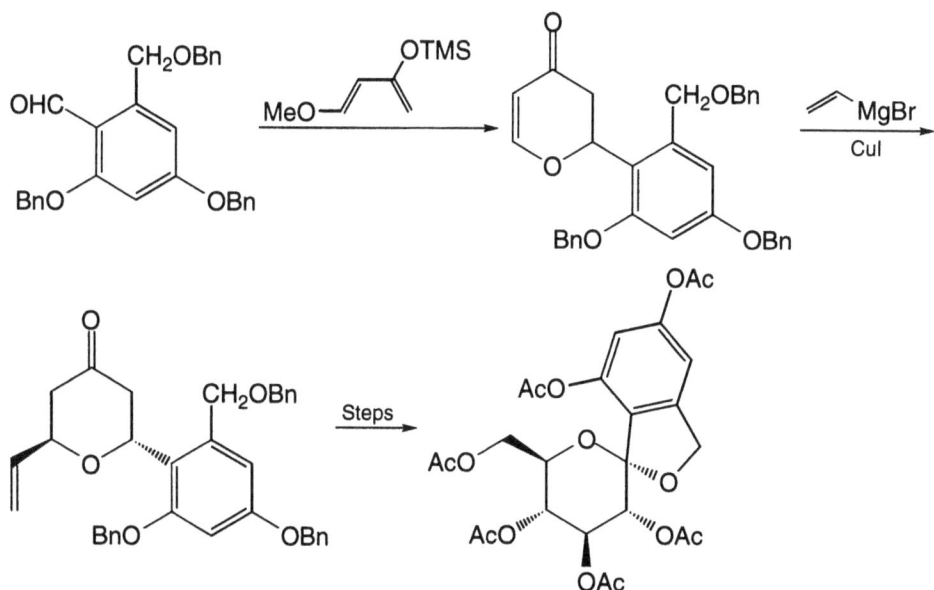

Where the previous example utilized carbonyl compounds as Diels-Alder dienes, complimentary technology is available incorporating carbonyl compounds as dienophiles. As shown in Scheme 7.5.3, this approach was explored by Danishefsky, *et al.*,[26] in the preparation of a papulacandin precursor. Utilizing the diene, shown, and the substituted benzaldehyde, Diels-

Alder methodology effected ring formation in a 92% yield. Subsequent Michael addition with vinylmagnesiumbromide provided the *C*-glycosidic precursor in 70% yield. This compound, after several synthetic transformations, was converted to the required peracetylated intermediate.

As illustrated in the previous two examples, the hetero Diels-Alder reaction provides a useful entry into the preparation of *C*-monosaccharides. However, as illustrated in Scheme 7.5.4, Danieshefsky, *et al.*,[2728] utilized this methodology in the preparation of a *C*-linked disaccharide. As shown, the aldehyde produced from a galactose derivative was reacted with a diene under chelation control. The ultimate result was a single isomer of a *C*-glycoside containing extremely versatile functionalities.

Scheme 7.5.4   *C*-**Disaccharides** *via*
Hetero Diels-Alder Reactions

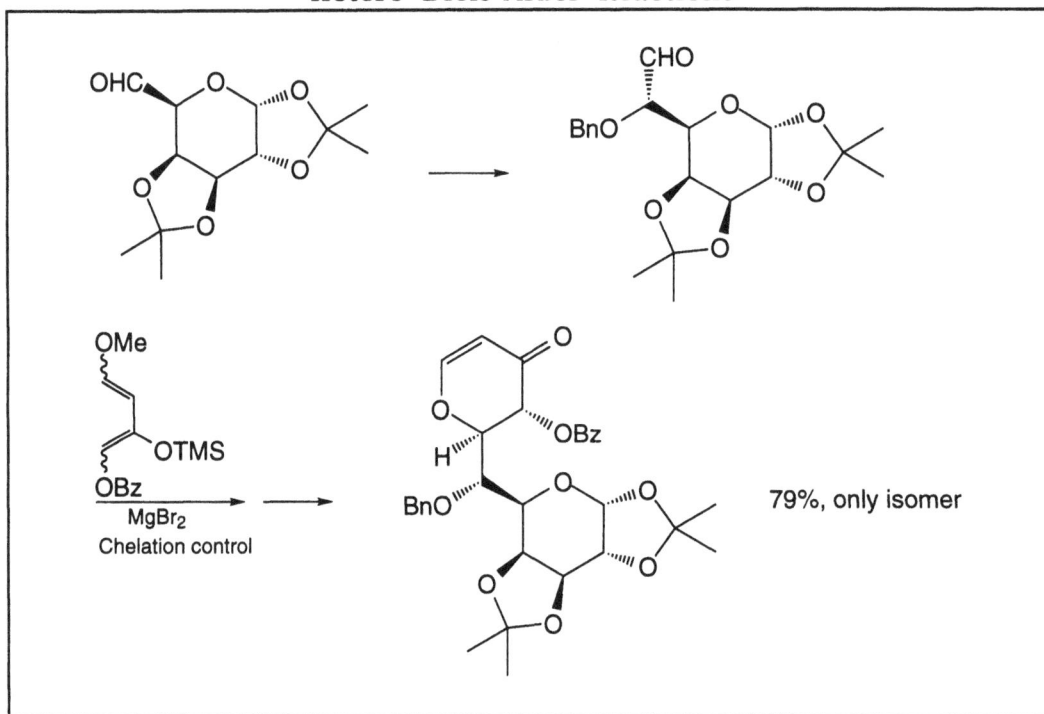

Diels-Alder methodology is well suited to the preparation of *C*-monosaccharides and higher carbohydrate analogs. In general, these reactions provide products with high degrees of stereochemical induction as well as exhibiting synthetically useful yields. Further examples of Diels-Alder reactions applied to the preparation of *C*-disaccharides will be elaborated upon in Chapter 8.

7.6   Other Methods for the Formation of Sugar Rings

The methods described thus far represent only a small subset of the available methodologies for the formation of *C*-furanosides and *C*-pyranosides. Other methods shown to be useful include cyclizations of halo olefins and ene-ynes. As shown in Scheme 7.6.1, Lee, *et al.*,[29] prepared halo olefins from suitable alcohols and acetylenic esters. Subsequent application of free radical conditions thus effected cyclizations. All reactions proceeded in yields exceeding 95%. Furthermore, where the formation of diastereomers was an issue, selective *cis* formation was observed. Thus the ease of preparation of the vinyl ether substrates required for these cyclization reactions makes this methodology an extremely useful addition to the technology surrounding the preparation of *C*-glycosides.

Scheme 7.6.1   **Radical Cyclizations**

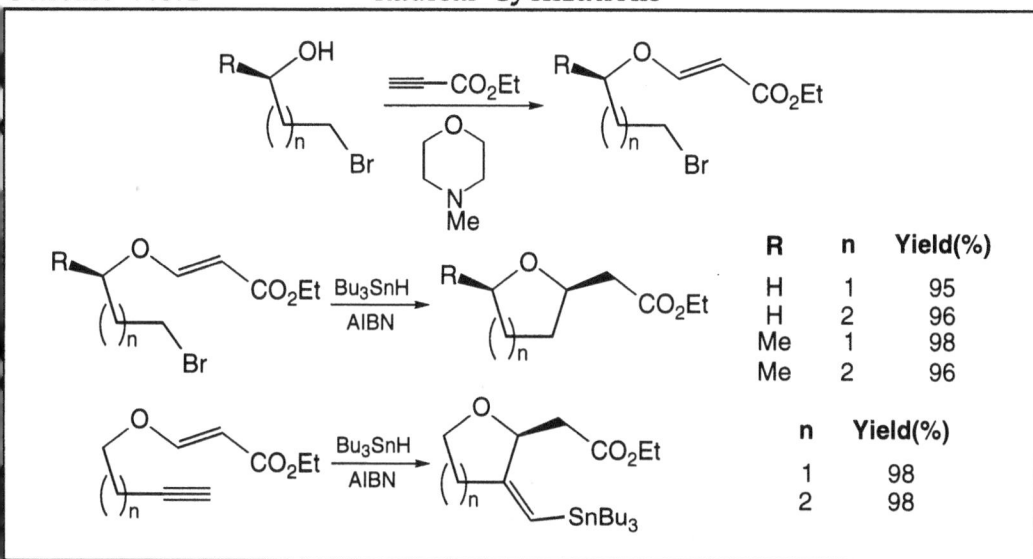

R	n	Yield(%)
H	1	95
H	2	96
Me	1	98
Me	2	96

n	Yield(%)
1	98
2	98

Scheme 7.6.2   **Pyran Modifications**

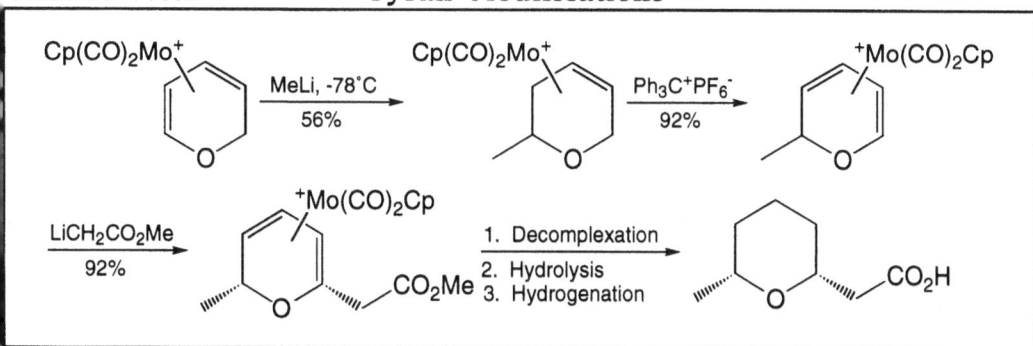

Where this chapter has examined the formation of *C*-glycosides *via* the formation of sugar rings, available methodologies allow the formation of *C*-glycosides by building off of pyran rings. As shown in Scheme 7.6.2, Hansson, *et al.*,[30] demonstrated the utility of reacting alkyllithium compounds with π complexes involving 2H-pyran and molybdenum. The initial addition of methyllithium provided a dihydropyran which, after oxidation back to a pyran, was ready for further elaboration. Following decomplexation, the illustrated disubstituted tetrahydropyran was isolated.

The lessons apparent in this chapter demonstrate that although some *C*-glycosides can be synthetically challenging targets, their availability from a variety of acyclic substrates compliments the chemistry made use of in the first six chapters. Many more methods are available for the preparation of *C*-glycosides and many more will be discussed in the final chapter as they apply to the preparation of *C*-di and trisaccharides.

7.7   References

1.   Vyplel, H.; Scholz, D.; Macher, I.; Schindlmaier, K.; Schutze, E. *J. Med. Chem.*, **1991**, *34*, 2759.
2.   Armstrong, R. W.; Teegarden, B. R. *J. Org. Chem.* **1992**, *57*, 915.
3.   Nicotra, F.; Panza, L.; Ronchetti, F.; Russo, G.; Toma, L. *Carbohydrate Res.* **1987**, *171*, 49.
4.   Pougny, J. R.; Nassr, M. A. M.; Sinay, P. *J. Chem. Soc. Chem. Comm.*, **1981**, 375.
5.   Liu, P. S. *J. Org. Chem.* **1987**, *52*, 4717.
6.   Qiao, L.; Vederas, J. C. *J. Org. Chem.*, **1993**, *58*, 3480.
7.   Lancelin, J. M.; Pougny, J. R.; Sinay, P. *Carbohydrate Res.* **1985**, *136*, 369.
8.   Freeman, F.; Robarge, K. D. *Carbohydrate Res.* **1985**, *137*, 89.
9.   Freeman, F.; Robarge, K. D. *Carbohydrate Res.* **1987**, *171*, 1.
10.   Boschetti, A.; Nicotra, F.; Panza, L.; Russo, G. *J. Org. Chem.*, **1989**, *54*, 1890.
11.   Lay, L.; Nicotra, F.; Panza, L.; Verani, A. *Gazzetta Chimica Italiana*, **1992**, *122*, 345.
12.   Carcano, M.; Nicotra, F.; Panza, L.; Russo, G. *J. Chem. Soc. Chem. Comm.* **1989**, 297.
13.   Frick, W.; Schmidt, R. R. *Liebig's Ann. Chem.* **1989**, 565.
14.   Casiraghi, G.; Cornia, M.; Rassu, G.; Sante, C. D.; Spanu, P. *Tetrahedron* **1992**, *48*, 5619.
15.   Schmidt, R. R.; Frick, W. *Tetrahedron* **1988**, *44*, 7163.
16.   Traxler, P.; Gruner, J.; Auden, J. A. L. *J. Antibiotics* **1977**, *30*, 289.
17.   Rosenblum, S. B.; Bihovsky, R. *J. Am. Chem. Soc.* **1990**, *112*, 2746.
18.   Schmid, W.; Whitesides, G. *J. Am. Chem. Soc.* **1990**, *112*, 9670.
19.   Wang, Y.; Babirad, S. A.; Kishi, Y. *J. Org. Chem.*, **1992**, *57*, 468.
20.   Burke, S. D.; Armistead, D. M.; Schoenen, F. J.; Fevig, J. M. *Tetrahedron*, **1986**, *42*, 2787.

21. Fleet, G. W.; Seymour, L. C. *Tetrahedron Lett.* **1987**, *28*, 3015.
22. Kaye, A.; Neidle, S.; Rees, C. B. *Tetrahedron Lett.* **1988**, *29*, 2711.
23. Kuszmann, J.; Podanyi, B.; Jerkovich, G. *Carbohydrate Res.* **1992**, *232*, 17.
24. Katagiri, N.; Akatsuka, H.; Haneda, T.; Kaneko, C. *J. Org. Chem.* **1988**, *53*, 5464.
25. Schmidt, R. R.; Frick, W.; Haag-Zeino, B.; Apparao, S. *Tetrahedron Lett.* **1987**, *28*, 4045.
26. Danishefsky, S.; Philips, G.; Ciufolini, M. *Carbohydrate Res.* **1987**, *171*, 317.
27. Danishefsky, S. J.; Pearson, W. H.; Harvey, D. F.; Charence, J. M.; Springer, J. P. *J. Am. Chem. Soc.,* **1985**, *107*, 1256.
28. Danishefsky, S. J.; DeNinno, M. P. *Angew. Chem. Int. Ed. Eng.,* **1987**, *26*, 15.
29. Lee, E.; Tae, J. S.; Lee, C.; Park, C. M. *Tetrahedron Lett.* **1993**, *30*, 4831.
30. Hansson, S.; Miller, J. F.; Liebskind, L. S. *J. Am. Chem. Soc.* **1990**, *112*, 9660.

8   Introduction

In this book, Chapter 1 served to introduce C-glycosides with respect to their nomenclature, availability from natural sources, and their use as potential clinical agents. Chapters 2 through 7 presented different technologies used in the preparation of a wide variety of C-glycosides. However, those intermediate six chapters also concentrated primarily on the simpler C-monosaccharides. In this chapter, the methods used in Chapters 2 through 7 will be applied to the preparation of the more complex C-di and trisaccharides. These compounds are characterized by the direct anomeric linking atom being carbon instead of oxygen. The examples are presented chronologically beginning with the earliest reported synthesis of a C-disaccharide.

8.1   Syntheses Reported in 1983

Scheme 8.1.1   Sinay C-Disaccharide Synthesis Part I

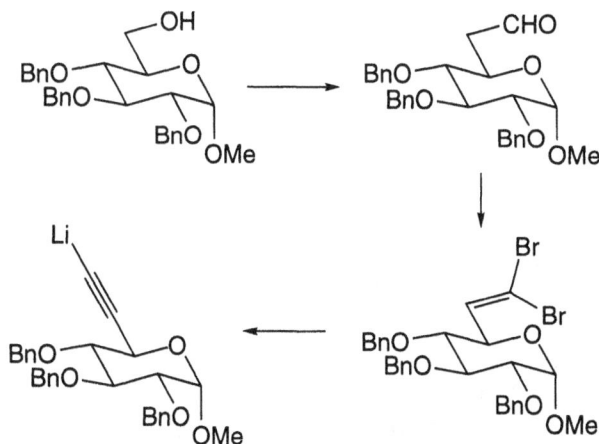

In 1983, the first synthesis of a C-disaccharide was reported by Sinay, *et al.*,[1] and involved the addition of an acetylenic anion to a sugar lactone. The synthesis, shown in Scheme 8.1.1, began by applying a Swern oxidation to the tri-O-benzyl methylglucoside shown. The resulting aldehyde was converted to an acetylene anion *via* the intermediate dibromo olefin.

The synthesis was completed, as shown in Scheme 8.1.2, by addition of the anion to the illustrated glucose lactone. As shown, this addition yielded the acetylene bridged glycoside. Deoxygenation was subsequently achieved utilizing triethylsilane in the presence of a Lewis acid. Catalytic hydrogenation subsequently reduced the acetylene and removed the benzyl protecting groups thus liberating C-gentiobiose.

Scheme 8.1.2   Sinay C-Disaccharide Synthesis Part II

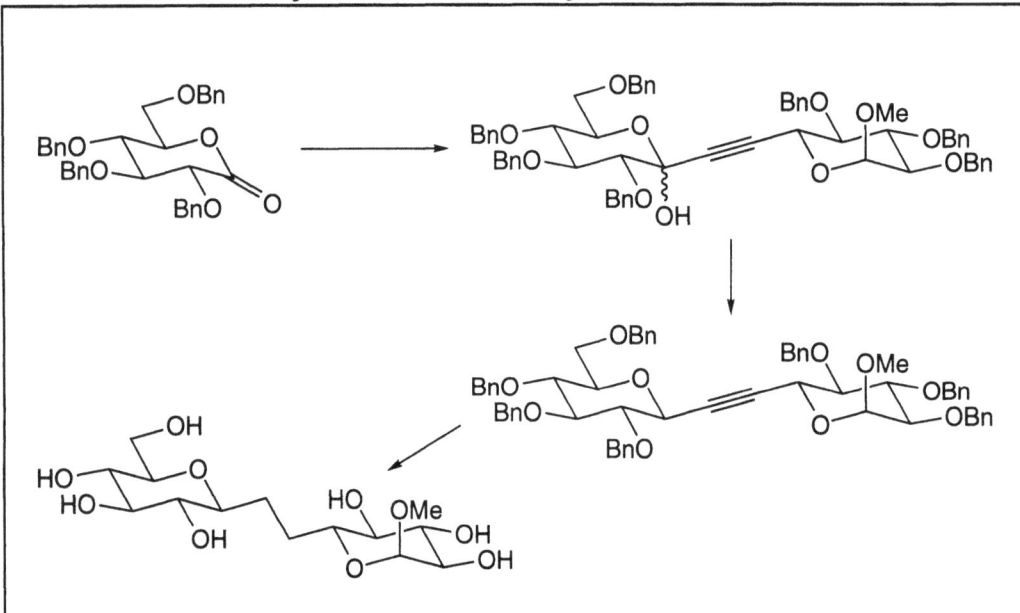

8.2   Syntheses Reported in 1984

Shortly after the Sinay report, additional reports of C-disaccharide syntheses became visible. In 1984, the reported technologies included the dimerization of nitroalkanes as well as the use of hetero Diels-Alder reactions. Beginning with nitroalkane dimerizations, Vasella, *et al.*,[2] initially studied the cross coupling of nitroalkanes and furanoside nitroglycosides discussed in section 2.6 and reviewed in Scheme 8.2.1. As demonstrated in the same report, the dimerization of furanoside nitroglycosides effectively produced C-difuranosides. The specific reaction, shown in Scheme 8.2.2, proceeded under basic conditions and the nitro group was removed on treatment with sodium sulfide.

Scheme 8.2.1   Cross Coupling of
Nitroalkanes and Nitroglycosides

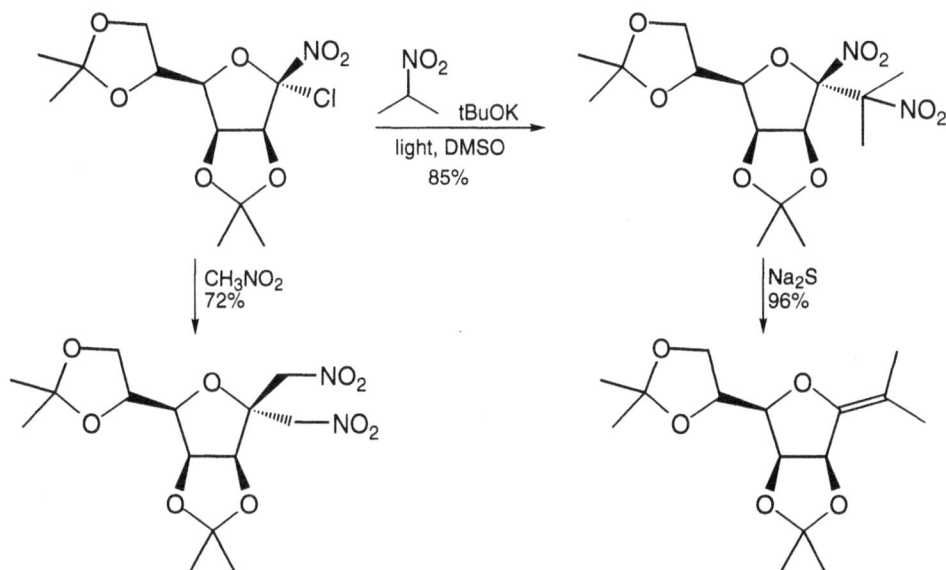

Scheme 8.2.2   Dimerization of Nitro Glycosides

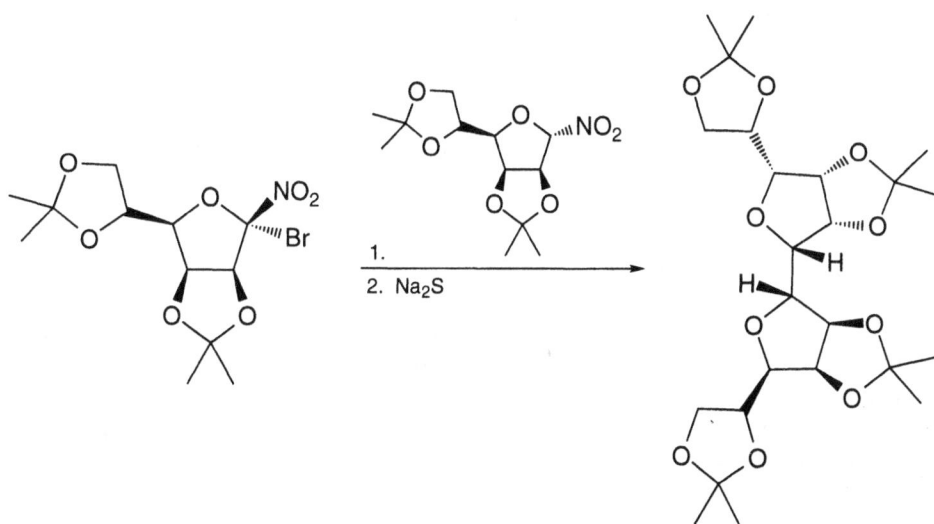

The use of hetero Diels-Alder reactions in the preparation of *C*-glycosides was introduced in Chapter 7. This cycloaddition reaction has shown substantial promise as a means of directly forming sugar rings from acyclic substrates. A. early as 1984, this technology was utilized in the preparation of *C*-disaccharides utilizing sugar-derived aldehydes. As shown in Scheme 8.2.3, Jurczak, *et al.*,[3] incorporated this methodology in the preparation of a sugar-substituted dihydropyran ring system. The reaction proceeded with the formation of a single isomer *via* an *endo* approach of the dienophile. The stereochemical preferences for these reactions were predictable from Felkin's models.[4,5]

**Scheme 8.2.3   Preparation of *C*-Disaccharides
                  Utilizing Hetero Diels-Alder Reactions**

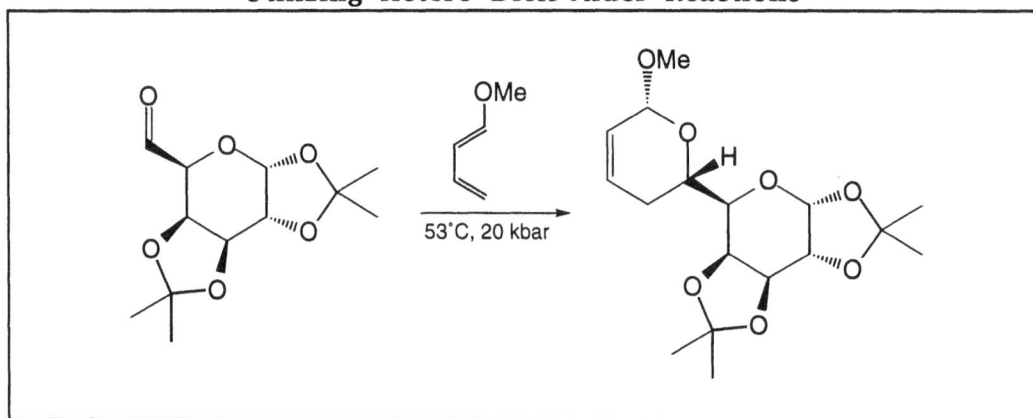

A similar approach to *C*-disaccharides was concurrently reported by Danishefsky, *et al.*,[6] thus positioning hetero Diels-Alder technology as a key strategy in the preparation of these complex molecules. As this chapter progresses, Diels-Alder methodology will become a recurring theme in increasingly more interesting synthetic strategies to biologically important targets.

8.3   Syntheses Reported in 1985

The examples presented in the previous two sections represent early approaches to the preparation of *C*-disaccharides. Building upon these early studies, new techniques became apparent and interesting reports surfaced in 1985. Aside from additional studies involving Diels-Alder methodology, Beau *et al.*,[7] reported the use of addition reactions between phenylsulfone anions and sugar-derived aldehydes as a viable method for the formation of *C*-disaccharides. As shown in Scheme 8.3.1, the sulfone associated with the addition product was cleaved on treatment with lithium naphthalide thus giving the final product.

Scheme 8.3.1   Sulfone Glycosides
and Sugar Aldehydes

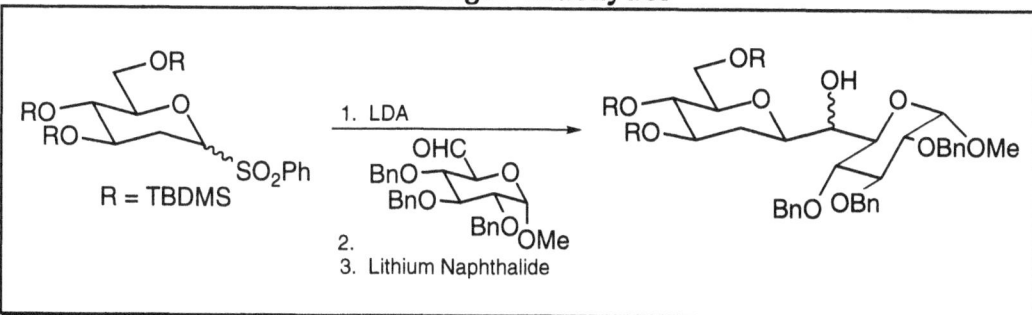

Returning to Diels-Alder methodology, Danishefsky, *et al.*,[8] demonstrated its use in the preparation of interesting *C*-disaccharides. The key synthetic reaction, illustrated in Scheme 8.3.2, proceeded in a chelation controlled manner giving a 79% yield of a *C*-disaccharide precursor.   The resulting dihydropyranone unit was recognized as suitable for subsequent conversion to various sugar derivatives.

Scheme 8.3.2   Hetero Diels-Alder

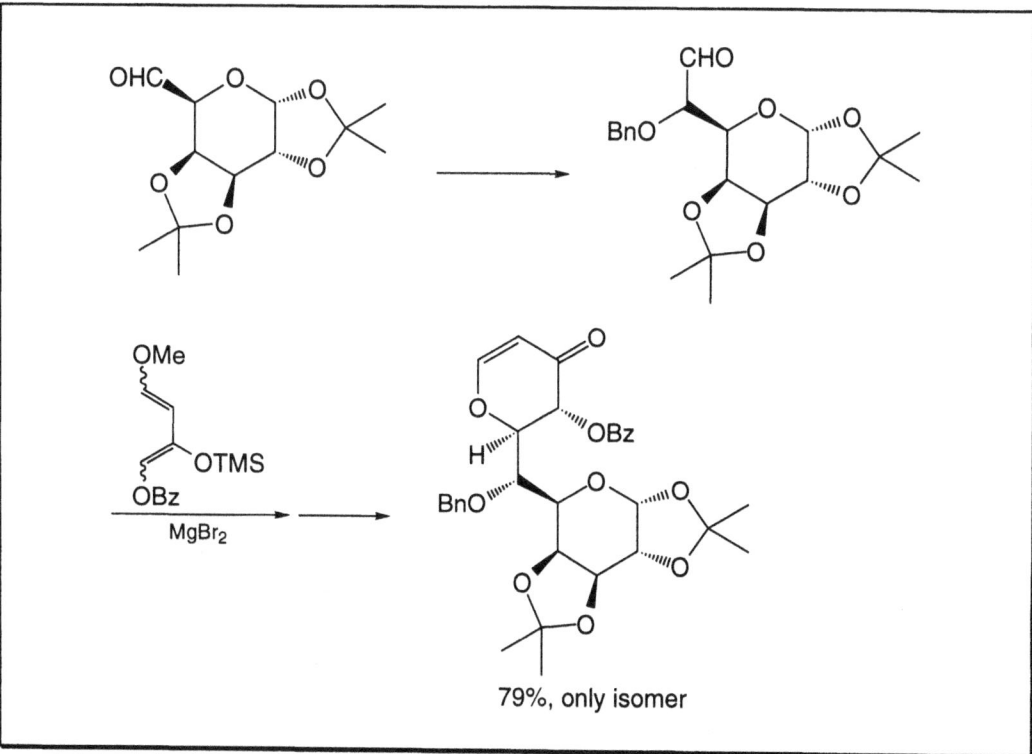

79%, only isomer

8.4 Syntheses Reported in 1986

Scheme 8.4.1   Metallations

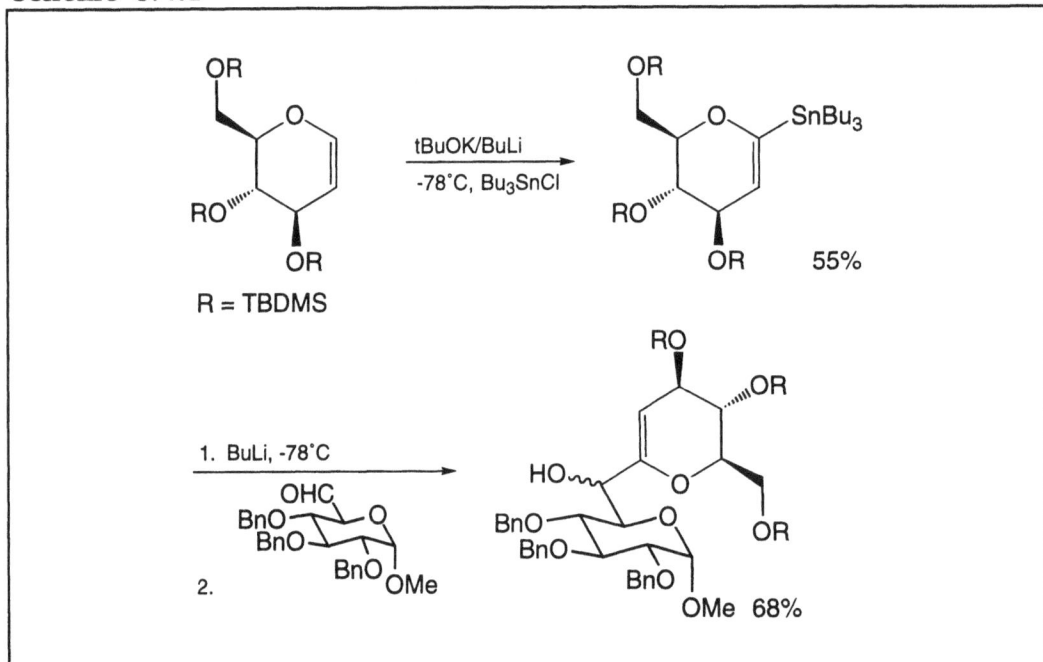

R = TBDMS

Scheme 8.4.2   1,3-Dipolar Cycloadditions

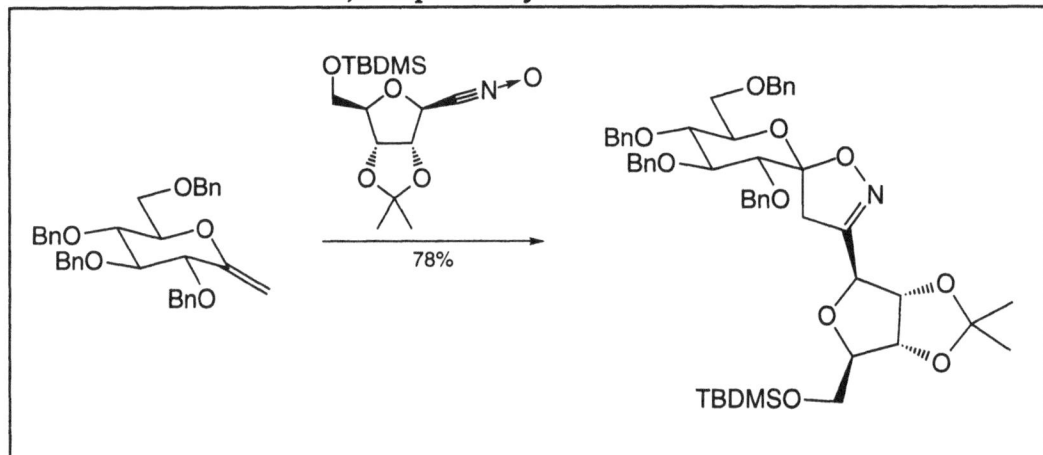

In 1986, as the technology surrounding the preparation of *C*-disaccharides progressed, the use of metallated glycals, radical reactions, and dipolar cycloadditions began to be used. With respect to metallated glycals, extensively discussed in Chapter 3, Hanessian, *et al.*,[9] stannylated the tri-*O*-silylated glucal shown in Scheme 8.4.1. The resulting product was then lithiated and added to a glucose aldehyde thus giving a mixture of diastereomeric alcohols at the newly formed *C*-glycosidic linkage.

Scheme 8.4.3   Radicals

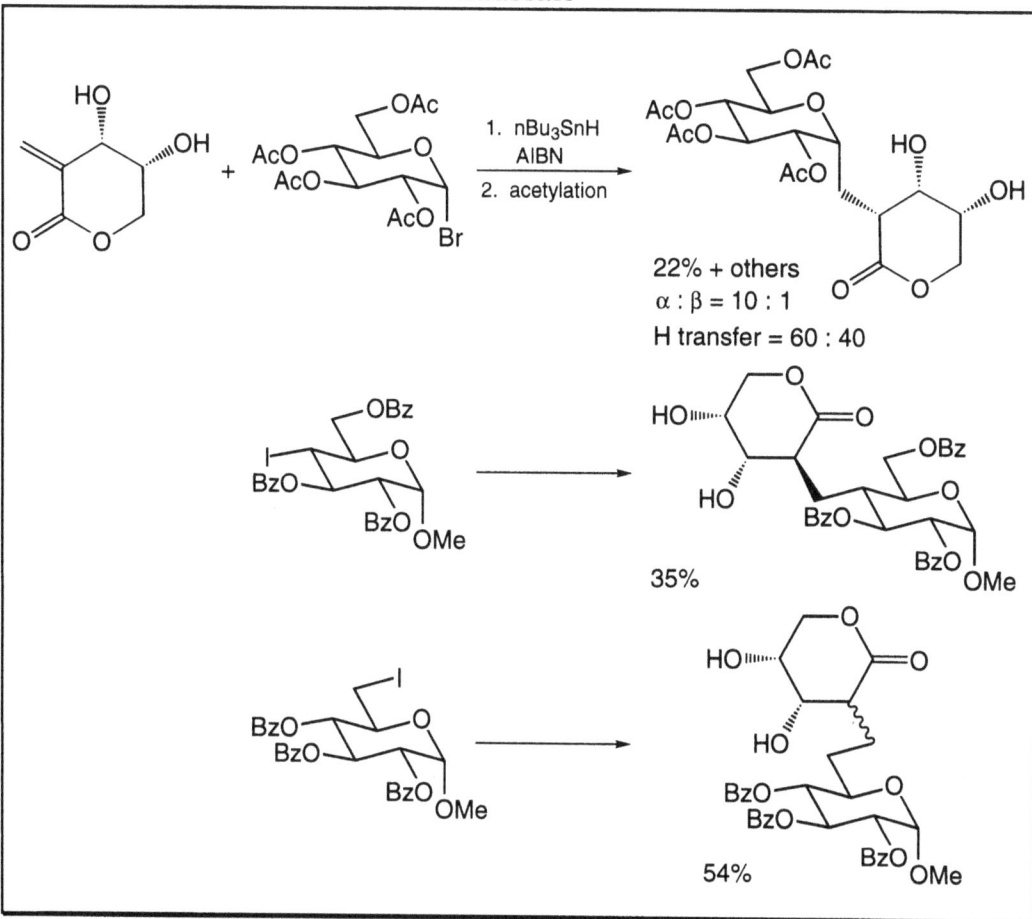

The 1,3-dipolar cycloaddition of exocyclic methylenes with nitrile oxides has already been addressed in section 2.13.[10] This work, shown in Scheme 8.4.2, is notable in that high stereoselectivity in the resulting isoxazoline can be obtained. This is presumably the consequence of the bulk of the nitrile oxide directing approach to the α face of the sugar methylene. Through the use of these reactions, *C*-disaccharides, bridged by isoxazoline rings have been made accessible.

Complementary to anion chemistry and 1,3-dipolar cycloadditions is the use of free radicals in the preparation of *C*-disaccharides. An early example of the use of this technology was reported by Giese, *et al.*,[11] and demonstrates the use of sugar halides as precursors to sugar radicals. Their subsequent addition to exocyclic methylenes present on sugar lactones is illustrated in Scheme 8.4.3. In all cases, yields were low to moderate but produced anomeric ratios in the range of 10 to 1.

8.5 Syntheses Reported in 1987

Scheme 8.5.1   Wittig/Mercury

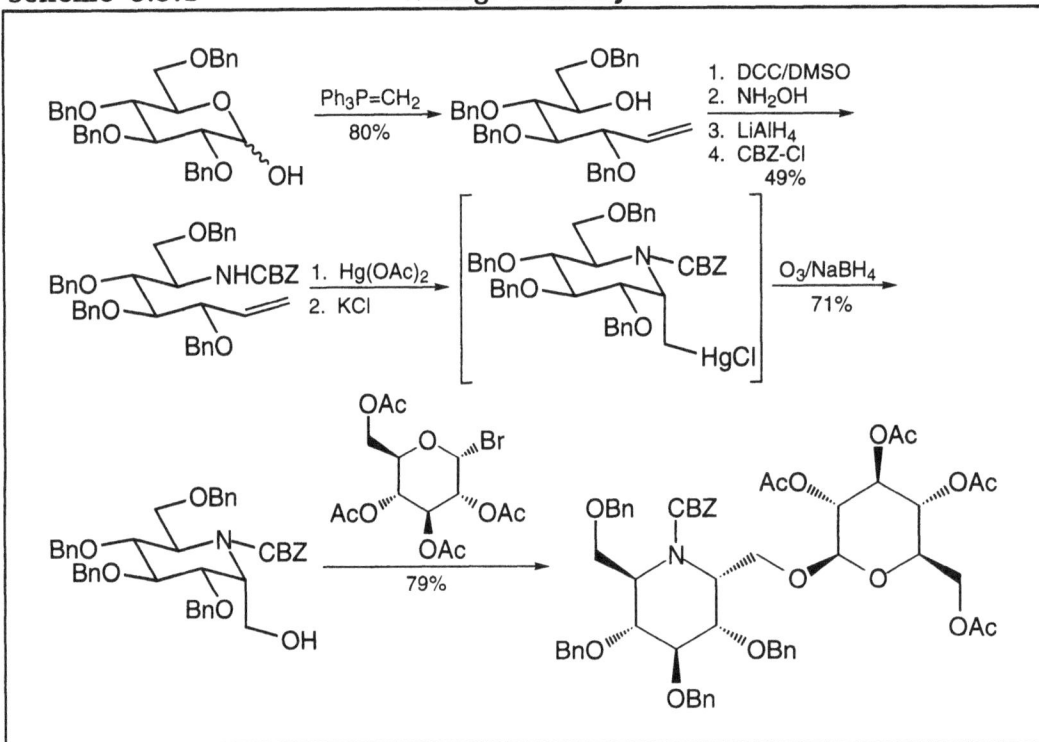

In a notable synthesis, reported in 1987, Liu, et al.,[12] illustrated the applicability of the Wittig reaction to sugars in the preparation of C-glycosides. The specific product of interest was a sugar derivative in which the ring oxygen was replaced with nitrogen. As illustrated in Scheme 8.5.1, treatment of 2,3,4,6-tetra-O-benzyl-D-glucopyranose with methylenetriphenylphosphorane effected formation of an 80% yield of the desired hydroxy olefin. Conversion of the free alcohol to a protected amine was accomplished in four steps with an overall yield of 49%. Final cyclization to the desired aza sugar was accomplished by first utilizing mercuric acetate and potassium chloride. Completion of the synthesis with ozone and sodium borohydride provided a three step yield of 71%. As the ultimate goal of this synthesis was the preparation of C-disaccharide inhibitors of α-glucosidases, the C-glycosidic aza sugar was coupled with 2,3,4,6-tetra-O-acetyl-D-glucopyranosyl bromide in a 79% yield.

Upon examining the structure of the final product, characteristics representative of both C and O-glycosides are apparent. This, however, does not weaken the importance of this report. The specific highlights utilizing Wittig and mercury methodology further demonstrates the utility of these reactions as general to the preparation of C-glycosides.

Figure 8.5.1   *C*-Isomaltose and
                          C-Gentiobiose Analogs

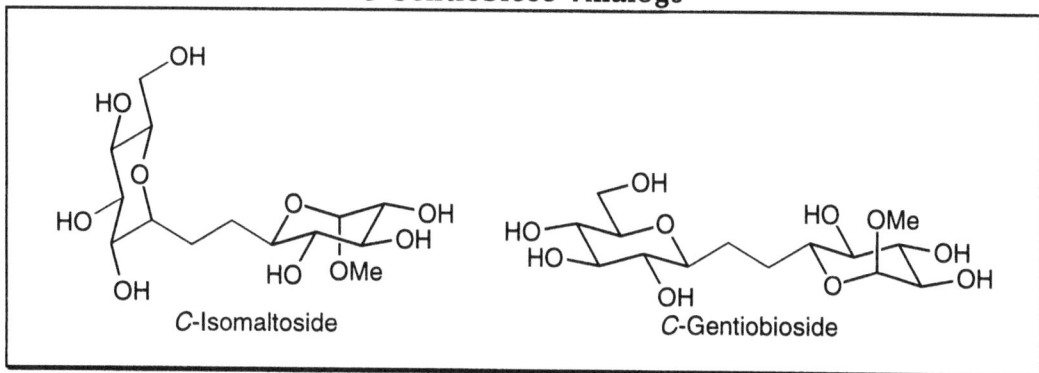

C-Isomaltoside   *C*-Gentiobioside

Additional work in 1987 centered around studies of the preferred conformations of various C-disaccharides. Preliminary work by Kishi, et al.,[13] was subsequently confirmed by the same group in 1991.[14] In these reports, utilizing C-glycosidic analogs of isomaltose and gentiobiose (Figure 8.5.1), it was determined that conformations of C-disaccharides can be predicted on the basis of steric interactions alone. This conclusion, based, in part, on similarities regarding the conformations and behaviors of conventional disaccharides in variable temperature proton NMR studies, was later extended to the conformations of C-trisaccharides introduced in Chapter 1.

8.6 Syntheses Reported in 1988

In 1988, methodologies utilized in the preparation of C-disaccharides included Michael additions, photochemical reactions, and 1,3-dipolar cycloadditions. Three examples illustrating the utility of these methodologies are presented in this section.

Beginning with conjugate additions, Giese, et al.,[15] utilized Michael-type additions as a direct means of coupling two sugar units. As shown in Scheme 8.6.1, the Michael acceptor was prepared from the phenylsulfoxide *via* treatment with base followed by formaldehyde and acid. The final C-glycosidation was accomplished by generating free radicals from 2,3,4,6-tetra-

O-acetyl-D-glucopyranosyl bromide. Subsequent 1,4-addition of this radical species yielded the desired methylene-bridged *C*-disaccharide with good stereoselectivity and a 45% yield for the overall sequence.

Scheme 8.6.1   **Michael Additions**

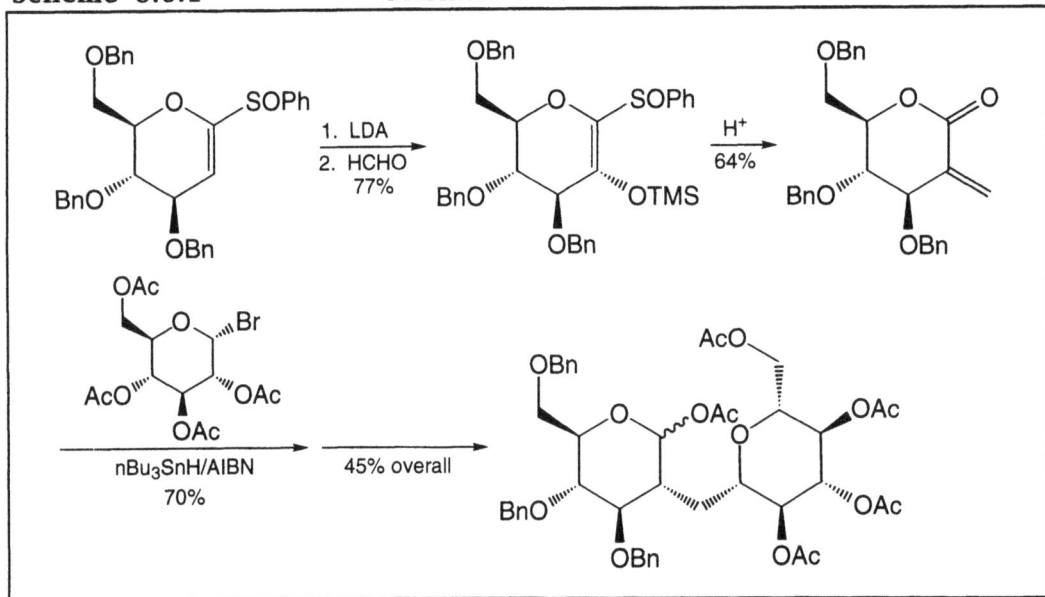

The use of free radicals is complimented by reactions catalyzed by light. Utilizing photochemical techniques, Giese, *et al.*,[16] effected the dimerization of a selenoglycoside to yield a *C*-disaccharide. The reaction, shown in Scheme 8.6.2, proceeded in poor yield with no stereoselectivity. However, through the use of tethers as well as varying the steric bulk on the protecting groups, it is reasonable to believe that this chemistry will ultimately find utility in a variety of applications.

Scheme 8.6.2   **Photochemistry**

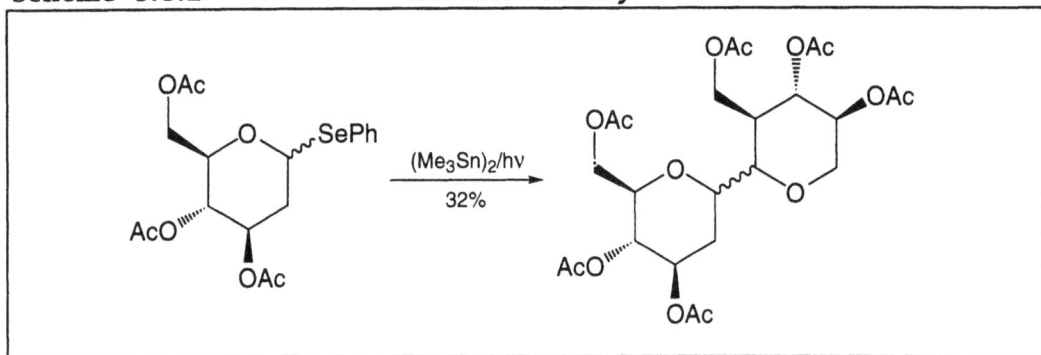

As an extension to the 1,3-dipolar cycloadditions presented in Scheme 8.4.2, Dawson, *et al.,*[17] demonstrated the formation and conversion of bridging isoxazolines to β-hydroxy ketones. A particular example, shown in Scheme 8.6.3, proceeded with good stereochemical control. However, although the chemical yield was 66%, the ratio of regioisomers formed in the reaction was 1:1. Subsequent reductive hydrolysis afforded the illustrated carbonyl bridged *C*-disaccharide.

Scheme 8.6.3   Dipolar Cycloadditions

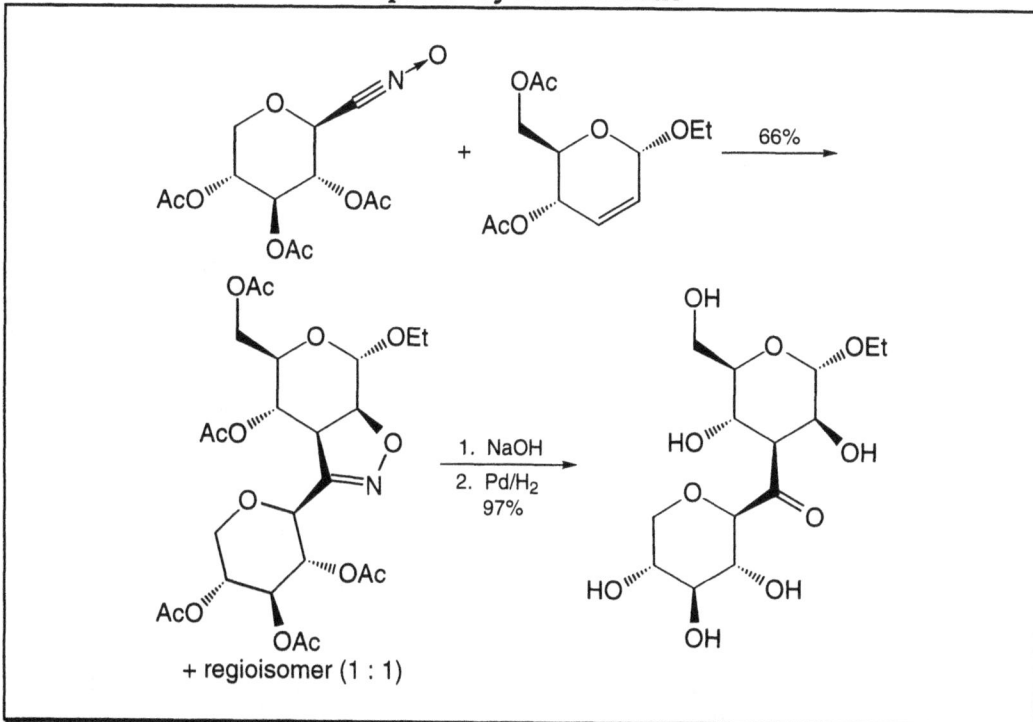

8.7   Syntheses Reported in 1989

In 1989, refinements and complementary additions to *C*-disaccharide technology surrounding acetylene anions were made. Additionally, technology relating to couplings with glycals bearing sulfoxides at C$_2$ was explored. Furthermore, examples of dimerization reactions, additions to difluoro olefins, and polyol cyclizations were reported. These technologies will be discussed in conjunction with furanose and pyranose sugars. Most notably, a synthesis of *C*-sucrose, bearing a furanose and a pyranose unit, will be examined.

Beginning with the use of acetylene anions, Armstrong, *et al.,*[18] prepared linear acetylene-bridged *C*-disaccharides. An example of the synthetic strategies is illustrated in Scheme 8.7.1 and begins with the ketone, shown. Stereoselective formation of the acetylene equivalent was achieved *via* formation of a primary alcohol followed by a Swern oxidation and a Wittig reaction with carbon tetrabromide. Although the intermediate aldehyde formed as a mixture of isomers, conversion to the desired stereoisomer was achieved on treatment with excess triethylamine.

Scheme 8.7.1   Acetylene Anions

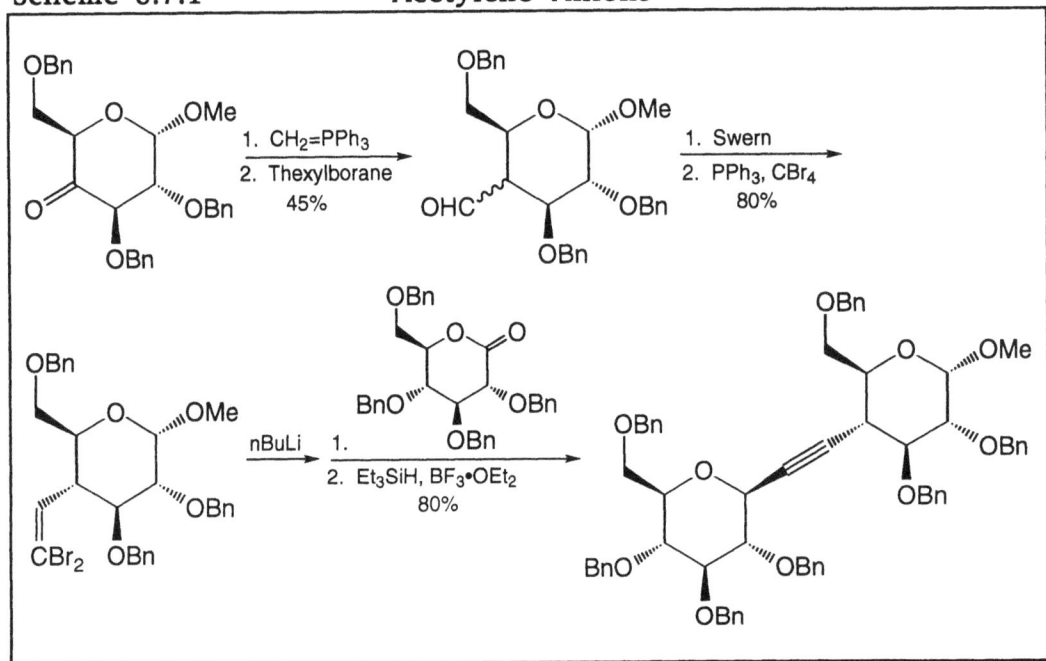

Regarding the final coupling to the *C*-disaccharide, the dibromo olefin was converted, *in situ*, to an acetylenic anion utilizing butyllithium. Addition of this species to tetra-*O*-benzyl gluconolactone followed by reduction of the hemiketal afforded the final product in approximately 80% yield for the two steps.

The use of 2-sulfoxide substituted glycals in the preparation of *C*-glycosides was extensively discussed in chapter 3. As shown in Scheme 8.7.2, Schmidt, *et al.*,[19] utilized this chemistry in the preparation of *C*-disaccharides. The key step involved deprotonation of the glycal followed by addition of the resulting anion to a sugar aldehyde. Final reductive cleavage of the sulfoxide provided a true *C*-disaccharide in good chemical yield and stereochemical purity.

As already alluded to, the dimerization of sugar units has provided some conceptually useful methods for generating *C*-disaccharides. As shown in Scheme 8.7.3, Boschetti, *et al.*,[20] effected the dimerization of two furanose units thus providing a one-step synthesis of a *C*-disaccharide. The reaction was initiated utilizing a Lewis acid and proceeded in 93% yield. Furthermore, the only site of multiple stereoisomers was the hemiketal. However, as demonstrated throughout this book, these species are easily reduced with substantial control over the stereochemistry of the relevant center. Thus, this important reaction sets substantial precedence in the applicability of this chemistry to the generation of increasingly more complex *C*-disaccharides and higher sugar analogs.

Scheme 8.7.2   **Glycal-2-Sulfoxides**

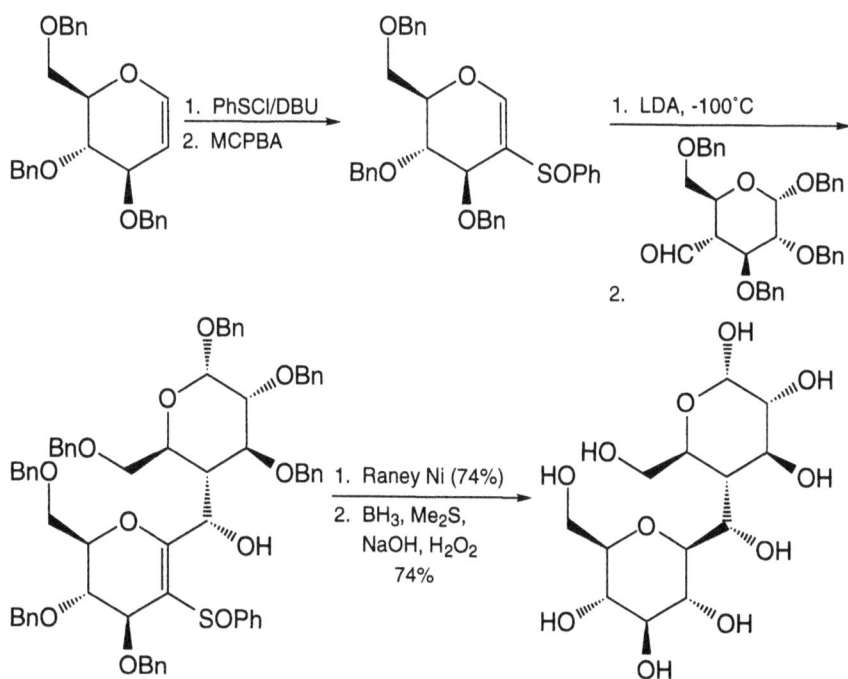

The use of free radicals in the generation of *C*-glycosides was extensively addressed in Chapter 5. Furthermore, Scheme 8.4.2 illustrates the use of free radicals in the preparation of *C*-disaccharides *via* Michael acceptors. An extremely complimentary addition to this technology is the use of difluoro olefins as free radical acceptors. As shown in Scheme 2.13.4, Motherwell, *et al.*,[21] utilized these species as substrates for *C*-glycosidations thus demonstrating their importance to free radical chemistry. In this same report, a difluoro olefin-derived from a furanose sugar was utilized in the formation of a *C*-disaccharide. The actual coupling, shown in Scheme 8.7.4, was initiated with tributyltin hydride and AIBN thus providing a 25% yield of the illustrated *C*-disaccharide as a single stereoisomer.

Scheme 8.7.3   Sugar Dimerizations

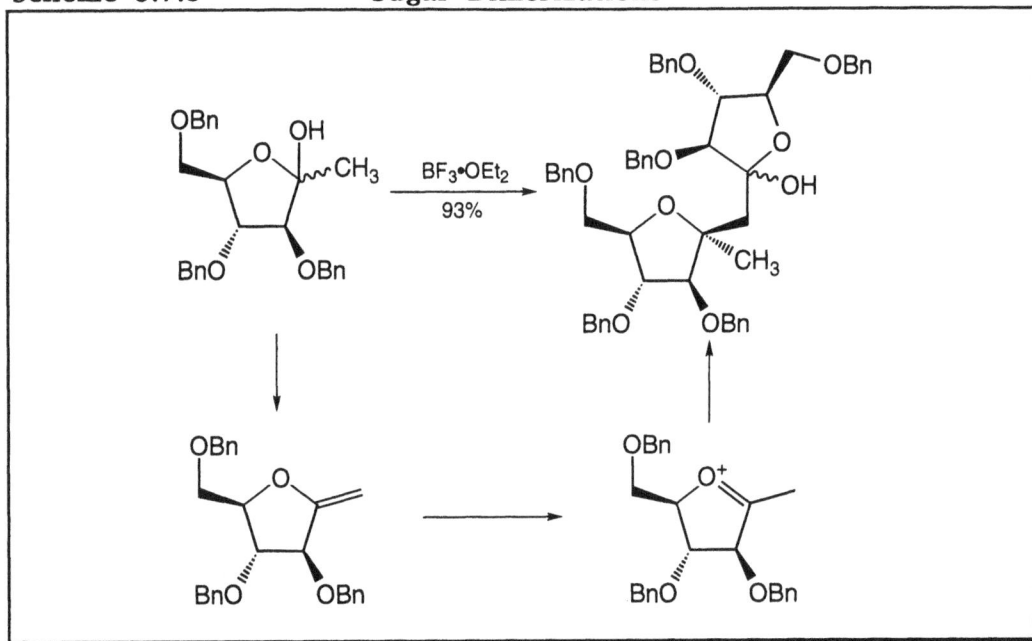

Scheme 8.7.4   Difluoro Olefins

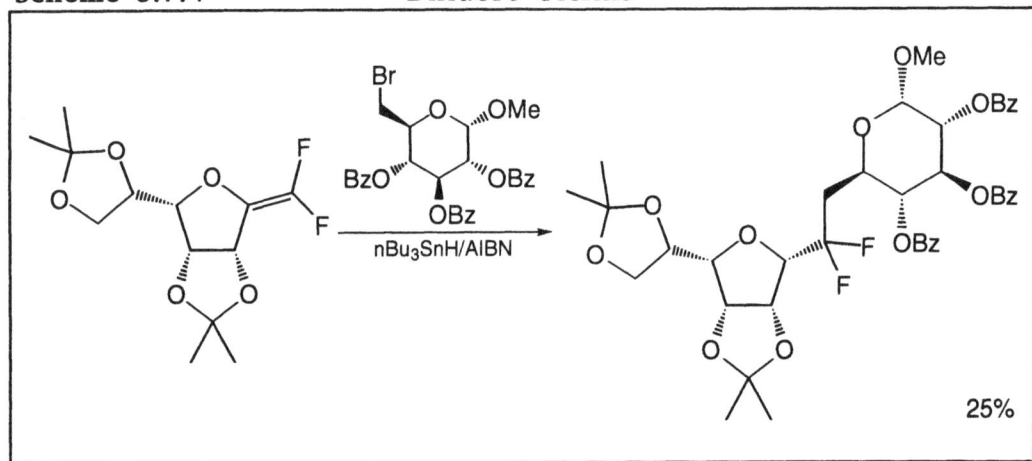

Among the most important C-disaccharide syntheses reported in 1989 is the preparation of C-sucrose. This analog of table sugar is interesting in that unlike natural sucrose, it cannot be metabolized to lower sugars. An actual synthesis of C-sucrose was reported by Nicotra, et al.,[22,23] and is illustrated in Scheme 8.7.5. As shown, the C-glucoside, prepared as discussed in Chapter 2, was converted to a metallated alcohol in 97% yield. Subsequent oxidation of the metal with iodine followed by oxidation of the alcohol to a ketone and treatment of the iodide with triphenylphosphine provided a 40% yield of the illustrated Wittig reagent. Condensation of the triphenylphosphorane with a protected glyceraldehyde gave the conjugated ketone, shown, in an 88% yield. The final conversion of this species to *C*-sucrose was achieved in six steps beginning with an osmylation of the olefin to a diol and ending with removal of the benzyl groups utilizing catalytic hydrogenation. Upon examination, the importance of this synthesis becomes apparent through recognition of a variety of chemical technologies described throughout this book. Furthermore, with respect to the food industry, sugar derivatives that possess the sweetness of natural sugars but lack susceptibility to metabolic processes are of substantial commercial value.

Scheme 8.7.5   *C*-Sucrose from a Polyol Glycoside

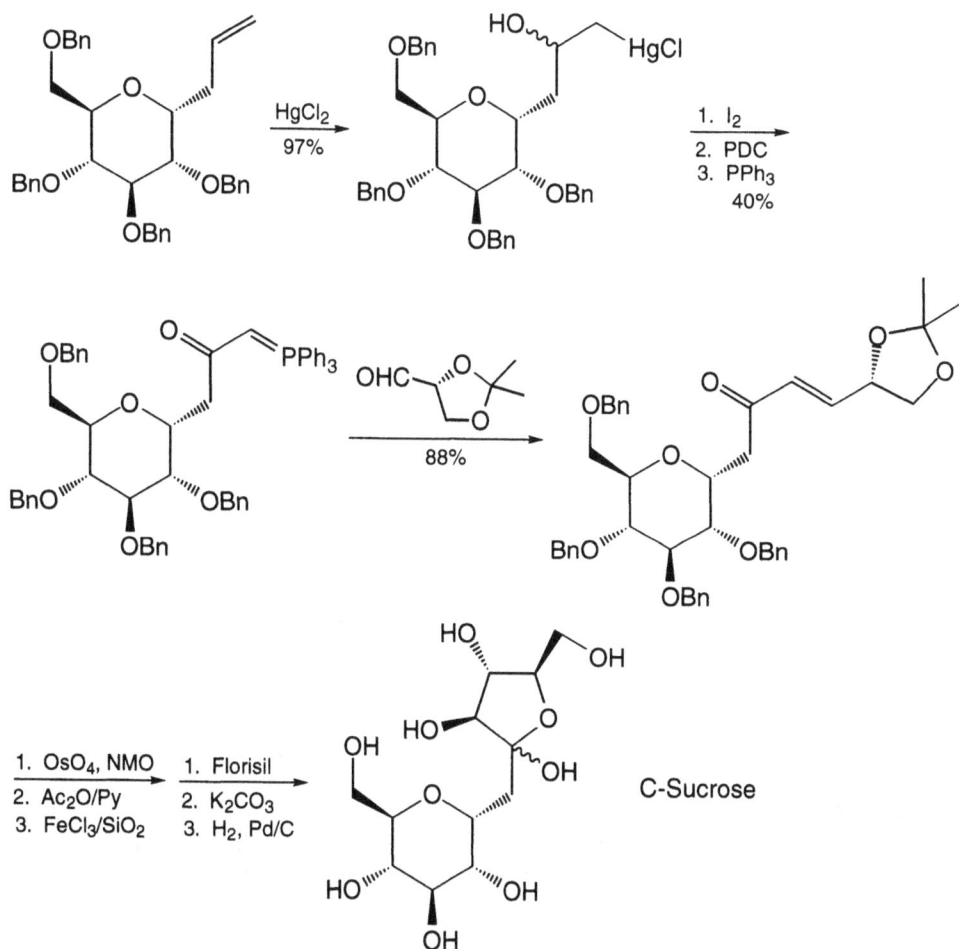

8.8   Syntheses Reported in 1990

In 1990, additional research into the conformational features of *C*-disaccharides was reported.[24] In addition to the insights provided by this study, significant developments in the synthetic methodology surrounding these compounds became apparent. Of notable interest are the use of *C*-nitroalkyl glycosides and palladium mediated coupling reactions.

Scheme 8.8.1   Palladium Mediated Couplings

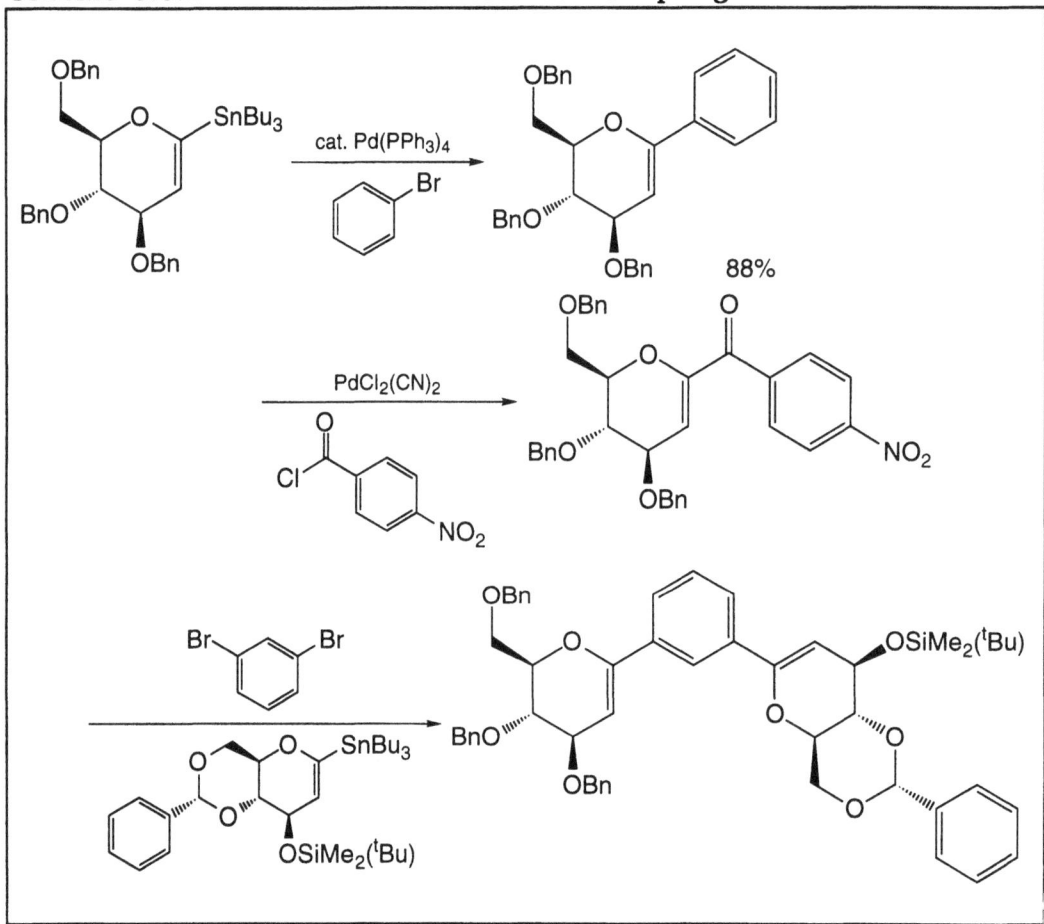

The use of palladium catalyzed couplings of 1-stannyl glycals with aryl halides was used by Dubois, *et al.*,[25] in the preparation of aryl-bridged *C*-disaccharides. As shown in Scheme 8.8.1, the high yielding reactions with arylhalides and acid chlorides were complimented by the use of 1,3-dibromobenzene. As illustrated, unlike the earlier examples, this compound acted as a linker thus joining two sugar units together. The resulting novel aryl-bridged C-disaccharide can be envisioned as capable of mimicking a trisaccharide. Furthermore, through the use of a variety of different sugar analogs, the ability to explore alternatives to a variety of naturally occurring trisaccharides becomes viable.

Scheme 8.8.2   **Nitroalkyl Couplings**

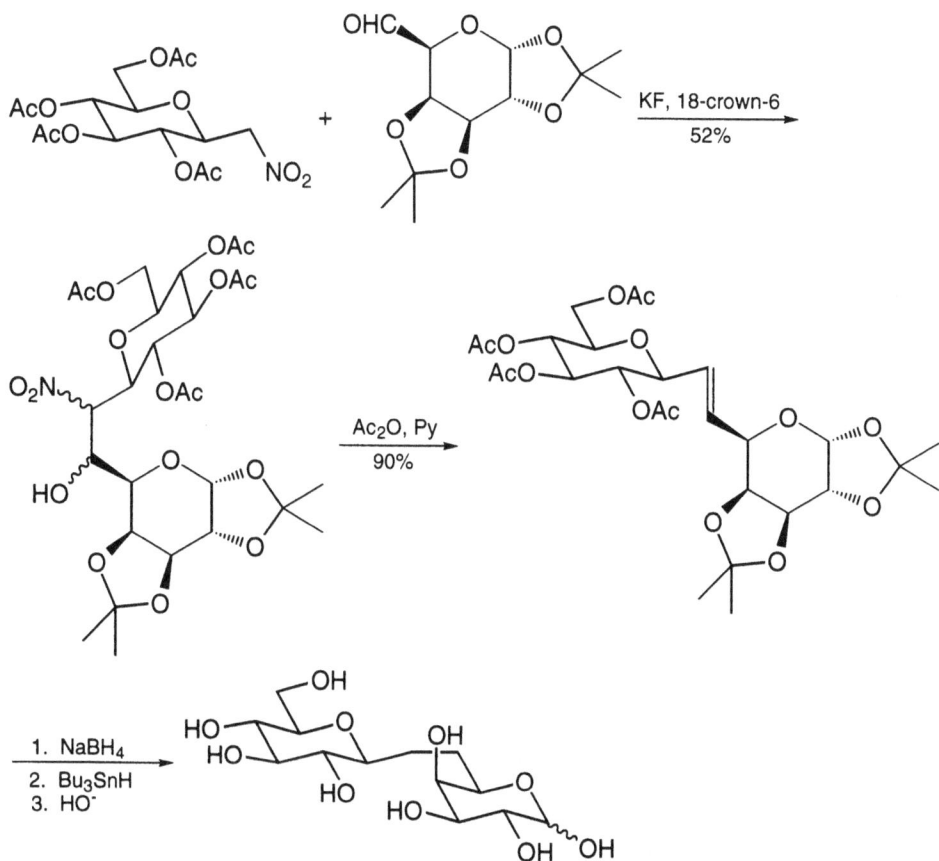

Chemistry involving the use of C-nitroalkyl glycosides has already been discussed in Chapter 2 and presented in Schemes 8.2.1 and 8.2.2. Complimentary to the dimerization reaction presented in Scheme 8.2.2 is the ability to apply similar chemistry to the unsymmetrical coupling of two different sugar units. Unlike the dimerization, the unsymmetrical couplings center around the addition of stabilized anions to aldehydes. As shown in Scheme 8.8.2, Martin, et al.,[26] reported the incorporation of this chemistry into the preparation of C-disaccharides. As illustrated, the actual coupling proceeded in a 52% yield with the subsequent elimination giving the olefin in a

90% yield. Final reduction of the double bond and removal of all protecting groups liberated the ethylene-bridged C-disaccharide. As shown in Scheme 8.8.3, the same report demonstrated the application of similar methodology to the preparation of methylene-bridged C-disaccharides via coupling with open-chain sugar analogs.

Scheme 8.8.3   **Nitroalkyl Couplings**

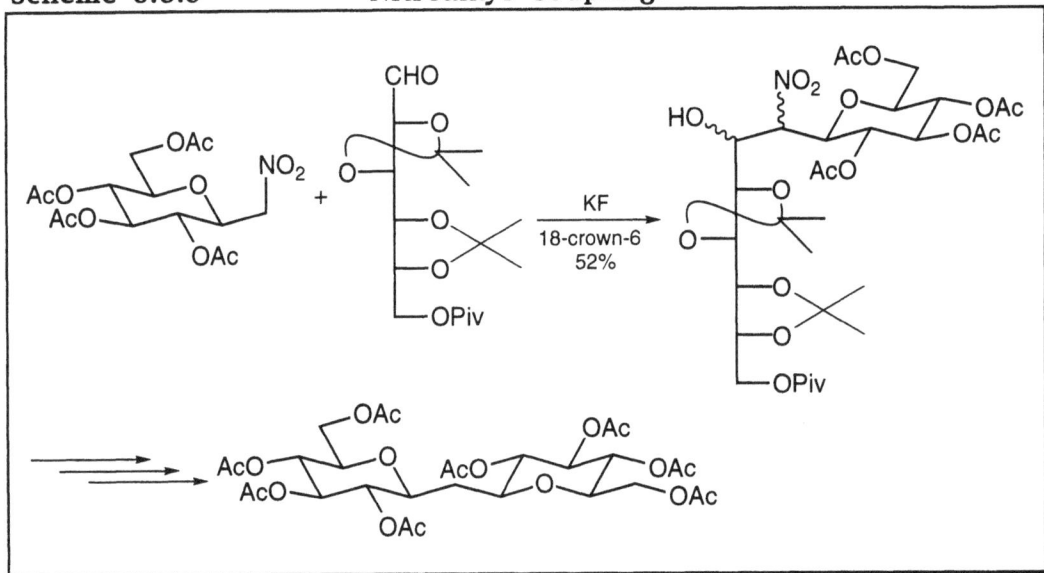

8.9 Syntheses Reported in 1991

In 1991, new syntheses of C-disaccharides incorporated Lewis acid mediated couplings with allylsilanes. Additionally, alkyllithium sugar derivatives were coupled to sugar lactones and electron deficient anions were added to sugar carbonates. Finally, complimentary to the anion chemistry used, sugar cuprates were utilized.

The use of allylsilanes in the formation of C-glycosides was extensively covered in chapter 2. This technology was adapted to the preparation of C-disaccharides by Stutz, et al.[27] As shown in Scheme 8.9.1, the illustrated glucose derivative was converted, in two steps, to the methyl ester. Treatment of the methyl ester with trimethylsilylmethyl Grignard followed by p-toluenesulfonic acid gave the allylsilane derivative shown. Combining this compound with tri-O-acetyl glucal in the presence of a Lewis acid gave the desired product in moderate yields. Thus, by combining well established Lewis acid mediated C-glycosidations, the preparation of a variety of C-disaccharides has been made available from relatively inexpensive and easily accessible starting materials.

Scheme 8.9.1   **Sugar Allylsilanes**

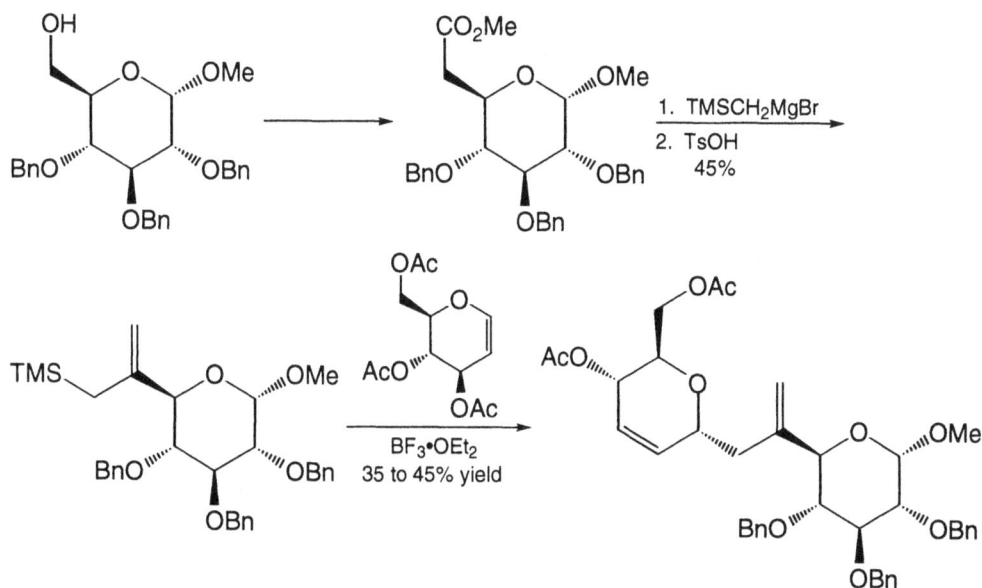

Where the use of Lewis acid chemistry demonstrates the utility of electrophilic substitutions in the preparation of *C*-disaccharides, methodologies involving nucleophiles are extremely complimentary. Already discussed in Chapter three and presented in numerous schemes throughout this chapter, sugar-derived nucleophiles provide excellent sources of *C*-glycosidation substrates. These observations were further exemplified by Schmidt, *et al.*,[28] in a study illustrating the generality of anionic additions to various sugar-derived lactones. As shown in Scheme 8.9.2, lithiated sugars were added to both glucose and galactose-derived lactones giving moderate yields of the coupled compounds. These products were subsequently deoxygenated under standard reductive conditions to give diastereomerically pure *C*-disaccharides. Although the yields of the additions were approximately 50%, the utility of these sequences is supported by the apparent control in the selective generation of β-*C*-glycosides.

Continuing with various anionic species utilized in the preparation of *C*-disaccharides, the importance of sugar-derived cuprates deserves discussion. As shown in Scheme 8.9.3, Prandi, *et al.*,[29] generated cuprates from stannyl glycosides *via* transmetallation and subsequent treatment with copper salts. On exposing these species to various epoxides in the presence of Lewis acids, *C*-glycosides were formed. An interesting observation is the stability of the stereochemistry of the formed cuprate thus allowing the formation of optically pure products. As shown, this chemistry was easily applied to the formation of glycal-derived cuprates. Interestingly, when glycals were used, the yields of

the reactions with epoxides appeared to increase as demonstrated by the
isolation of an 80% yield of the illustrated C-disaccharide. Furthermore, in
retaining the glycal skeleton, in conjunction with the methoxy glycoside of the
added unit, further selective chemical manipulations may possibly generate
substantially more complex structures.

Scheme 8.9.2   **Lithiated Sugars**

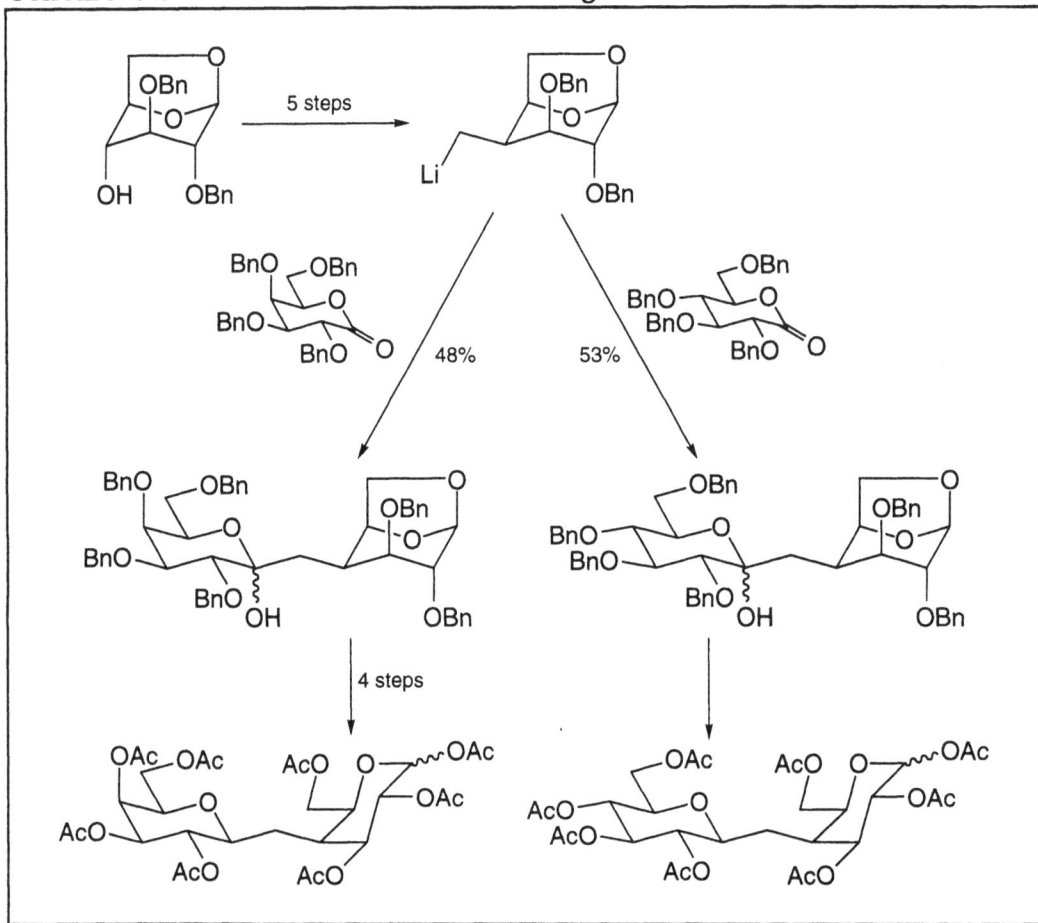

Although anionic species have proven useful in the preparation of C-
glycosides and C-disaccharides, substantial versatility is available from the use
of neutral palladium mediated couplings discussed in Chapter four and
presented in Scheme 8.8.1. Such reactions, applied to non-aromatic systems,
were demonstrated by Engelbrecht, et al.,[30] in the reaction of electron deficient
species with glycosidic carbonates. The reactions, shown in Scheme 8.9.4,
incorporate diethyl malonate as the nucleophilic species. Palladium was used to
effect the coupling of this compound to unsaturated sugars. When the more

electron deficient cyano ester was used, a double coupling was observed thus forming a symmetrical *C*-disaccharide. Of notable interest is the apparent control over the stereochemistry of the addition by varying the steric bulk surrounding the glycosidic carbonate. Furthermore, areas where further derivitization may be useful can be envisioned following decarboxylation of the methylene-bridge of the *C*-disaccharide.

Scheme 8.9.3   **Sugar Cuprates**

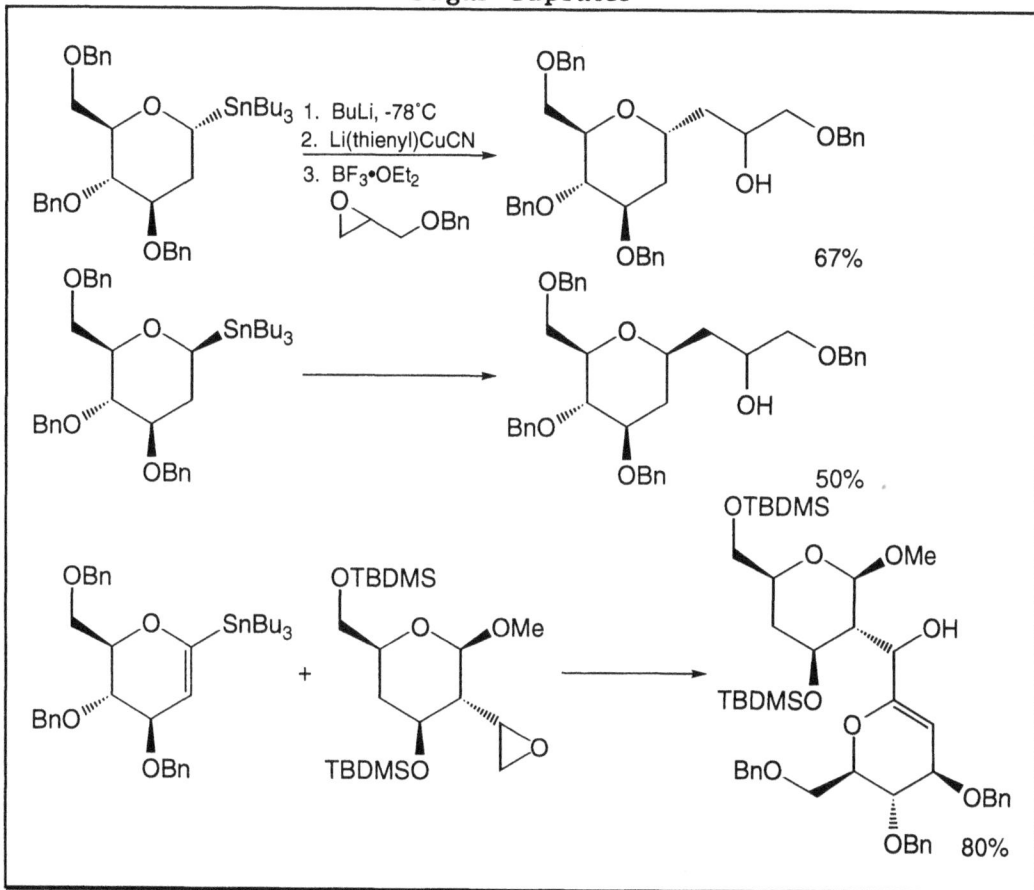

As a final example to close this section, further advances in free radical chemistry will be explored. Unlike the examples presented earlier in this chapter, the radical acceptors used by Bimwala, *et al.*,[31] were derived from sugar synthons. Specifically, as shown in Scheme 8.9.5, treatment of the bicyclic ketone, shown, with *tert*-butyl-dimethylsilyltriflate followed by the Eschenmoser salt provided the desired conjugated ketone.[32-34] The nature of the actual coupling reaction, which proceeded in a 48% yield, will be further elaborated upon in the next section.

Scheme 8.9.4   Glycosidic Carbonates

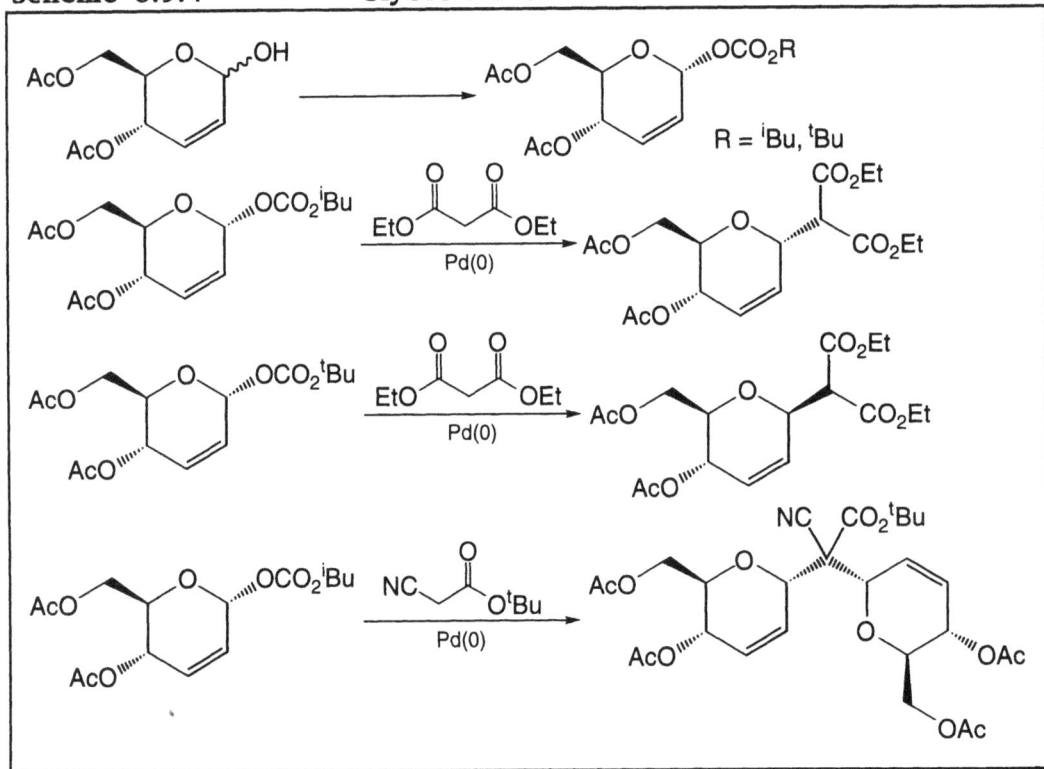

Scheme 8.9.5   Glycosidic Radicals
                              Added to Michael Acceptors

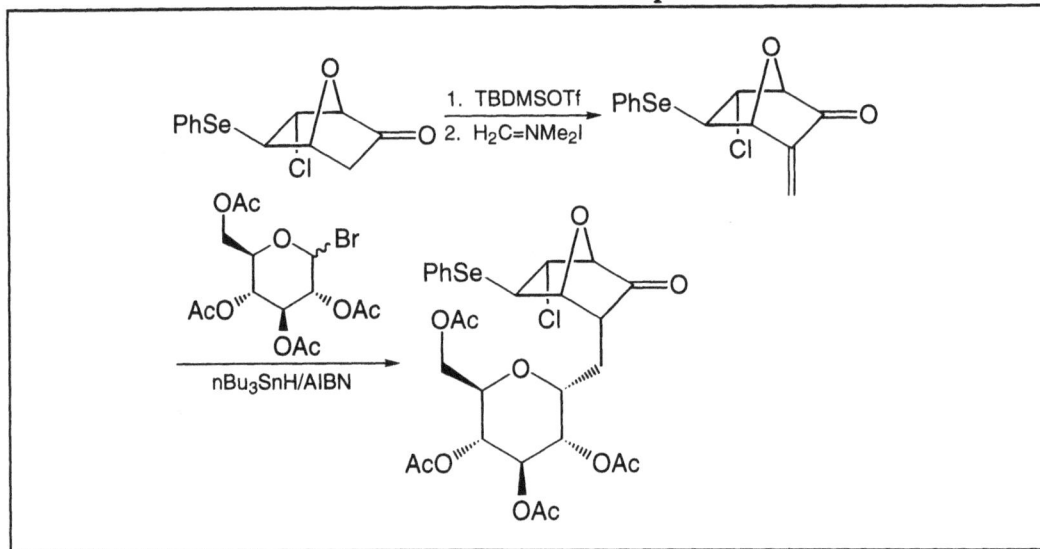

8.10 Syntheses Reported in 1992

Scheme 8.10.1
Glycosidic Radicals
Added to Michael Acceptors

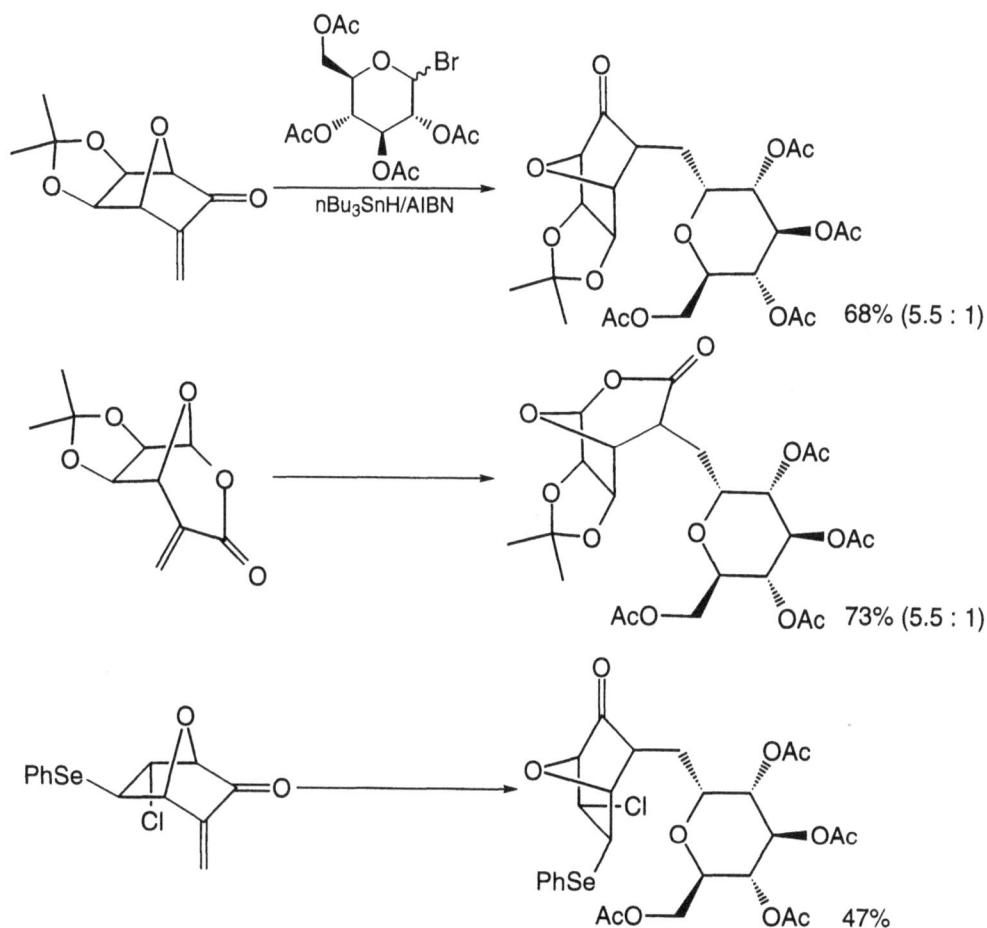

In 1992, the reported preparations of *C*-disaccharides included the utilization of free radical reactions. Additionally, the use of various polyol cyclizations were discussed. In the application of radical chemistry to the synthesis of *C*-disaccharides, Bimwala, *et al.*,[35] expanded upon his earlier work illustrated in Scheme 8.9.5. As shown in Scheme 8.10.1, glycosidic radicals were added α,β-unsaturated carbonyl compounds producing yields ranging from 47% to 73%. Furthermore, the illustrated *C*-disaccharide precursors exhibited diastereomeric ratios of approximately 5.5 : 1.

Scheme 8.10.2   Bromocyclizations

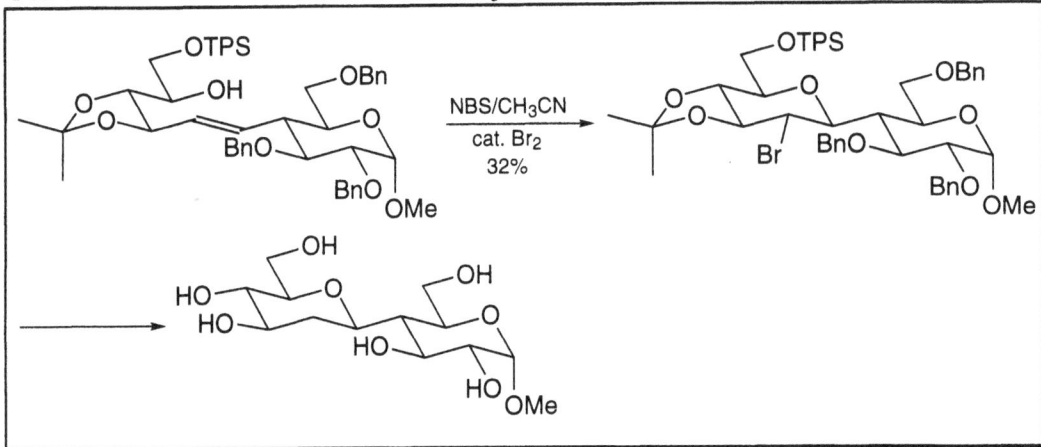

NBS/CH$_3$CN
cat. Br$_2$
32%

Scheme 8.10.3   Route A Starting Material

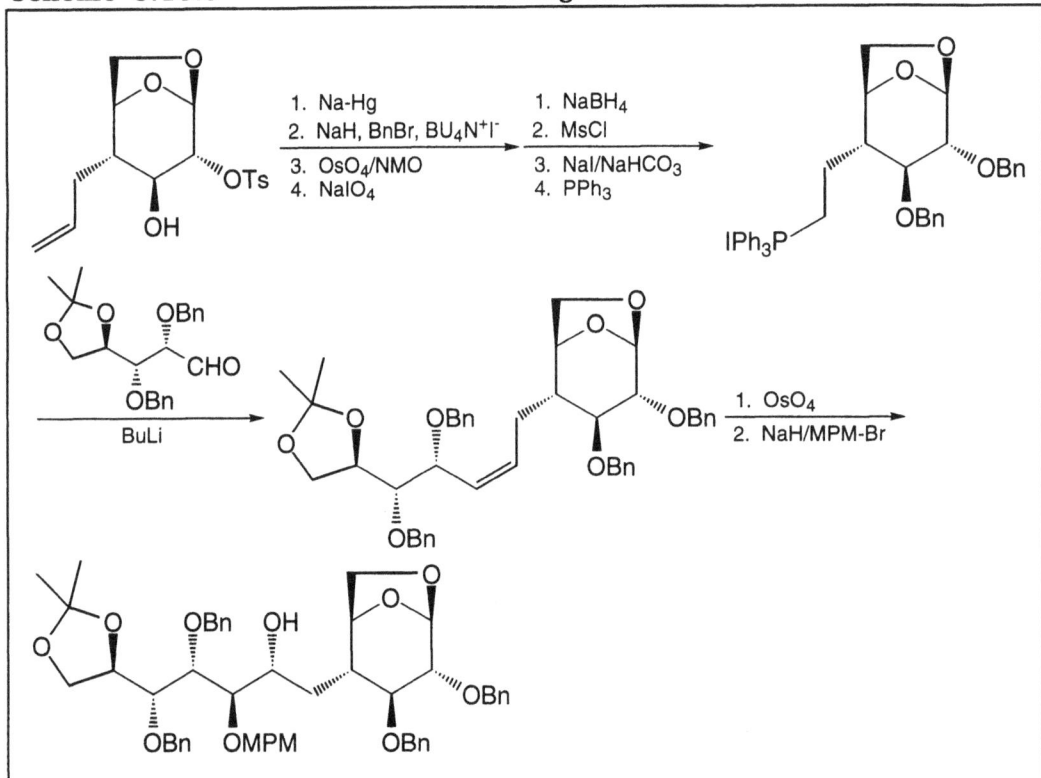

1. Na-Hg
2. NaH, BnBr, BU$_4$N$^+$I$^-$
3. OsO$_4$/NMO
4. NaIO$_4$

1. NaBH$_4$
2. MsCl
3. NaI/NaHCO$_3$
4. PPh$_3$

BuLi

1. OsO$_4$
2. NaH/MPM-Br

The cyclization of polyols to tetrahydropyran rings was discussed, in context with the formation of *C*-glycosides, in Chapter seven.   Through extensions of this chemistry, such cyclizations have been adapted to the preparation of higher *C*-polysaccharides. As an introductory example, a report by Armstrong, *et al.*,[36] is presented in Scheme 8.10.2. As illustrated, the use of bromine as an agent capable of effecting ring closure of certain hydroxyolefins provides a means of forming *C*-disaccharides from suitably derivitized sugars. Although the yield is relatively low, the illustrated reaction provides a novel and stereoselective strategy for ring closures.

The above mentioned bromocyclization is an excellent example of the formation of *C*-disaccharides through the cyclization of appropriately protected polyols. Utilizing different methodologies, Kishi, *et al.*,[37] effected similar reactions on sugar-derived polyols. The general strategies were illustrated in Schemes 7.3.6, 7.3.7, and 7.3.8 and are discussed, in detail, in the following paragraphs.

Scheme 8.10.4   **Route A Completion**

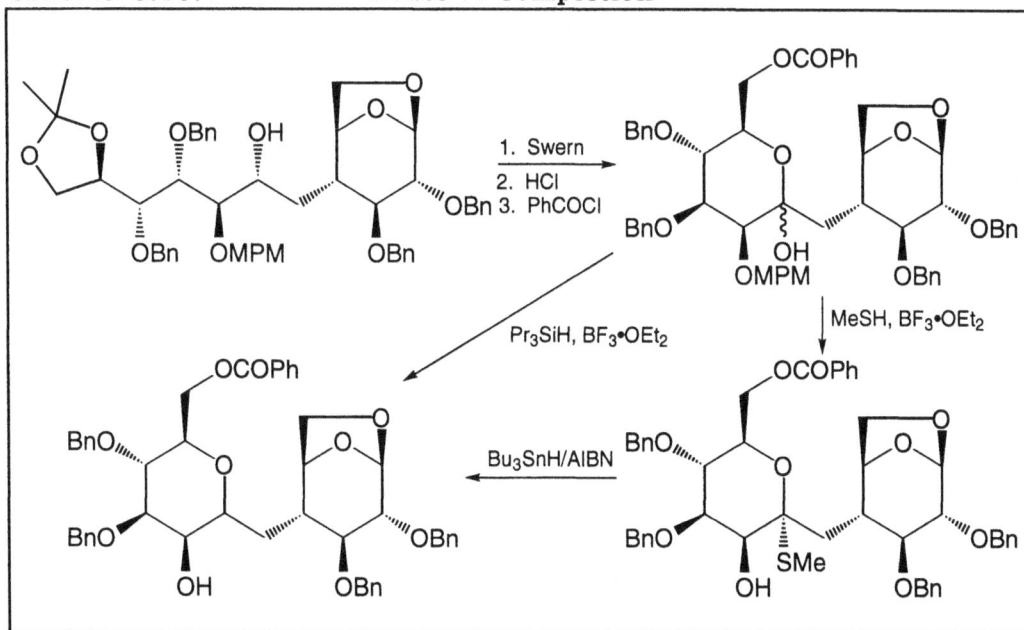

The strategy eluded to in Scheme 7.3.6 is elaborated upon in Scheme 8.10.3 and involves the allyl sugar derivative shown. Conversion of the allyl group to the phosphonium salt is accomplished in eight steps. Wittig olefination with the illustrated protected aldehyde provides the *cis* olefin which is dihydroxylated under asymmetric conditions and selectively protected as the PMB ether.

In Scheme 8.10.4, the above described protected polyol is converted to the illustrated hemiacetal *via* a Swern oxidation followed by spontaneous cyclization. The hemiacetal is converted to the final product through two routes. The first, being a direct strategy involves the Lewis acid mediated deoxygenation. The second route provides the same compound through the two step process involving formation of a thiomethylglycoside followed by removal of the thiomethyl group under radical conditions.

The strategy eluded to in Scheme 7.3.7 is elaborated upon in Scheme 8.10.5 and involved the starting material prepared in Scheme 8.10.3. Cyclization of the polyol is effected on removal of the acetonide under acidic conditions followed by the oxidative cleavage of the resulting diol. Spontaneous formation of the hemiacetal is followed by protection as the *p*-nitrobenzyl ether. Although two diastereomers are formed in this sequence, the minor product is easily converted to the major under mild basic hydrolysis of the *p*-nitrobenzyl ether followed by reprotection of the isomerized hemiacetal. This compound is then converted to the allene, shown, on treatment with propargyltrimethylsilane in the presence of trimethylsilyl triflate.

Scheme 8.10.5   Route B Starting Material

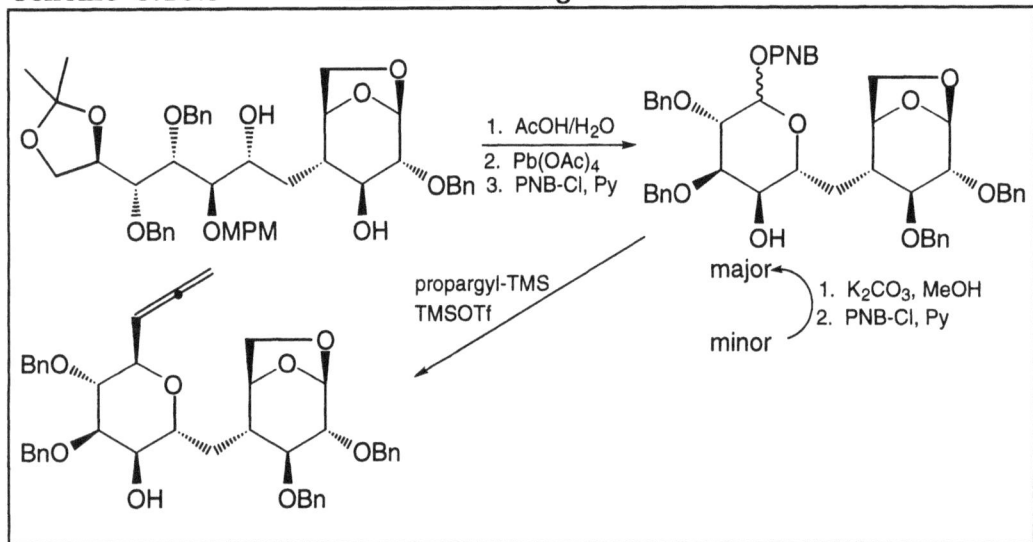

In Scheme 8.10.6, the above formed allene disaccharide is selectively converted to two diastereomeric disaccharides. This is accomplished through the inversion of the unprotected alcohol *via* Swern oxidation followed by borane reduction. The beauty of this demonstration is apparent through the remarkable stereochemical versatility achieved through the choice of the method used to effect polyol cyclization.

The final strategy, eluded to in Scheme 7.3.8, further exemplifies the range of *C*-disaccharides available from a single polyol when utilizing different methods of cyclizations. As shown in Scheme 8.10.7, the easily prepared aldehyde is coupled to the phosphonium salt described in Scheme 8.10.3. The resulting *cis* olefin was asymmetrically dihydroxylated and the resulting diol was protected as the *p*-methoxybenzylidine ketal. Ring opening of the ketal gave the PMB ether and the resulting free hydroxyl group was protected as the corresponding acetate. Cleavage of the acetonide followed by tosylation of the primary alcohol and treatment with sodium hydride gave the epoxide. Cyclization with ring opening of the epoxide was effected on removal of the PMB group followed by treatment with acid. The resulting *C*-disaccharide was then functionalized to represent the products described in the previous two approaches.

Scheme 8.10.6   **Route B Completion**

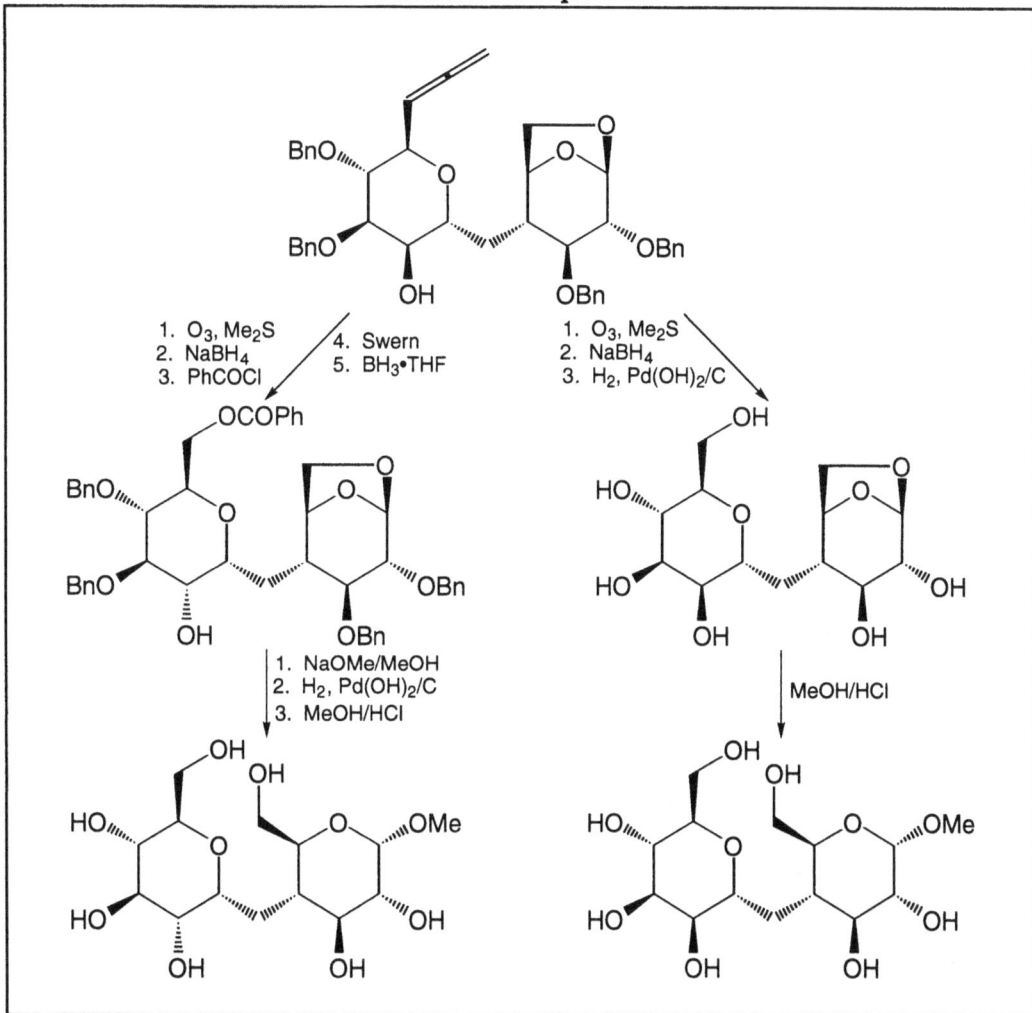

Scheme 8.10.7   Route C

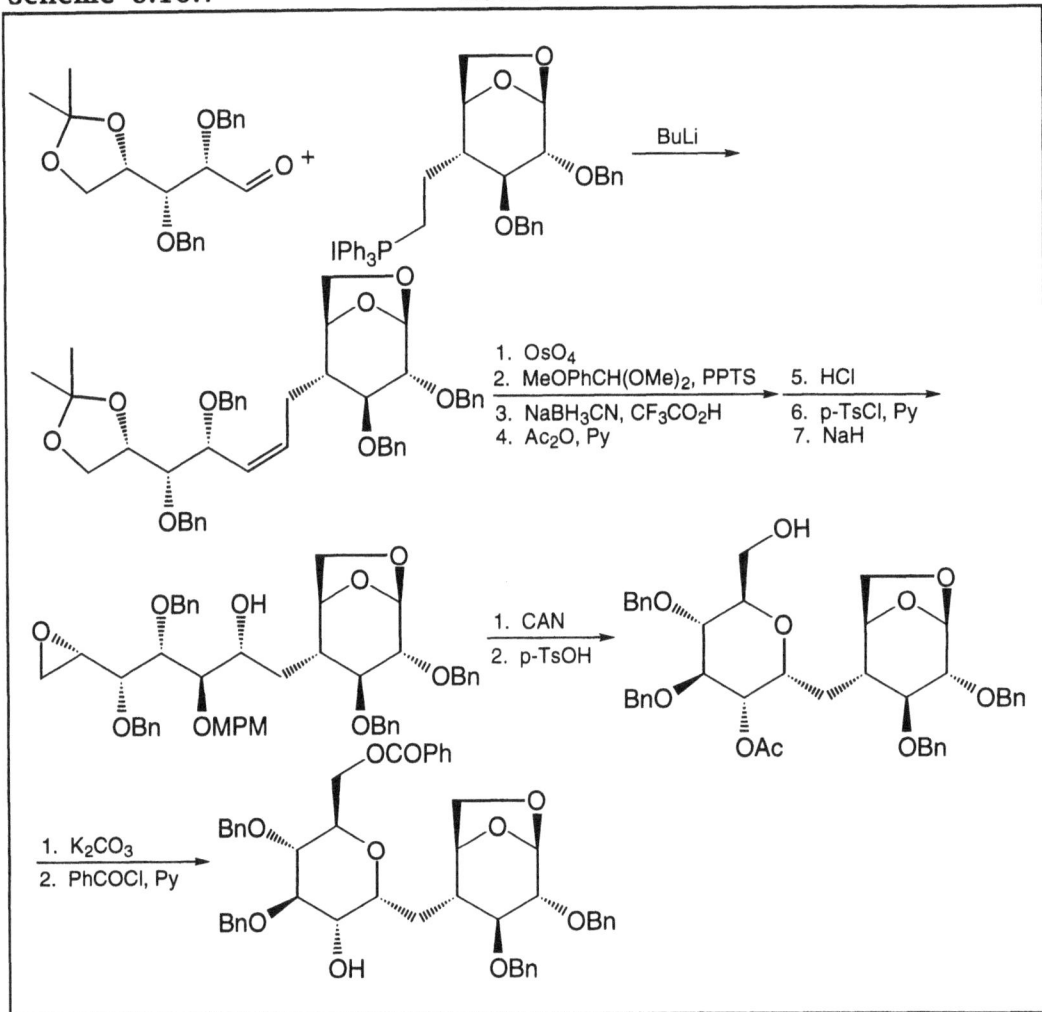

All the work described in Schemes 8.10.3 through 8.10.7 are nicely complimentary to an earlier preparation of C-disaccharides by Kishi, et al.[38] This report, published in 1991, differs from that just described in that the products are ethylene-linked C-disaccharides instead of being methylene-linked. The synthesis began, as shown in Scheme 8.10.8, with the illustrated hydroxymethyl C-glucoside. Swern oxidation followed by a Wittig reaction and subsequent treatment with base gave the acetylenic glycoside. Elaboration to the vinyl iodide was accomplished on treatment with iodine followed by potassium azodicarboxylate. Coupling to the aldehyde, shown, was then effected utilizing $NiCl_2/CrCl_2$ in DMSO.

Scheme 8.10.8   **Ethylene-Bridged**
                                     ***C*-Disaccharides -- Part 1**

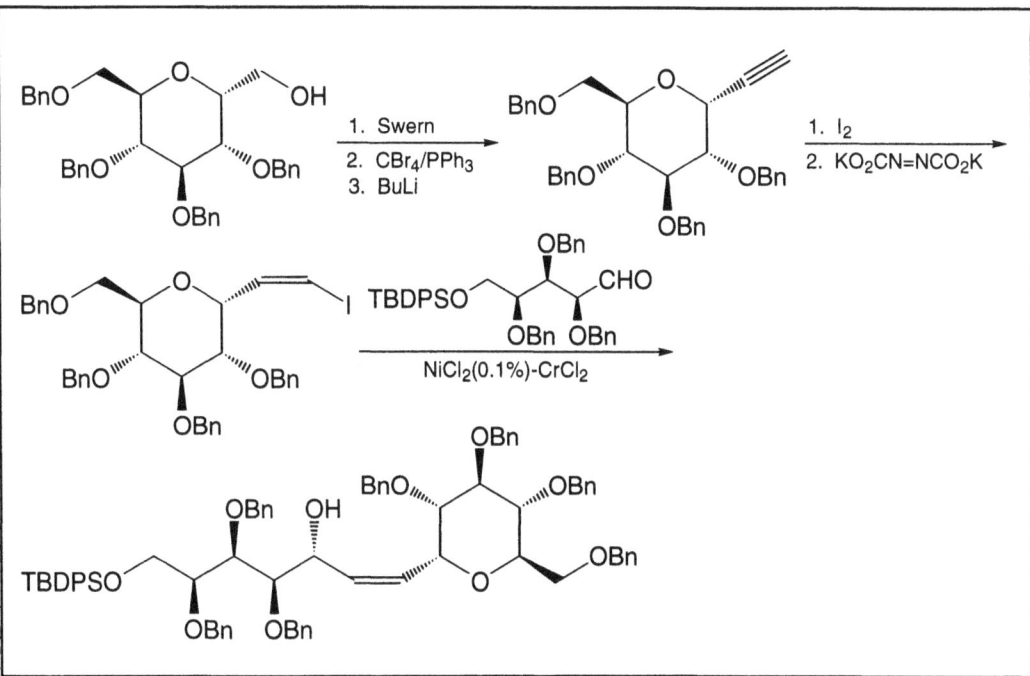

1. Swern
2. CBr₄/PPh₃
3. BuLi

1. I₂
2. KO₂CN=NCO₂K

NiCl₂(0.1%)-CrCl₂

Scheme 8.10.9   **Ethylene-Bridged**
                                     ***C*-Disaccharides -- Part 2**

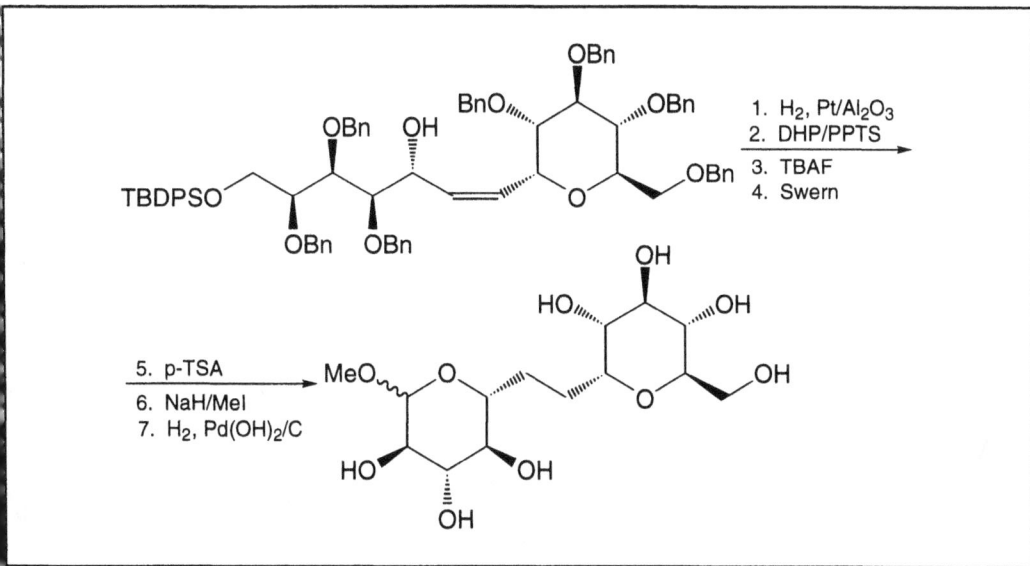

1. H₂, Pt/Al₂O₃
2. DHP/PPTS
3. TBAF
4. Swern

5. p-TSA
6. NaH/MeI
7. H₂, Pd(OH)₂/C

Completion of the synthesis was accomplished, as shown in Scheme 8.10.9, beginning with the selective catalytic hydrogenation of the olefin followed by protection of the secondary alcohol and removal of the silyl protecting group. Cyclization of the protected polyol was subsequently effected by oxidation of the alcohol under Swern conditions followed by removal of the THP protecting group. After spontaneous formation of the hemiacetal, conversion to the O-methylglycoside resulted from treatment with sodium hydride and methyl iodide. Finally, all protecting groups were removed under catalytic conditions to give the O-methylglycoside-ethylene-bridged-C-disaccharide.

8.11 Syntheses Reported in 1993

Scheme 8.11.1   Tethered Radicals

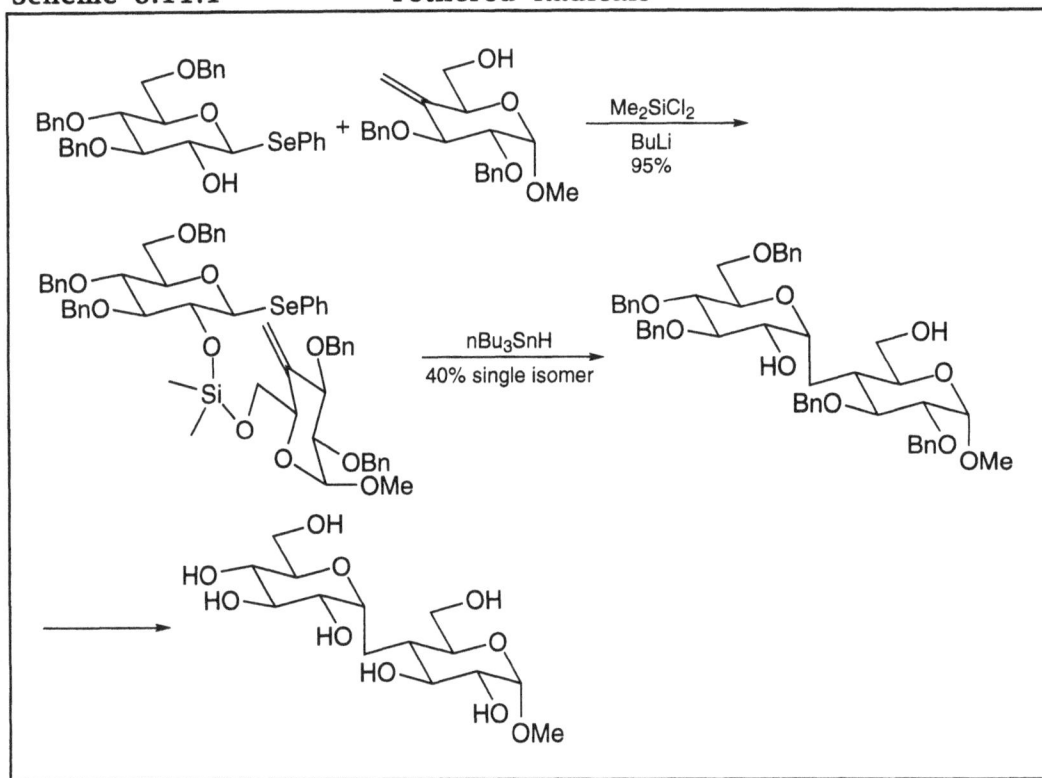

Trends in the state of C-disaccharide chemistry continued to change, and, in 1993, acetylene anions were revisited. Additionally, the use of tethers directing the approach of radical species to olefins gained much notoriety. An application of this technology, reported by Sinay, et al.,[39,40] is shown in Scheme 8.11.1 and begins with the selenoglucoside, shown. Coupling of this species to the olefinic sugar derivative was effected using a dimethyldichlorosilane linker.

Treatment of the resulting coupled compound with tributyltin hydride effected the formation of a 40% yield of the *C*-disaccharide as a single isomer. Removal of the protecting groups yielded the α-*C*-maltoside shown.

Scheme 8.11.2   Acetylene-Bridged *C*-Disaccharides

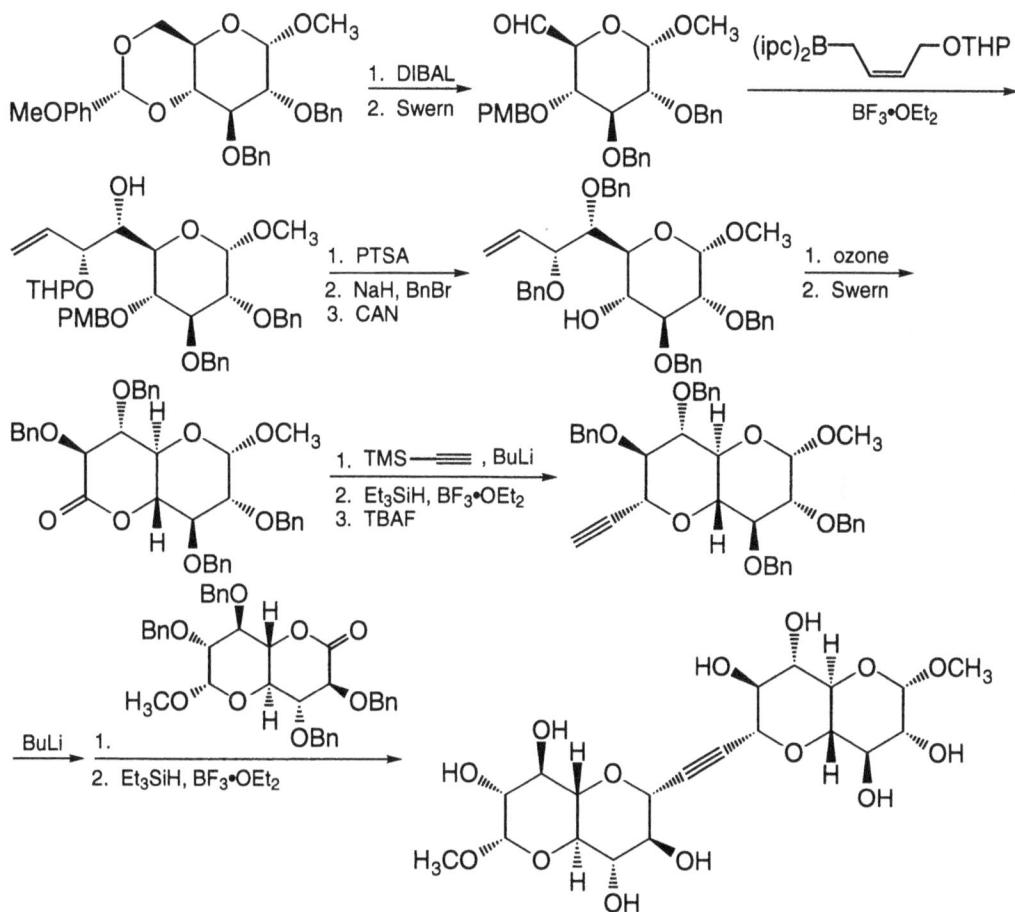

Utilizing the addition of acetylene anions to sugars, Armstrong, et al.,[41] prepared novel acetylene-bridged *C*-disaccharides. His synthesis, shown in Scheme 8.11.2, begins with the *p*-methoxybenzylidine protected sugar shown. Treatment of this compound with DIBAL followed by the application of the Swern oxidation to the resulting alcohol gave the aldehyde. Allylboration gave the extended side chain which was adjusted through protecting group manipulations giving the alcohol. Ozonolysis of the olefin followed by Swern oxidation of the spontaneously formed hemiacetal gave the sugar-fused sugar lactone.   Addition of the propargyltrimethylsilane anion followed by deoxygenation and removal of the silyl group gave the acetylene glycoside. Repeating this process with the sugar lactone gave the novel acetylene linked dipyranyl C-disaccharide after removal of the protecting groups.   A novel extension of this technology resulted in the acetylene-bridged glucose oligomer shown in Figure 8.11.1.

Figure 8.11.1   Acetylene Bridged C-Disaccharides

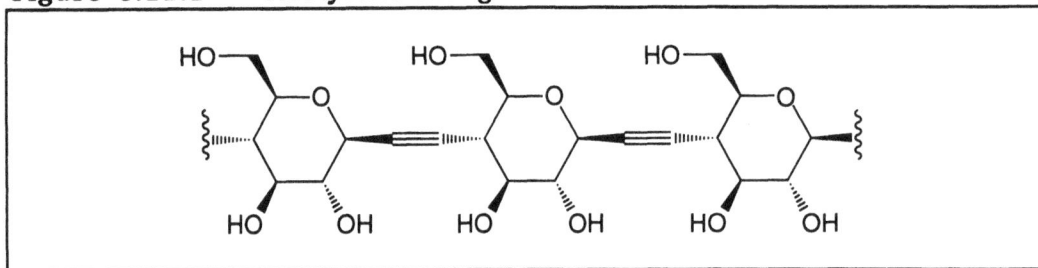

Scheme 8.11.3   C-Sucrose

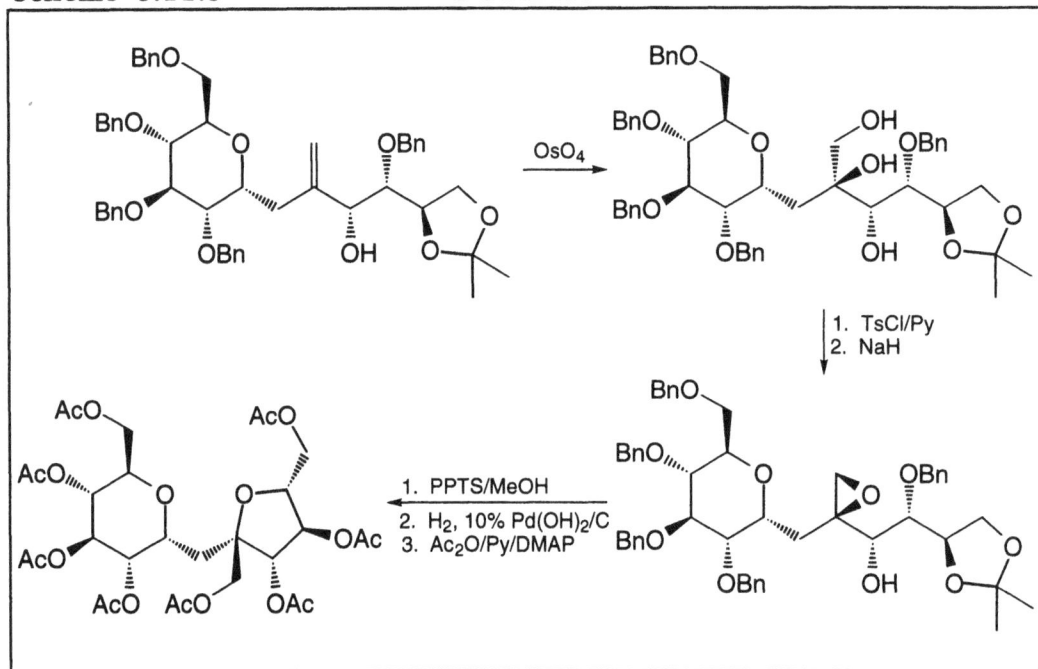

Additional work, originally reported in 1988 and elaborated upon in 1993, presents a synthesis of epi-C-sucrose and C-sucrose complimentary to that presented in Scheme 8.7.5. This chemistry, reported by Kishi, et al.,[42,43] involves the cyclization of a polyol. However, unlike the previously described preparation (Scheme 8.7.5), the key cyclization involves the tandem ring opening of an epoxide.

As shown in Scheme 8.11.3, C-sucrose was prepared from the illustrated olefin. Osmylation of the olefin followed by tosylation of the primary alcohol and treatment with base gave the epoxide. The acetonide was then hydrolyzed and the benzyl groups were removed. Spontaneous cyclization of the furanose ring was effected on removal of the acetonide. In all, an efficient synthesis of C-sucrose was accomplished in five steps from the starting olefin.

Figure 8.11.2   Blood Group Determinants

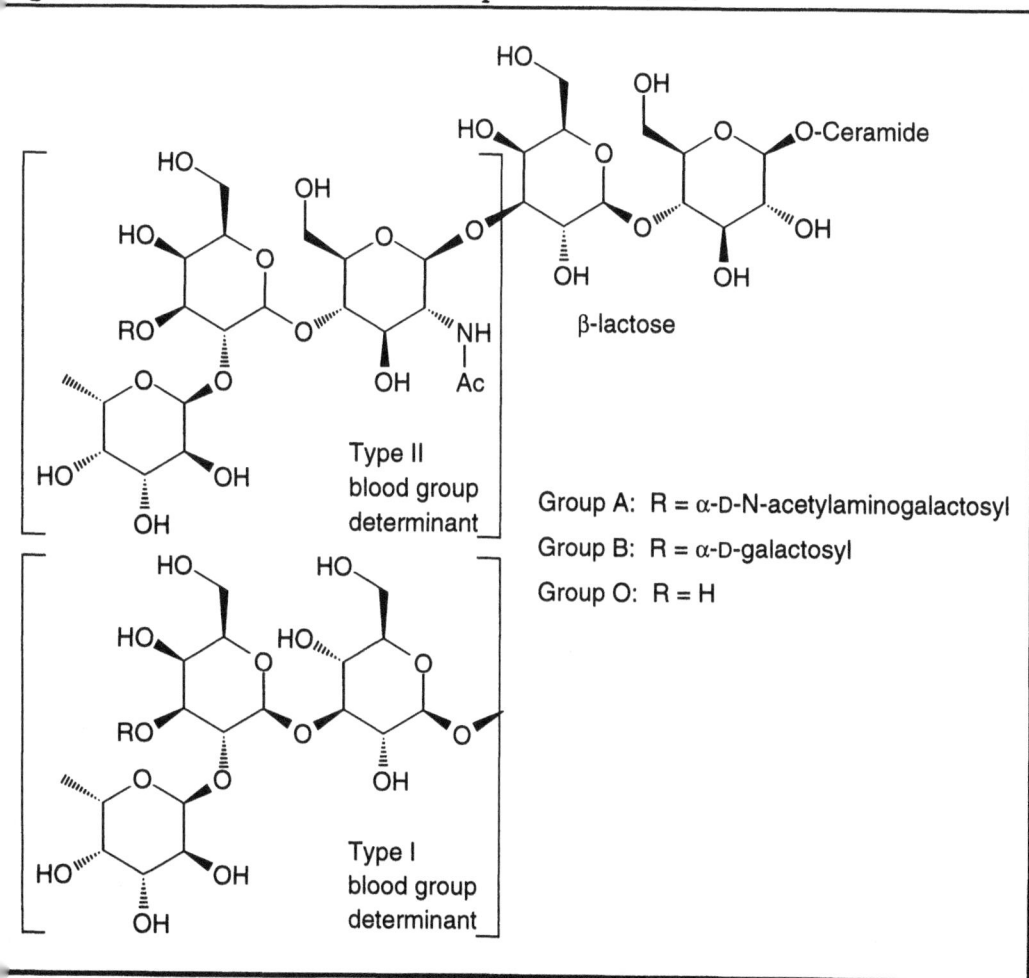

β-lactose

Type II blood group determinant

Group A: R = α-D-N-acetylaminogalactosyl

Group B: R = α-D-galactosyl

Group O: R = H

Type I blood group determinant

So far, this chapter has primarily focused on the synthesis of C-disaccharides. However, notable preparations of C-trisaccharides have been reported. In one particular example, Kishi, et al.,[44,45] targeted the synthesis of the blood group determinants illustrated in Figure 8.11.2. In this rather extensive synthesis, the initial starting material, prepared as shown in Scheme 8.11.4, was derived from the illustrated sugar analog. Conversion to the methyl ketone was accomplished *via* Swern oxidation to the aldehyde followed by addition of methyl Grignard and an additional Swern oxidation. The coupled product was prepared *via* aldol condensation with the aldehyde. This compound was then cyclized on removal of the silyl protecting group. The resulting hemiacetal was deoxygenated under previously described conditions and the alcohol was oxidized to a ketone rendering the diastereomeric mixture irrelevant.

Scheme 8.11.4   **Starting Material A**

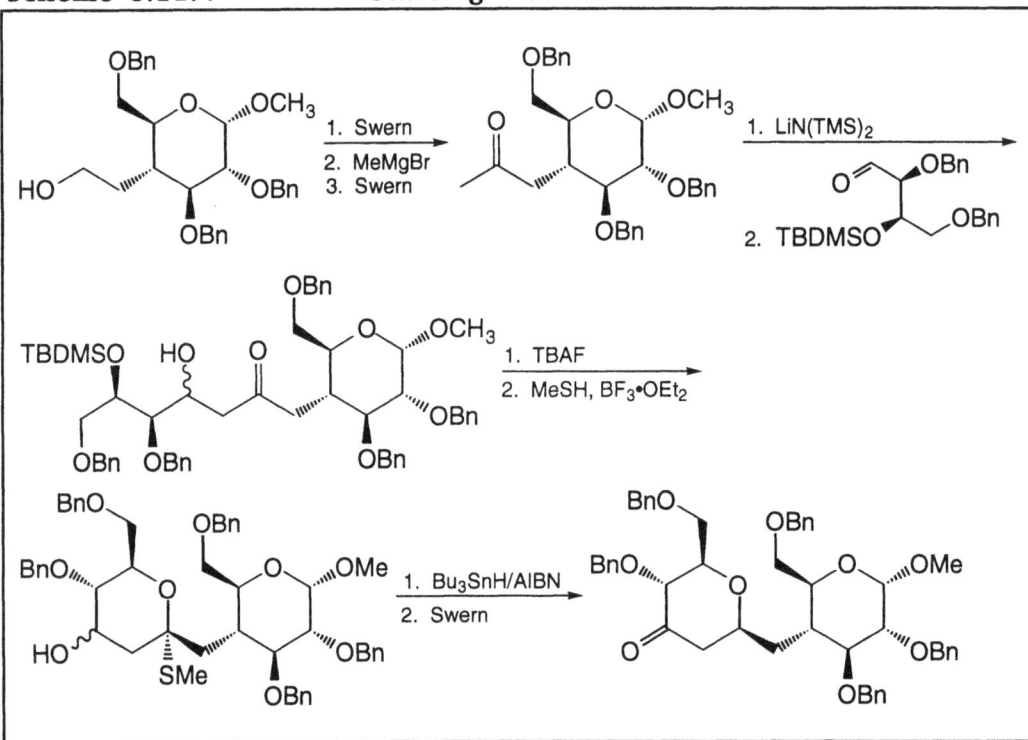

Scheme 8.11.5   **Starting Material B**

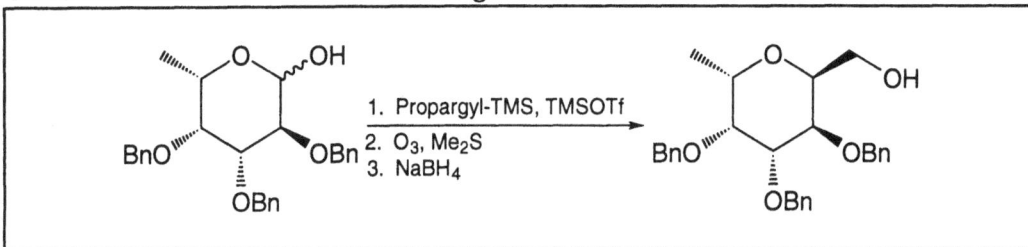

Scheme 8.11.6   A + B

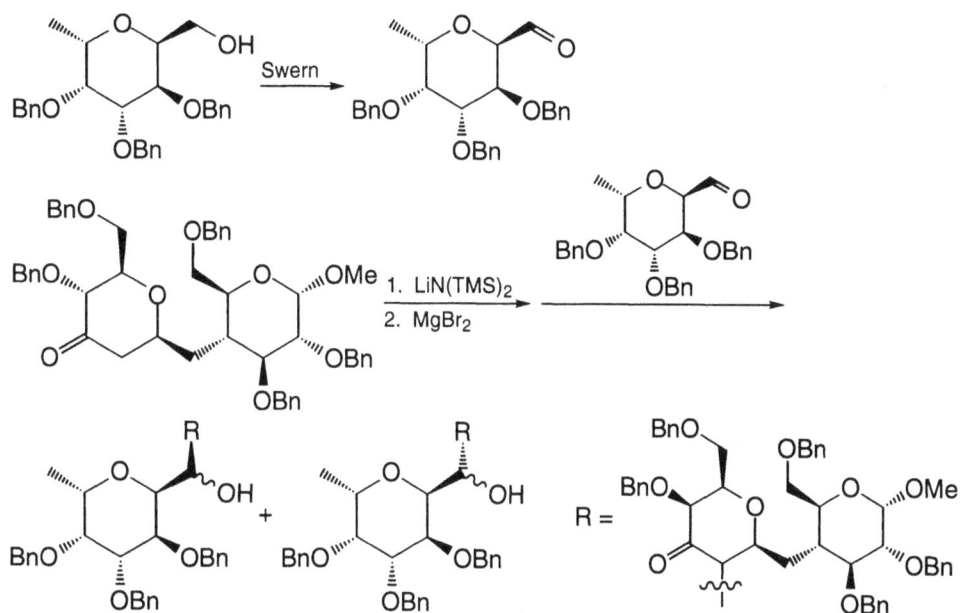

Scheme 8.11.7   Conclusion of Synthesis

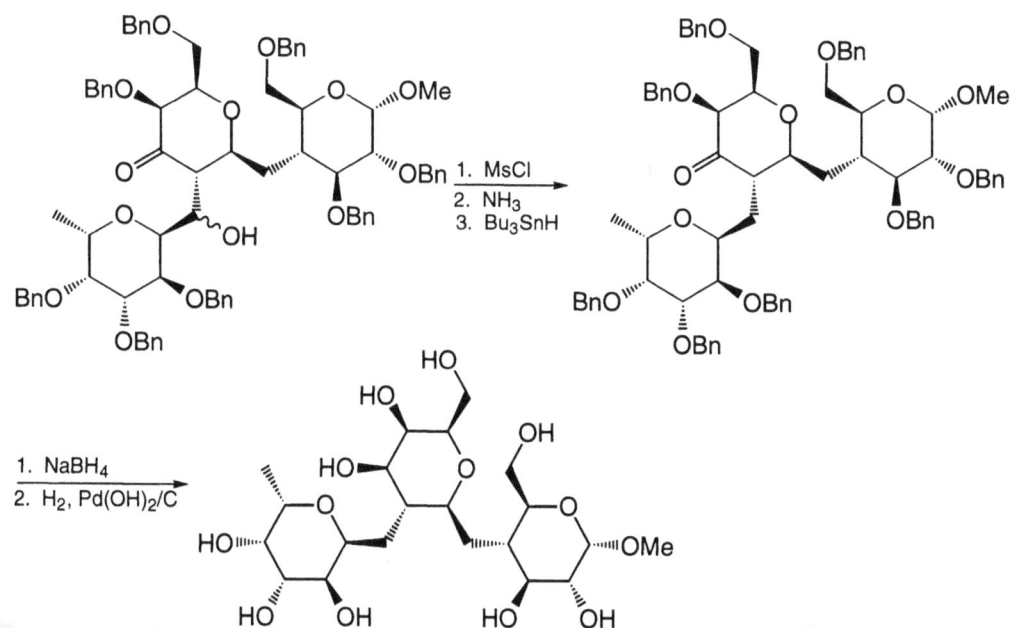

The next starting material, prepared as shown in Scheme 8.11.5, is the result of Lewis acid mediated *C*-glycosidation of the illustrated fucose derivative. Ozonolysis of the resulting allene followed by reduction gave the desired alcohol.

Starting materials A and B were subsequently coupled as shown in Scheme 8.11.6 *via* oxidation of the alcohol to an aldehyde followed by an aldol condensation. The resulting products were separated and the diastereomeric mixture of alcohols were carried on as shown in Scheme 8.11.7.

The conclusion Kishi's synthesis of *C*-glycosidic analogs of the blood group determinants stems from the diastereomeric mixture of alcohols described in Scheme 8.11.6. As shown in Scheme 8.11.7, the alcohol groups were mesylated and subsequently treated with ammonia in tetrahydrofuran. The resulting enones were reduced on treatment with tributyltin hydride thus yielding the illustrated $C_{2'}$-equatorial ketone as a single isomer. Completion of the synthesis was accomplished on reduction of the ketone with sodium borohydride followed by removal of the benzyl groups under catalytic conditions.

8.12 Syntheses Reported in 1994

Scheme 8.12.1   Hydroxyethylene-Bridged *C*-Dipyranosides

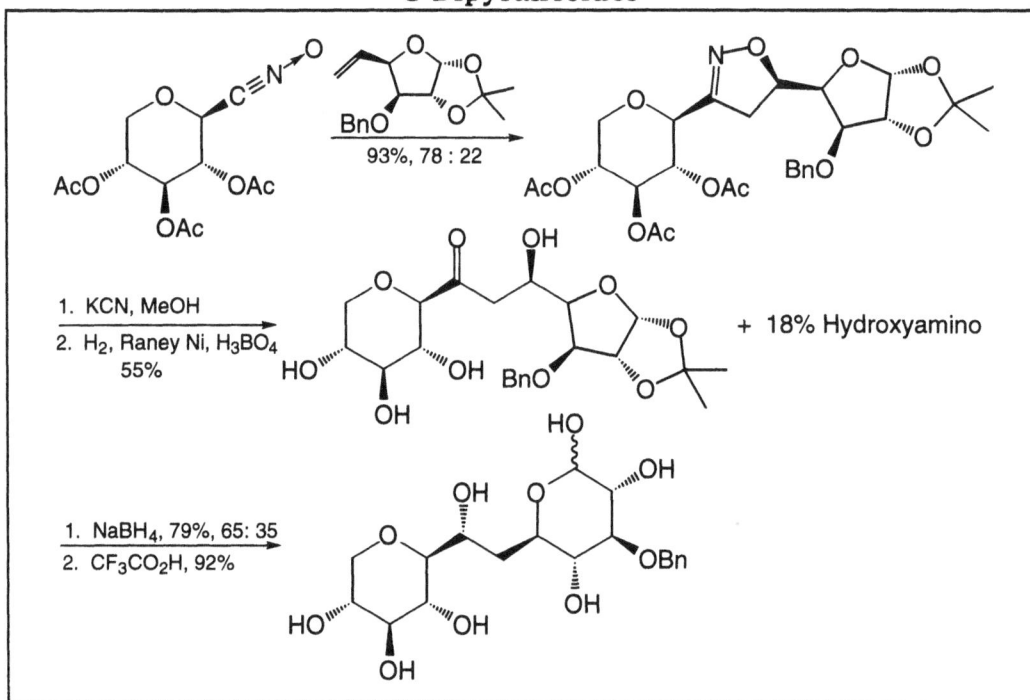

Although no preparations of *C*-trisaccharides were reported in 1994, advances in the technologies applied to the synthesis of *C*-disaccharides yielded several interesting papers. Among the chemistry reported is an extension of the isoxazoline chemistry discussed in Schemes 8.4.2 and 8.6.3. As shown in Scheme 8.12.1, Paton, *et al.,*[46] effected isoxazoline formation utilizing pyranose-derived nitrile oxides and furanose sugars bearing vinyl substituents. As expected, the 1,3-dipolar cycloadditions provided isoxazoline-bridged *C*-disaccharides. The illustrated reaction favored the product shown and exhibited a 93% yield with a 78 : 22 ratio of isomers at the newly formed stereogenic center. However, where hydrolysis of the isoxazolines prepared in Scheme 8.6.3 yielded carbonyl-bridged *C*-disaccharides, hydrogenation of Paton's isoxazolines afforded hydroxyketones. The illustrated example proceeded in 55% yield with an additional 18% accounted for as the corresponding hydroxyamino analog. As shown, reduction of the isolated hydroxyketone produced a 65 : 35 mixture of diols in 79% combined yield favoring the R configuration. Exposure of the major isomer to trifluoroacetic acid effected rearrangement of the furanose component thus providing a novel method to the preparation of hydroxyethylene-bridged *C*-dipyranosides.

Scheme 8.12.2   **Tethered Radicals**

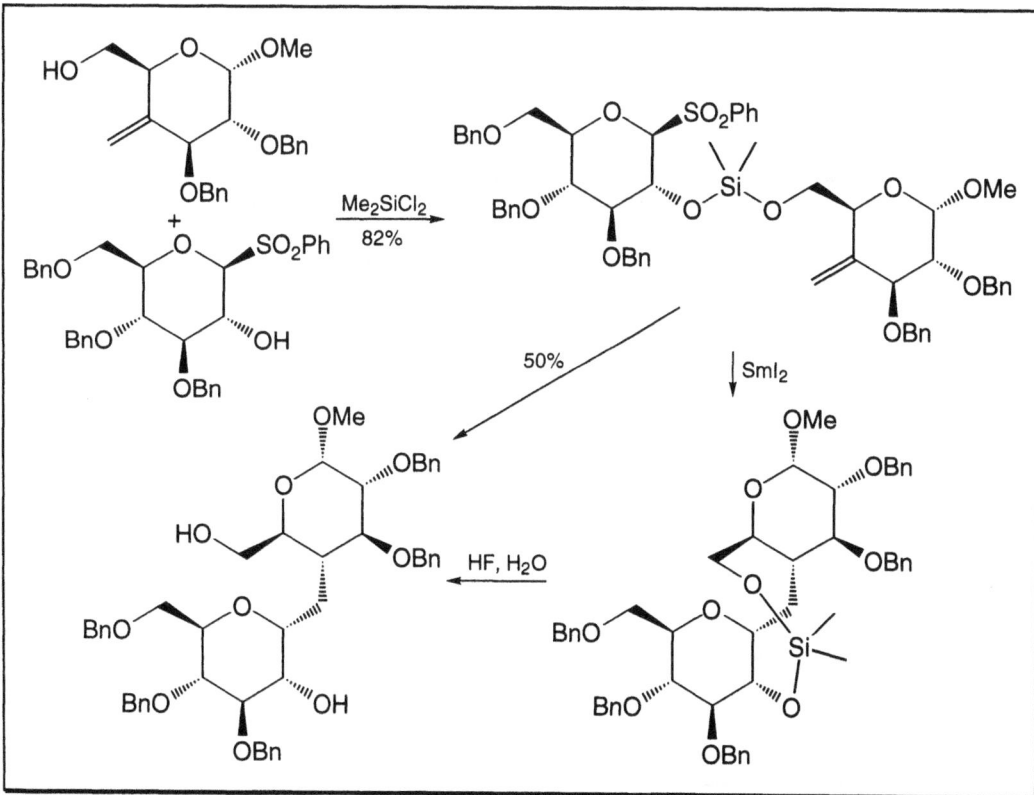

As discussed in the previous section, the use of dimethylsilyl tethers as a means of directing the delivery of one sugar to another under free radical conditions has gained much notoriety. As illustrated in Scheme 8.11.1, the source of the radicals were selenoglycosides and the radicals were generated utilizing tributyltin hydride. In 1994, Sinay, et al.,[47] extended this chemistry to include phenylsulfones as sources of glycosidic free radicals generated on treatment with samarium diiodide. As shown in Scheme 8.12.2, the tethered compound was obtained in 82% yield. Subsequent formation of the *C*-glycosidic linkage was achieved on treatment with samarium diiodide. Completion of the synthesis was accomplished on removal of the tether utilizing HF in water. The combined yield for the final two steps was 50% thus demonstrating a high yielding and potentially general method for the preparation of *C*-disaccharides.

8.13 Trends of the Future

Figure 8.13.1   Oxabicyclo[4.1.0]heptenes

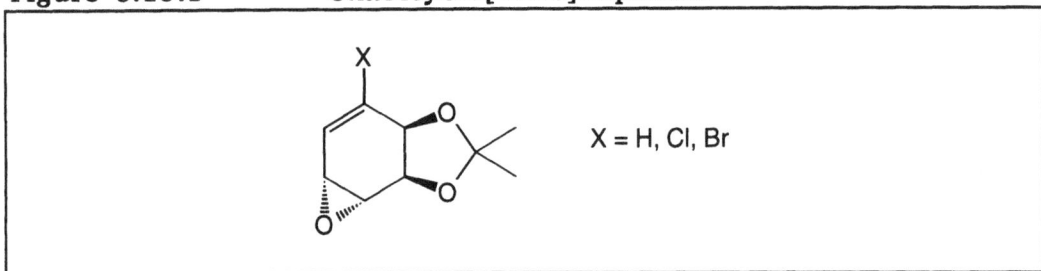

X = H, Cl, Br

**Scheme 8.13.1   Preparation of
Oxabicyclo[4.1.0]heptenes**

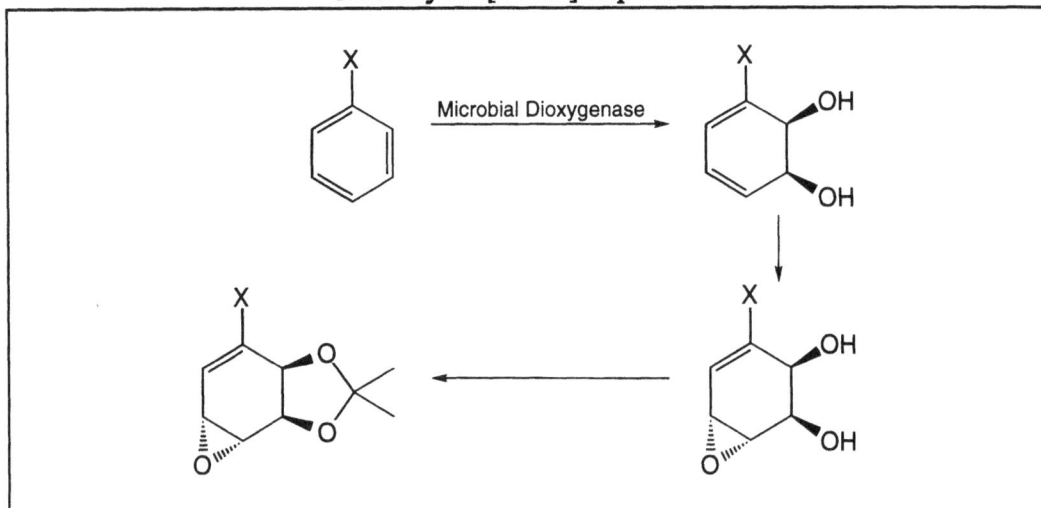

Throughout this book, the evolution of the chemistry surrounding the preparation of *C*-glycosides was presented. Examples were shown which not only presented the actual *C*-glycosidations but also utilized them is the preparation of increasingly more complex target molecules. As it would be presumptuous to assume that we now know all we need in order to prepare any *C*-glycoside of choice, one final example is presented.

In mid 1994, Hudlicky, *et al.,*[48] published a report describing the preparation of new synthons for *C*-disaccharides. The actual compounds of interest are the oxabicyclo[4.1.0]heptenes illustrated in Figure 8.13.1. Unlike the chemistry presented thus far, the compounds represented in Figure 8.13.1 were prepared utilizing the biocatalytic oxidation of haloarenes illustrated in Scheme 8.13.1.[49,50]

Scheme 8.13.2   Addition of Carbon Nucleophiles to Oxabicyclo[4.1.0]heptenes

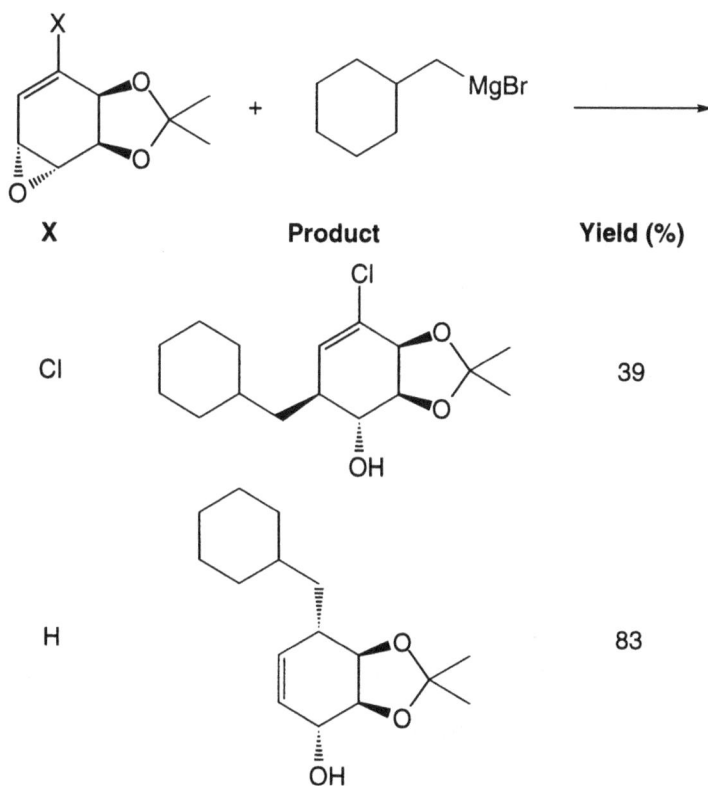

With the preparation of the oxabicyclo[4.1.0]heptenes established, their use as synthons for *C*-disaccharides was explored. As shown in Scheme 8.13.2, addition of carbon nucleophiles such as cyclohexylmethylmagnesiumbromide provided good yields of various addition products. In contrast to the good results obtained with Grignard reagents, the use of organocuprates provided substantially lower yields with measurable quantities of side products. However, as can be deduced from the illustrated results, use of Grignard reagents derived from protected sugars may potentially lead to new *C*-glycoside analogs capable of exhibiting interesting and useful pharmacological properties.

The ability to prepare *C*-glycosides has benefited numerous areas of carbohydrate research from the study of conformational stabilities to the potential development of stable carbohydrate analogs suitable for clinical use. As this field continues to mature, we can look back upon the chemistry developed thus far and truly appreciate the excitement associated with the discovery of future technologies and potential applications.

8.14 References

1. Rouzaud, D.; Sinay, P. *J. Chem. Soc. Chem. Comm.* **1983**, 1353.
2. Aebischer, B.; Meuwly, R.; Vasella, A. *Helv. Chim. Acta* **1984**, *67*, 2236.
3. Jurczak, J.; Bauer, T.; Jarosz, S. *Tetrahedron Lett.* **1984**, *25*, 4809.
4. Cherest, M.; Felkin, H.; Prudent, N. *Tetrahedron Lett.* **1968**, 2201.
5. Cherest, M.; Felkin, H. *Tetrahedron Lett.* **1968**, 2205.
6. Danishefsky, S. J. Maring. C. J.; Barbachyn, M. R.; Segmuller, B. E. *J. Org. Chem.* **1984**, *49*, 4564.
7. Beau, J. M.; Sinay, P. *Tetrahedron Lett.* **1985**, *26*, 6189.
8. Danishefsky, S. J.; Pearson, W. H.; Harvey, D. F.; Charence, J. M.; Springer, J. P. *J. Am. Chem. Soc.* **1985**, *107*, 1256.
9. Hanessian, S.; Martin, M.; Desai, R. C. *J. Chem. Soc. Chem. Comm.* **1986**, 926.
10. RajanBabu, T. V.; Reddy, G. S. *J. Org. Chem.* **1986**, *51*, 5458.
11. Giese, B.; Witzel, T. *Angew. Chem. Int. Ed. Eng.* **1986**, *25*, 450.
12. Liu, P. S. *J. Org. Chem.* **1987**, *52*, 4717.
13. Goekjian, P. G.; Wu, T.-C.; Kang, H.-Y.; Kishi, Y. *J. Org. Chem.* **1987**, *52*, 4823.
14. Goekjian, P. G.; Wu, T.-C.; Kang, H.-Y.; Kishi, Y. *J. Org. Chem.* **1991**, *56*, 6422.
15. Giese, B.; Hoch, M.; Lamberth, C.; Schmidt, R. R. *Tetrahedron Lett.* **1988**, *29*, 1375.
16. Giese, B.; Ruckert, B.; Groninger, K. S.; Muhn, R.; Lindner, H. J. *Liebigs Ann. Chem.* **1988**, 997.
17. Dawson, I. M.; Johnson, T.; Paton, R. M.; Rennie, R. A. C. *J. Chem. Soc. Chem. Comm.* **1988**, 1339.
18. Daly, S. M.; Armstrong, R. W. *Tetrahedron Lett.* **1989**, *30*, 5713.
19. Schmidt, R.; Preuss, R. *Tetrahedron Lett.* **1989**, *30*, 3409.

20. Boschetti, D.; Nicotra, F.; Panza, L.; Russo, G.; Zucchelli, L. *J. Chem. Soc. Chem. Comm.* **1989**, 1085.
21. Motherwell, W. B.; Ross, R. C.; Tozer, M. J. *Synlett* **1989**, 68.
22. Carcano, M.; Nicotra, F.; Panza, L.; Russo, G. *J. Chem. Soc. Chem. Comm.* **1989**, 642.
23. Lay, L.; Nicotra, F.; Pangrazio, C.; Panza, L.; Russo, G. *J. Chem. Soc. Perkin Trans. 1* **1994**, *3*, 333.
24. Neuman, A.; Longchambon, F.; Abbes, O.; Pandraud, H. G.; Perez, S.; Rouzaud, D.; Sinay, P. *Carbohydrate Res.* **1990**, *195*, 187.
25. Dubois, E.; Beau, J. M. *J. Chem. Soc. Chem. Comm.* **1990**, 1191.
26. Martin, O. R.; Lai, W. *J. Org. Chem.* **1990**, *55*, 5188.
27. de Raddt, A.; Stulz, A. E. *Carbohydrate Res.* **1991**, *220*, 101.
28. Preuss, R.; Schmidt, R. R. *J. Carbohydrate Chem.* **1991**, *10*, 887.
29. Prandi, J.; Audin, C.; Beau, J. M. *Tetrahedron Lett.* **1991**, *32*, 769.
30. Engelbrecht, G. J.; Holzaphel, C. W. *Heterocycles* **1991**, *32*, 1267.
31. Bimwala, R. M.; Vogel, P. *Tetrahedron Lett.* **1991**, *32*, 1429.
32. Black, K. A.; Vogel, P. *J. Org. Chem.* **1986**, *51*, 5341.
33. Fattori, D.; de Guchteneere, E.; Vogel, P. *Tetrahedron Lett.* **1989**, *30*, 7415.
34. Roberts, J. L.; Borromeo, C. D.; Poulter, C. D. *Tetrahedron Lett.* **1977**, 1621.
35. Bimwala, R. M.; Vogel, P. *J. Org. Chem.* **1992**, *57*, 2076.
36. Armstrong, R. W.; Teegarden, B. R. *J. Org. Chem.* **1992**, *57*, 915.
37. Wang, Y.; Babirad, S. A.; Kishi, Y. *J. Org. Chem.* **1992**, *57*, 468.
38. Goekjian, P.; Wu, T.-C.; Kang, H. Y.; Kishi, Y. *J. Org. Chem.* **1991**, *56*, 6423.
39. Xin, Y. C.; Mallet, J. M.; Sinay, P. *J. Chem. Soc. Chem. Comm.* **1993**, 864.
40. Vauzeilles, B.; Cravo, D.; Mallet, J. M.; Sinay, P. *Synlett* **1993**, 522.
41. Sutherlin, D. P.; Armstrong, R. W. *Tetrahedron Lett.* **1993**, *34*, 4897.
42. O'Leary, D. J.; Kishi, Y. *J. Org. Chem.* **1993**, *58*, 304.
43. Dyer, U. C.; Kishi, Y. *J. Org. Chem.* **1988**, *53*, 3383.
44. Haneda, T.; Goekjian, P. G.; Kim. S. H.; Kishi, Y. *J. Org. Chem.* **1992**, *57*, 490.
45. Kishi, Y. *Pure & Appl. Chem.* **1993**, *65*, 771.
46. Paton, R. M.; Penman, K. J. *Tetrahedron Lett.* **1994**, *35*, 3163.
47. Chenede, A.; Perrin, E.; Rekai, E. D.; Sinay, P. *Synlett* **1994**, *6*, 420.
48. Hudlicky, T.; Tian, X.; Konigsberger, K.; Rouden, J. J. Org. Chem. **1994**, *59*, 4037.
49. Hudlicky, T.; Fan, R.; Tsunoda, T.; Luna, H.; Andersen, C.; Price, J. D. *Isr. J. Chem.* **1991**, *31*, 229.
50. Carless, H. A. *Tetrahedron Lett.* **1992**, *33*, 6379.

Index

Authors' Biographies

Daniel E. Levy received his Bachelor of Science from the University of California at Berkeley where he pursued research programs under the direction of Professor Henry Rapoport. While pursuing his Ph.D. at the Massachusetts Institute of Technology, under the direction of Professor Satoru Masamune, Dr. Levy studied and compiled his thesis on the total synthesis of calyculin A. Since completing his Ph.D., Dr. Levy has been studying the chemistry surrounding *C*-glycosides at Glycomed, Inc.

Cho Tang received his Bachelor of Science from the Chinese University of Hong Kong after which, he pursued his Ph.D. at the University of Chicago under the direction of Professor William Wulff. In this capacity, Dr. Tang studied and compiled his thesis on the subject of chromium carbene complexes. After completing a postdoctoral appointment at Columbia University, under the direction of Professor Gilbert Stork, Dr. Tang became involved in drug design and drug discovery at Hoffman-La Roche. Upon moving to Glycomed, Inc., Dr. Tang became involved in numerous programs centering around the chemistry of *C*-glycosides. Dr. Tang is currently studying tyrosine kinase inhibitors at Sugen, Inc.